EXPANDED IDEAS IN BIOTECHNOLOGY
TECHNIQUES AND THEORY

EXPANDED IDEAS IN BIOTECHNOLOGY
TECHNIQUES AND THEORY

Dr. Khushboo Chaudhary

Delve
PUBLISHING

www.delvepublishing.com

Expanded Ideas in Biotechnology: Techniques and Theory

Dr. Khushboo Chaudhary

Delve Publishing
4164 Lakeshore Road
Burlington, ON L7L 1A4
Canada
www.delvepublishing.com
Email: orders@arclereducation.com

ISBN: 978-1-77956-359-0 (Hardcover)

ABOUT THE AUTHOR

Dr Khushboo Chaudhary is presently working as a Technical Officer II in THSTI, Faridabad, Haryana, India and having 11 year of research experience. Previously, she worked on "Improvement of Phytoremediation efficiency of Fluoride". She has published several research papers in international and national journals. She has got best paper and poster presentation award by Indian Society of Genetics and Biotechnology Research and Development and received the president appreciation awards also in the International conference. She has got best poster award by ISSGPU Central Institute Research on Goats, Makhdoom. She has published several gene bank in NCBI Pubmed. She has published research article in virology journal. Khushboo is likely to be a co-author in several of the publications and have submitted/likely to submit. She is a coauthor in J. Virological Methods we submitted couple of weeks ago. Besides, she has been a co-author in 5 of the abstracts we presented/likely to present in various conferences of National and International scientific societies such as Indian Society of Virology, American Society of Virology, Indian Society of Sheep and Goat Production and Utilization etc.

TABLE OF CONTENTS

LIST OF FIGURES

LIST OF TABLES

GLOSSARY

Agarose Gel Electrophoresis: It is a process in which a matrix composed of a highly purified form of agar is used to separate large DNA and RNA molecules.

Allele: One of pair of variant forms of a gene occurring at a given locus of a gene.

Amplification: There are formation of many copies of DNA segment of PCR.

Annealing: The process of healing and slowly cooling of double stranded DNA to allow the formation of hybrid DNA or DNA-RNA molecules.

Antibiotic: A biological substances produced by one organism inhibiting the growth of another organism.

Antibody: A protein synthesized by B lymphocytes that recognizes a specific site on an antigen.

Antigen: A compound that induces the production of antibodies.

Antiparallel orientation: Arrangement of two strands of DNA molecule oriented in opposite directions so that the 5' PO4 end of one strand is aligned with 3'OH end of the complementary strand.

Antisense DNA: The sequence of Chromosomal DNA that is transcribed.

Antisense RNA: An RNA sequence complementary to all or part of a functional RNA.

Antisense therapy: In vitro treatment of genetic disease by blocking translation of a protein with DNA or RNA sequences that is complementary to a specific mRNA.

Artificial seeds: A synthetic seed consisting of a somatic embryo surrounded by nutrient medium which is protected by a thin membrane of chemical.

Autoclave: An instrument used for sterilization of glassware and culture media.

Auxins: A class of plant growth regulators that stimulate cell division, cell elongation, apical dominance and root initiation. They are used in plant cell and tissue culture, e.g., 2,4-D, IAA, NAA, etc.

Axenic culture: A pure of culture single propagule of an organism not contaminated by any other microorganism.

Axillary bud: A bud produced in the axis of leaves.

Bioaccumulation: The concentration of a chemical agent, e.g., DDT in the increasing amount in the organisms of a food chain.

Biodegradation: The gradual breakdown of a compound to its constituents by a living organism.

Biodiversity: The variability among the living organism from all sources, soil, water, air, extreme habitat or associated with organisms.

Biofertilizers: The commercial preparation of microorganisms by using which nitrogen and phosphorus level and growth of plants increased.

Bioremediation: The process of using living organisms to remove contaminants, pollutants or unwanted substances from soil or water.

Biosensor: An electronic device that uses biological molecules or cells to detect specific compounds.

Biotic stress: Stress resulting from living organisms which can harm the other plants such as virus, bacteria, fungi, weeds, harmful insects.

Callus: An organized mass of plant cells capable of cell division and growth in vitro.

cDNA Clone: A double stranded DNA complement of mRNA synthesized in vitro by using reverse transcriptase and DNA polymerase.

cDNA library: A collection of cDNA clones generated in vitro from the mRNA sequences of a single tissue or cell population.

Cell culture: The cell culture of growing cells in vitro in liquid medium.

Cell line: The cell line is derived from a primary culture and selected at the time of the first subculture or derived from cell population with particular attributes.

Chimera: Recombinat DNA molecules containing sequences from different organisms.

Clone: A collection of genetically identical cells or organisms derived from a common ancestor where all members have similar genetic composition.

Consensus sequences: The nucleotide sequences that is present in majority of genetic signals that perform a specific function.

Cosmid: A plasmid vector which consists of the COS site of phage lambda (λ) and one or more selectable markers such as an antibiotic resistance gene.

Crown gall: A tumor occurring at the base of certain plants due to infection of the plant by *Agrobacterium tumifaciens*.

Cytokinins: The class of plant growth regulators which causes cell division, cell differentiation, shoot differentiation and breaking of apical dominance, e.g., BAP, Kinetin and zeatin.

Denaturation: The separation of DNA double stranded molecule into single stands.

Differentiation: The process of biochemical and structural changes by which cells become specialized in form and function.

DNA construct: A suitable DNA which has been prepared for cloning purpose.

DNA polymerase: A n enzyme that catalyzes the phosphodiester bond in the formation of DNA.

DNA probe: A radiolabeled (32P) DNA segment used to hybridize the base between the probe and a complementary base sequences in a DNA sample.

Electrophoresis: The method of separation of molecule such as DNA, RNA or protein based on their relative migration applying a strong electric field.

Embryo transfer: Implantation of embryos from donor animals or generated by in vitro fertilization into the uterus of the recipient animals.

Excision: Enzymatic removal of DNA segment from a chromosome.

Exonucleases: An enzyme the digest DNA or RNA from the ends of strand; 5' exonuclease requires a free 5' end and degrades the molecule in 5'-3 direction; 3' exonuclease requires a free 3' end and degrades the DNA IN 3'-5 direction.

Explant: The tissue taken from a plant or seed and transferred to a culture medium to establish a tissue culture system or regenerate the plant.

Fermentation: The process by which microorganism turn raw materials such as glucose into products such as alcohol.

Fusion product: A single protein molecule encoded by parts of two or more genes combined together as one unit.

Gene: A hereditary unit or the segment of DNA coding a specific protein.

Genetic fingerprinting: A technique of analyzing DNA of an individual to reveal the pattern of repetition of particular nucleotide sequences through the genome.

Hybridoma: A unique fused cells or protoplasts into a matrix.

In vitro: A ny process carried out in sterile cultures or measurement of biological processes outside the intact organism such as enzyme reaction.

In vivo: The natural condition in which organism live.

Leader sequences: It is the non-translated sequence at the 5' end of mRNA that precedes the initiation codon.

Lysogeny: The phenomenon in which prophage survives within a host bacterium as a part of host genome or as extrachromosomal element and does not cause lysis.

Micropropagation: The small piece of tissue such as meristem grown in culture to produce large number of plants.

Monoclonal antibodies: It is derived from a single clone cells which recognize only one kind of antigen.

Nod genes: Genens encode nodulins.

Oncogene: A gene that cause cancer.

Operon: A cluster of gene that are coordinately regulated.

Palindromic sequences: The complementary DNA sequences that are the same when each stand is read in the same direction, e.g., (5'-3). These sequences act as recognition sites for type II restriction endonuclease.

Patent: A legal document issued by a government that permits the holder the exclusive right to manufacture, use of sell an invention for a defined period, which varies country to country.

Plasmid: Extrachromosomal, self-replicating circular double stranded DNA molecules containing some non-essential genes.

Primer: A short oligonucleotide that hybridizes the template strand and give 3' OH end for the initiation of nucleic acid synthesis.

Recombinant: An individual whose genes on a chromosomes results from one or more cross over events.

Secondary metabolites: The metabolites produced as end products of primary metabolism and not involved directly in metabolic activity.

Splicing: The gene manipulation where the one DNA molecule is attached to one another.

Totipotency: A property of normal cells that they have the genetic potential to give rise to a complete individuals except terminally specialized cells.

Transfection: DNA transfer into a eukaryotic cell.

Transposon: A DNA segment that can Jump from locus and join a new locus on the DNA molecule by involving the enzyme transposase.

Vectors: The vehicles for transferring DNA from one cell to another or DNA molecule that can carry inserted DNA and be perpetuated in a host cell.

Western blot: A technique in which protein is transfer from an electrophoresis gen to a cellulose or nylon support membrane following electrophoresis.

Yeast artificial chromosomes: A vector is used for hundreds to kilobases long used for cloning of DNA fragment.

Zooblot: The cloned DNA from one species to DNA from the other organism to determine the extent to which the cloned DNA is evolutionary conserved.

PREFACE

The book "**Expanded Ideas in Biotechnology** ***Techniques and Theory***" has been written to serve as a textbook for undergraduate and postgraduate students. This book covers a range of syllabus, making it easy for students to go through the content. It is prepared for students to gain a fair knowledge of "Biotech" with an easy presentation of language for proper understanding of each topic. The chapters are arranged to provide a sequential knowledge of biotechnology. Examples to illustrate different points have been chosen from various universities and institutions in the country.

This book would benefit not only undergraduate and postgraduate students of biology but also research scholars from different universities. The scientists and authors quoted or referenced in the book are fully acknowledged at the end. I have been greatly helped by several books in preparing this book, and I have expressed my acknowledgments in the references. If I have forgotten any references, please disregard these mistakes. Comments and suggestions are welcome from students, teachers, and friends for corrections and improvements, which will be greatly acknowledged.

Last but not least, I would be thankful to readers for pointing out errors and most welcome suggestions in the next book.

—Khushboo Chaudhary

CHAPTER 1

BIOTECHNOLOGY

Contents

The term biotechnology was coined in 1919 by a Hungarian engineer, Karl Ereky, to describe a process for the large-scale production of pigs. According to him, all types of work are biotechnology by which products are produced from raw materials using living organisms. By the end of the 20th century, biotechnology emerged as a new discipline of biology integrating with technology.

There was no sudden sprout of this discipline, but some of the methods for the production of products were developed centuries back. Therefore, biotechnology is concerned with the exploitation of biological components for the production of useful products. There are many explanations about biotechnology which are as follows:

- Biotechnology is defined by different organizations in different ways. It has been broadly defined as "the development and □tilization of biological processes, forms, and systems for obtaining maximum benefits to man and other forms of life."
- Biotechnology is "the science of applied biological process" (Biotechnology: A Dutch Perspective, 1981).
- Biotechnology is the "controlled use of biological agents such as microorganisms or cell□lar components for beneficial □se" (U.S. National Science Foundation).
- Biological agents: we can call biological agents such as microorganisms, enzymes, plant and animal cells. The meaning of these three definitions and others given by many organizations are more or less similar.

The definition of genetic engineering has been given by Smith (1996) as "the formation of new combinations of heritable material by the insertion of nucleic acid molecules produced by whatever means outside the cell, into any virus, bacterial plasmid, or other vector system so as to allow their incorporation into host organisms in which they do not naturally occur but in which they are capable of continued propagation." Biotechnology is widely used in pharmacy to create more efficient and less expensive dr□gs. Recombinant DNA technology is □sed for the prod□ction of specific enzymes, which enhance the rate of production of a particular range of antibodies in the organism. Hormones such as somatostatin, insulin, and the human growth hormone can be synthesized easily and cheaply.

The first h□man hormone to be synthesized by genetic engineering was somatostatin. Somatostatin is a brain hormone originating from the hypothalamus. It acts to inhibit the release of human growth hormone and insulin and is related to the treatment of diabetes, pancreatitis, and other conditions. Genentech, a California-based company, has produced human growth hormone (hGH) from genetically engineered bacteria. Human insulin is the first genetically engineered pharmaceutical product, developed by Eli Lilly and Company in 1982. Bovine Somatotropin (BST) is produced for a large quantity of milk production in cows. Antibiotics are chemical substances produced by several microorganisms. Recombinant DNA technology has helped in the increased production of antibiotics; for example, the rate of penicillin produced at present is about 150,000 units/ml against about 10 units/ml in the 1950s. Antibiotics produced using such technology have very specific effects and cause fewer side effects. Currently, scientists are working on vaccines for fatal illnesses such as AIDS, hepatitis, malaria, flu, and even some forms of cancer. Interferon, an antiviral protein, is prepared from mammalian cells by recombinant DNA technology. By cloning cDNA to genes for human interferon, it has been found that there are a large number of interferons differing in amino acid sequences and properties. A large number of interferons are prepared in yeast cells by the fermentation process (Mandels and Stenberg, 1976).

1.1. BIOTECHNOLOGY CENTURY

Biotechnology is the technology applied to biology, molecular biology, genetics, and many other subfields of biology. Biotechnology utilizes cellular and biomolecular processes to create technologies and products that help improve our lives and nature. By making useful food, such as bread and cheese, and preserving dairy products, we have been doing this for many years now. Recent biotechnology develops breakthrough products and technologies to fight diseases, reduce environmental harm, feed the hungry, use less and cleaner energy, and have safer, cleaner, and more efficient industrial manufacturing processes (Agarwal, 2000). So far, more than 250 biotechnology healthcare products and vaccines have been made available to patients, many for previously untreatable diseases. More than 13.3 million farmers around the world use agricultural biotechnology to increase yields, prevent damage from insects and pests, and reduce the damage done to the environment due to farming. Additionally, more than 50 biorefineries are being built across North America to test and refine technologies to produce biofuels and chemicals from renewable biomass, which can help reduce greenhouse gas emissions (Boffey, 1987).

Biotechnology is the third wave in biological science and represents an interface of basic and applied sciences, where the gradual and subtle transformation of science into technology can be witnessed. Biotechnology is defined as the application of scientific and engineering principles to the processing of materials by biological agents to provide goods and services. Biotechnology comprises a number of technologies based on increasing understanding of biology at the cellular and molecular level. The Bible already provides numerous examples of biotechnology. Namely, it deals with the conversion of grapes to wine, dough to bread, and milk to cheese. The oldest biotechnological processes are found in microbial fermentations, as borne out by the Babylonian tablet dated circa 6000 B.C. explaining the preparation of beer (Smith, 1966). The Sumerians were able to brew as many as 20 types of beer in the third millennium B.C.

In about 4000 B.C., leavened bread was produced with the aid of yeast. During the Vedic period (5000–7000 B.C.), Aryans had been performing daily *Agnihotra* or *Yajna.* In *Ayurveda,* the production of 'Asava' and 'Arista' using different substrates and flowers of mahua (*Madhuca indica*) or dhataki (*Wodfordia fructicosa*) has been well characterized till today since the Vedic period. One of the materials used in *Yajna* is animal fat (i.e., ghee), which is a fermented product of milk.

The term 'biotechnology' was described in a Bulletin of the Bureau of Biotechnology published in July 1920 from the office of the same name in Leeds Yorkshire (Crueger, 1984).

There are numerous sub-fields of biotechnology:

- **Red Biotechnology:** It is biotechnology applied to medical processes. Some examples are the designing of organisms to produce antibiotics, and the engineering of genetic cures to cure diseases through genomic manipulation.
- **White Biotechnology:** It also known as gray biotechnology, is biotechnology applied to industrial processes. An example is the designing of an organism to produce a useful chemical. White biotechnology tends to consume fewer resources than traditional processes when used to produce industrial goods.
- **Green Biotechnology:** It is biotechnology applied to agricultural processes. An example is the designing of an organism to grow under specific environmental conditions or in the presence (or absence) of certain agricultural chemicals. One hope is that green biotechnology might produce more environmentally friendly solutions than traditional industrial agriculture. An example of this is the engineering of a plant to express a pesticide, thereby eliminating the need for external application of pesticides. Whether or not green biotechnology products such as this are ultimately more environmentally friendly is a topic of considerable debate.
- **Blue Biotechnology:** The term blue biotechnology has also been used to describe the marine and aquatic applications of biotechnology, but its use is relatively rare.

1.2. BROADLY BIOTECHNOLOGY CAN BE DIVIDED INTO TWO MAJOR BRANCHES

- **Non-Gene Biotechnology:** It deals with whole cells, tissues, or even individual organisms
- **Gene Biotechnology:** It involves gene manipulation, cloning, etc.

Non-gene biotechnology is a more popular practice, and plant tissue culture, hybrid seed production, microbial fermentation, and production of hybridoma antibodies or immunochemicals are widespread biotechnology practices.

For centuries, humans have used microorganisms to produce foods and drinks without understanding the microbial processes underlying their production. In recent years, the understanding of the biosynthetic pathways and regulatory control mechanisms used by microorganisms for the production of several metabolites has increased by developing the knowledge of the biochemistry of industrially important organisms. Notable biotechnologies for food processing include fermentation technology, enzyme technology, and monoclonal antibody technology.

Beneficial microbes participate in fermentation processes, producing many useful metabolites such as enzymes, organic acids, solvents, vitamins, amino acids, antibiotics, growth regulators, flavors, and nutritious foods (Sasson, 1984). Some leading food bioprocessing technologies are dairy processing, and alcohol and beverage processing. Production of alcoholic beverages includes wine, beer, whiskey, rum, sake, etc., utilizing microorganisms like *Clostridium acetobutylicum, Leuconostoc mesenteroides, Aspergillus oryzae, Saccharomyces cerevisiae, Rhizopus sp.,* and *Mucor sp.* Biotechnologically produced organic acids like citric acid, acetic acid, gluconic acid, D-Lactic acid, and fumaric acid also have very high market value.

The application of biotechnology can result in (a) new ways of producing existing products with the use of new inputs, and (b) new ways of producing new products. Examples of the former include the production of gasoline from ethanol which in turn is produced from sugar; the production of insulin using recombinant DNA technology; the production of hepatitis B vaccine using recombinant DNA technology and the extraction of copper using mineral leaching bacteria. The alternative inputs are oil for gasoline, porcine pancreases for insulin, human blood for hepatitis vaccine, and the conventional mining techniques for copper. Examples of the latter include possible medicinal substances that are produced in minute quantity in the human body and which cannot be synthesized such as insulin, interleukin or Tissue Plasminogen Activator (TPA). A wide variety of microorganisms are now being employed as tools in biotechnology to produce useful products or services. Raw materials can be converted to useful finished products both by ordinary chemical processes and by biological means. Generally, the costs of chemical conversion are quite high as the reactions require high temperatures or pressure. In contrast, biological alternatives, using microbes or cultured animal or plant cells, operate at physiologically normal conditions of temperature, pressure, pH, etc. During the next few decades, biotechnology would have overtaken chemical technology and many such chemicals which are today produced chemically would be made through biotechnology (Dubey 2009).

Enzyme technology is an area of considerable current interest and development. Enzymes are biological catalysts and have been used for many years as isolated agents, particularly in food, e.g., rennin, papain, and invertase. These enzymes have increasingly replaced plants and animal enzymes. For example, amylase from *Bacillus* and *Aspergillus* has substituted those of malted wheat and barley in brewing, baking, and biscuit-making, as well as in the textile industry, etc. Today, enzyme technologies have four distinct areas of application: cosmetics, therapy, the food and feed industry, and diagnostic purposes. One very important recent application is the production of foodstuffs from non-traditional raw materials. For instance, the development of the sweetener high fructose corn syrup (HFCS), also called isoglucose. Another recent application is the use of phytase in animal feed.

Nowadays, interest in the traditional fermentation technology for food processing has greatly increased because of the emphasis placed on plant materials as human foods. Single-cell protein (SCP) is the term generally accepted to mean the microbial cells (algae, bacteria, actinomycetes, and fungi) grown and harvested for animal or human food. During World War II, when there were shortages in proteins and vitamins in the diet, the Germans produced yeasts and a mold (*Geotrichum candidum*) in some quantity for food. Research on SCP has been stimulated by a concern over the eventual food crisis or food shortages that will occur if the world's population is not controlled. Many scientists believe that the use of microbial fermentations and the development of an industry to produce and supply SCP are possible solutions to meet a shortage of protein if and when the amount of protein produced or obtained by agriculture and fishing becomes insufficient (Ferber 2004).

The roots of molecular biology were established only after the British biophysicist Francis Crick and the American biochemist James Watson, in 1953, proposed the structure of DNA (deoxyribonucleic acid) molecule, which is well known as the chemical bearer of genetic information of most organisms. We really began understanding and utilizing molecular biotechnology (or gene biotechnology) only after recombinant DNA technology was developed in the 1970s. Daniel Nathans (in 1971) of Johns Hopkins University utilized the restriction enzyme to split the DNA of the monkey tumor virus, Simian Virus (SV_{40}). Recombinant DNA technology, often referred to as genetic engineering or gene manipulation, involves the extraction of a particular gene of interest from one organism and then the insertion of the gene into other organisms.

Genetic manip□lation may be defined as the extracell□lar (i.e., *in vitro*) creation of new forms or arrangements of DNA in such a way as to allow the incorporation or continued propagation of altered genetic conditions in nature. Among the first scientists to attempt genetic manip□lation was Pa□l Berg of Stanford University who, in 1971, along with his co-workers, opened the DNA molecule of SV_{40} and spliced it into a bacterial chromosome, constructing the first recombinant DNA molec□le.

Genetic engineering techniques are useful tools for genetic research. They can help to gain insight into the structure, function, and regulation of genes. They also help to prepare physical maps of viral genomes. Maps of several viruses have been made available, such as SV_{40}, Polyoma virus, and adenovirus. Another goal in genetic engineering is to design a super bug that can degrade most of the major hydrocarbon components of petroleum. Different strains of *Pseudomonas putida* contain a plasmid with genes coding for enzymes that digest a single family of hydrocarbons. By crossing the various strains of this bacterium, a super bug has been created. The multiplasmid bacterium is able to grow on a diet of crude oil. The super bug has the potential to clear up oil spills.

One of the best-known applications of genetic engineering is that of the creation of genetically modified organisms (GMOs). There are potentially momentous biotechnological applications of GM, for example, oral vaccines produced naturally in fruit, at very low cost. This represents, however, a spread of genetic modification to medical p□rposes and opens an ethical door to other uses of the technology to directly modify human genomes. A genetically modified food is a food prod□ct derived in whole or part from a genetically modified organism (GMO) s□ch as a crop plant, animal, or microbe s□ch as yeast. Genetically modified foods have been available since the 1990s. The principal ingredients of GM foods currently available are derived from genetically modified soybean, maize, and canola. Between 1996 and 2001, the total surface area of land cultivated with GMOs had increased by a factor of 30, from 17,000 km^2 (4.2 million acres) to 520,000 km^2 (128 million acres). The value for 2002 was 145 million acres (587,000 km^2) and for 2003 was 167 million acres (676,000 km^2). Future applications of GMOs include bananas that produce human vaccines against infectious diseases such as Hepatitis B, fish that mat□re more q□ickly, fr□it and n□t trees that yield years earlier, and plants that produce new plastics with unique properties. Now scientists have transformed *Tobacco Mosaic Virus* (TMV) to infect host plants and produce immunizing proteins rather than debilitating leaf shrivel, turning greenhouse tobacco into a biofactory for plague vaccine.

Genetic diseases could be treated through the use of genetic engineering. Defective genes in an organism cause genetic disorders. If a defective gene co□ld be identified and located in a partic□lar gro□p of cells, it co□ld be replaced with a functional one. The transgenic cells are then implanted into the organism, resulting in a cure for the disorder. Cloning is a relatively new sector of biotechnology, but it promises answers to very important problems related to surgery. Tissues and organs could be cloned for surgical purposes. If scientists could isolate stem cells and then direct their development, they would be able to create any kind of tissue, organ, or even a whole part of a body (Schlegel 1986; Amirato 1983).

Another revolutionizing tool of biotechnology is DNA fingerprinting. DNA fingerprints are □sef□l in several applications of h□man health care research, as well as in the j□stice system. DNA fingerprinting is □sed to diagnose inherited disorders in both prenatal and newborn babies in hospitals around the world. These disorders may incl□de cystic fibrosis, hemophilia, H□ntington's disease, familial Alzheimer's, sickle cell anemia, thalassemia, and many others. Early detection of such disorders enables the medical staff to prepare themselves and the parents for proper treatment of the child. In some programs, genetic co□nselors □se DNA fingerprint information to help prospective parents □nderstand the risk of having an affected child. DNA fingerprint information can also help in developing c□res for inherited disorders. DNA fingerprints help to link suspects to biological evidence such as blood or semen stains, hair, or items of clothing found at the scene of a crime and help in solving crimes. Another important □se of DNA fingerprints in the co□rt system is to establish paternity in custody and child support litigation. The U.S. armed services have j□st beg□n a program to collect DNA fingerprints from all personnel for □se later, in case they are needed to identify casualties or persons missing in action or for s□spect verification.

Due to the revolutionary development of biotechnology during the last couple of decades, agriculture has drastically advanced. Sensational achievements were made in both plant cultivation and animal husbandry. Plants have been improved in four different ways:

- Enhanced potential for more vigorous growth and increasing yields.
- Increased resistance to natural predators and pests, including insects and disease-causing microorganisms.
- Production of hybrids exhibiting a combination of superior traits derived from two different strains or even different species.

- Selection of genetic variants with desirable qualities such as increased protein value, increased content of limiting amino acids, which are essential in the human diet, or smaller plant size, reducing vulnerability to adverse weather conditions.

Another important area of biotechnology is the improvement of livestock. Improvements in disease control, the efficiency of reprod□ction, yields of livestock products (such as meat, milk, wool, eggs), and the composition of livestock products (such as leaner meat), as well as the feed value of low-quality feeds (such as straw), are some of the applications of biotechnology.

One of the major scientific revol□tions of the twentieth cent□ry was the breaking of the genetic code and the development of tools that enabled scientists to probe the molecules of life with incredible precision. Now, in the 21st century, these developments in biology are being combined with the use of ever-increasing computer power to help us face the challenges that the new century brings.

Biotechnology has a promising future. In the future, biotechnology will be credited for some revolutionary technology. Recent advances in bioenergy, bioremediation, synthetic biology, DNA computers, virtual cells, genomics, proteomics, bioinformatics, and bio-nanotechnology have made biotechnology even more powerful. The recent discovery of the conduction of electricity by DNA and its behavior as a superconductor has opened a new realm in modern science. In the future, biotechnology will have a profound impact on the world economy. Biotechnology is a golden tool to solve some of the key global problems like global epidemics, fatal diseases, global warming, rising petroleum fuel crisis, and above all poverty. For all the positive effects of biotechnology, there are some possible side effects. Nobody knows what ecological hazards could be caused by transgenic organisms. Some even speculate that some transgenic organisms could fall into the wrong hands to develop bioweapons. The opposition to genetic engineering says that "the science is very young and needs a lot more research." The path from a test t□be to the field is not a straight highway. Both intellect□al and financial reso□rces sho□ld be realized before discoveries pave their way to industrial applications. In conclusion, biotechnology has also proved to be extremely productive and innovative, and the 21st century should be the century of biotechnology (Bull, 1983; Cruger, 1984; Boffey, 1987; Cummings, 2004; Ferber, 2004; Dubey, 2009).

1.3. HISTORY OF BIOTECHNOLOGY

The origin of biotechnology is as old as human civilization. The development of biotechnology can be studied considering its growth that occurred in two phases:

- Traditional biotechnology (old); and
- New biotechnology (modern).

1.4. ORIGIN AND DEFINITION: OLD VS. NEW BIOTECHNOLOGY

The origin of biotechnology can be traced back to prehistoric times when microorganisms were already used for processes like fermentation. Although a molecular biologist may consider the cloning of DNA to be the most important event in the history of biotechnology, it was rediscovered in the 1970s for the third time during the present century. In the 1920s, *Clostridium acetobutylicum* was used by Chaim Weizmann for converting starch into butanol and acetone, the latter being an essential component of explosives during World War I. This raised hopes for commercial production of useful chemicals through biological processes and may be considered as the first rediscovery of biotechnology in the present century. Similarly, during World War II (in the 1940s), the production of penicillin (as an antibiotic discovered by Alexander Fleming in 1929) on a large scale from cultures of *Penicillium notatum*, was the second rediscovery of biotechnology. This marked the beginning of an era of antibiotic research. The third rediscovery of biotechnology is its recent reincarnation in the form of recombinant DNA technology, which led to the development of a variety of gene technologies and is considered to be the greatest scientific revolution of the century. Thus, even though biotechnology had its origin in the past, it assumed special significance only in the 1970s and 1980s.

Biotechnology is the product of the interaction between the science of biology and technology. This relationship between science and technology has been observed to be complex, so that not only has science influenced technology, but technology has also influenced science. Because of this complex relationship and its major impact on human welfare, it is believed that biotechnology in the future may become a major force for human existence. Already, the products of biotechnology (including diagnosis, prevention, and cure of diseases; new and cheaper chemical products, e.g., pharmaceutical drugs and new food sources; devices for environmental protection and energy conservation, etc.) are playing a very important role in employment, productivity, trade, economics, and the quality of human life throughout the world. Biotechnology is such an important area of study that attempts have been made to define it. It has been recognized that a complete definition of biotechnology is difficult due to the wide range of its usage (Mendel, 1988; Umar, 2003). Fermentation by some microorganisms, formation of yogurt (curd) and cheese, production of penicillin from certain fungi, and processes of baking and brewing are often included in describing what is called old biotechnology. Another example of biological processes involving the techniques of recombinant DNA and polymerase chain reaction (PCR) and bioprocessing, which became possible only through research in molecular biology, has been described as new biotechnology. However, in recent literature, no effort has been made to distinguish between old and new biotechnology. The following are some available definitions of biotechnology, and one of them may be really adequate.

- Biotechnology is "the application of biological organisms, systems, or processes to manufacturing and service industries" (British biotechnology).
- Biotechnology is "the integrated use of biochemistry, microbiology, and engineering sciences in order to achieve the technological application of the capabilities of microorganisms, cultured tissue cells, and parts."
- Biotechnology is "a technology using biological phenomena for copying and manufacturing various kinds of useful substances."
- Biotechnology is "the controlled use of biological agents, such as microorganisms or cell□lar components, for beneficial □se."

1.5. TRADITIONAL BIOTECHNOLOGY

It is kitchen technology developed by our ancestors using fermenting bacteria. It is as old as human civilization. The Aryans performed daily *Agnihotra* or *Yajna* during the *Vedic* period (5000-7000 BC). The material used in Yajna is animal fat (ghee), which is a fermented product of milk.

Similarly, the divine "soma" had been offered to God (a fermented microbial product used as a beverage). Sumerians and Babylonians (6000 BC) were drinking beer. Egyptians were baking leavened bread by 4000 BC. The preparation of curd, ghee, wine, beer, and vinegar, etc., were part of kitchen technology. Despite all developments, the preparation of curd, ghee, vinegar, and alcoholic beverages, idli, jalebi, dosa (Table 1.1) have become an art in the kitchens of all Indians.

Traditional biotechnology refers to conventional biotechnology that has been used for many centuries. Beer, cheese, wine, and many other foods have been produced using traditional biotechnology. Thus, traditional biotechnology includes processes that are based on traditional methods.

It includes the process that is based on the natural capabilities of microorganisms. It has established a huge and expanding world market. In monetary terms, it represents a major part of all biotechnology financial profits.

There is one story related to the butter "*Makhan Chor*" (butter stealing) by Lord Sri Krishna during the Mahabharat period. The butter has been produced following the same kitchen method or art.

Table 1.1. *Fermented Foods Prepared in Different Parts of India*

Types of Foods	Substrate	Regions	Quality and Uses
Wari	Black gram	North	The spongy cake used as a snack
Vadai (wada)	Black Gram	North	Deep-fried cake is used as a snack
Tharra	Mahua	North	Sweet alcoholic beverages obtained through distillation
Tari	Date Palm	East	Sweet alcoholic beverage
Sinki	Radish tap root	Himalaya	Sun-dried sour soup pickles
Srikhand	Milk	West	Concentrated sweetened preparation
Rabdi	A mixture of butter-milk/ wheat/ barley/ pearl millet	West	The cooked paste is used as a Staple food
Paneer	Milk	North	Soft cheese used as fried curry
Papad	Black gram flour (besan)	North	Circular wafers used as a snack
Nan	Wheat Flour (Maida)	North	Leavened flat-baked bread is used as a staple food
Mishti dahi	Milk	East	Thick sweet gel
Mesu	Bamboo shot	Himalaya	Sour pickles
Khalpi	Cucumber	Himalaya	Sour pickles
Khaman	Bengal gram	West	Spongy cake, used in breakfast
Jalebi	Wheat Flour (Maida)	North	Crispy, deep-fried used as sweet confectionery
Idli	Rice and Black gram	South	Steamed Spongy cake used in Breakfast

Gundrum	Leafy Vegetables	Himalaya	Sun-dried, sour taste, used as-soup or pickles
Dhokla	Bengal-gram	West	The spongy cake used as a snack
Dahi	Milk	North	Sour, thick gel with Whey
Dosa	Rice and black gram	South	Spongy, shallow fries used as a staple food
Chhurpi	Milk	Himalaya	Cheese like mild sour, soft mass used as curry
Ambali	Millet and rice	South	Steamed cake used in snack
Bhatura	Wheat flour (Maida)	North	Flat-fried, leavened bread used with chhola

The Egyptians during the 2000 BC period used to prepare vinegar from crushed dates by keeping it for a longer time. But the crushed dates produce intoxicants at first. In Egypt, Mesopotamia, and Palestine d□ring 1500 BC, the art of production of wine from crushed grapes, and beer from germinated cereals (malt) using bread leaven (a mass of yeast) was established. In Indian Ay□rveda, the prod□ction of "Asava" □sing different s□bstrates and flowers of mahua (*Madhuca indica)* or dhataki (*Woodfordia fruticosa*) has been well characterized till today since the Vedic period. In these methods, various substrates are transformed into a number of products. Hence, the odor, color, and taste of final prod□cts are changed. Moreover, the □se of salts for the preservation of different foods has been earlier developed even in Europe. The preservation methodology is still of the mummies of Egypt are noteworthy. The possible m□mmification has been done thro□gh dehydration of a dead body followed by the use of a mixture of salts, largely sodium carbonate. Traditional technology is an art rather than a science.

The cause of fermentation could be discovered after observing microorganisms using a microscope by Antony van Leeuwenhoek (1673–1723) in Delft (Holland). A significant contrib□tion was made by chemists to the process and the products of fermentation during the 18th century. The milky precipitate was obtained when the gas evolved from fermentation was passed through the limewater in 1757. It was called the limewater test.

A similar gas comes out from the burning of charcoal. Henry Cavendish demonstrated that the gas evolved from brown sugar in water treated with yeast and was absorbed by sodium hydroxide solution. Scheele & Priestley in 1772–1774 confirmed the identity of oxygen gas present in air. Using analytical techniques for carbon estimation, Antoine Lavoisier gave the chemical basis of alcoholic fermentation. A Frenchman, Nichola Appert in 1810 described the method of food preservation. In the same year, Peter Durand also gave the use of tin containers for food preservation. It was done by putting an air-tight vessel containing food material in boiling water. It increased the importance of the canning industry. Lack of oxygen in such a closed and heated vessel was reported by Gay Lussac. He concluded that oxygen is required for the initiation of alcoholic fermentation, but not for the further progress of fermentation.

Charles B. Astier introduced the concept that air is the carrier of all kinds of germs after 1830. Theodore Schwann, after a series of experiments, demonstrated that the development of fungus (sugar fungus) on fruit juice ca□ses fermentation. He was the first to observe and describe yeast in the growth process. Charles Cagiard-Latour (1838) observed yeast budding using a microscope with 300–400 power magnification.

Justus von Liebig, a well-known chemist, proclaimed in 1839 that all the activity of yeast cells was the result of chemical and physical reactions occurring in the medium. Microbes, also known as microorganisms, are invisible creatures that are too small to be seen with the naked eye, requiring a microscope for study. They are diverse, present everywhere, and exist in a variety of environments, from mountains and volcanoes to deep-sea vents and hot springs. *Thermus aquaticus* is an example of an unusually heat-stable bacterium isolated from hot springs, from which the heat-resistant enzyme *Taq* polymerase was derived.

Microorganisms can be found everywhere, including in the air, water, the food we eat, and even in our bodies. In fact, we come into contact with countless microorganisms every day. They play a vital role in the earth's ecology, decomposing dead plant and animal bodies and recycling chemical elements for use by living plants and animals. Some are commercially used in the production of important chemicals such as alcohol, organic acids, and antibiotics. They are also responsible for food fermentations, enhancing quality and shelf life. The use of microbes to produce various products such as different foods and chemicals is known as biotechnology. By using recombinant DNA, microbes can produce essential substances like proteins, vaccines, and enzymes. Microbes include bacteria, archaea, fungi, algae, protozoa, viruses, viroids, and prions. Bacteria and archaea are classified as prokaryotes, a term introd□ced by Edorand Chatton for cells in which the nuclear material is not surrounded by a haploid.

The fungi, algae, and protozoa are eukaryotic organisms whose genomic DNA is contained in a nucleus surrounded by a membrane. Others don't have reg□lar cell str□ct□res and are classified separately (Aneja, 2014). Microbes range in size from ultramicroscopic viruses 20 nm in diameter to large protozoans 5 mm or more in diameter. In simple words, the largest microbes are as much as 250,000 times the size of the smallest ones. The study of all organisms in the microscopic range is called microbiology. The naming of organisms, called the binomial system of nomenclature, was devised by Carolus Linnaeus in 1735. In this system, each living organism is assigned two names – the genus and the specific epithet, both of which are □nderlined or italicized, e.g., *Escherichia coli*. Microorganisms are relevant to all of us in different ways. Sometimes the infl□ences of microorganisms in h□man life are beneficial, while other times they are detrimental or harmful to human beings. For example, microorganisms are required for the production of cheese, beer, bread, yogurt, alcohol, wine, antibiotics, vitamins, vaccines, and many other important products that are useful in human life.

Microorganisms are essential components of our ecosystem. Fungi and bacteria play an important role in the recycling of organic and inorganic materials through their involvement in the carbon, nitrogen, and sulfur cycles. They also serve as a source of nutrients at the base of all ecological food chains and webs. A recent study by Professor Jonathan Rhodes and his colleagues at the University of Liverpool in North West England reveals that common edible mushrooms contain a protein lectin that can inhibit cancer cell multiplication. This discovery could potentially lead to new targets for therapy. *Taxomyces andreanae*, an endophytic fungus found in the U.S., is being utilized to produce taxol, an antitumor diterpenoid used in the treatment of certain cancers. Taxol was originally derived from the bark of the Pacific yew (*Taxus brevifolia*). In 1347, the plague, also known as the Black Death, struck Europe and within 4 years claimed the lives of 25 million people, which was one-third of the population. Subsequently, the disease resurfaced multiple times, ultimately wiping out 75% of the European population by 1431. This disease is believed to have had a profound impact on European culture. German bacteriologist Robert Koch (1843–1910) first demonstrated in 1876 that anthrax is ca□sed by the microbe *Bacillus anthracis*. In the past 30 years, there has been a rise in new and emerging infectious agents such as the AIDS virus (HIV – human imm□nodeficiency vir□s), *Escherichia coli*, and several other previously known diseases that are resurfacing. Recent studies have shown correlations between gastric ulcers and the bacterium *Helicobacter* that invades the stomach, between type 1 diabetes and certain coxsackie viruses, and between coronary artery disease and cytomegalovirus infection.

Microbes play a role in the decomposition of waste material and help other animals digest grass such as cattle, sheep, and termites, so they can also balance nature. Microbes affect human health. The environment provides the development of biodegradable products and safe drinking water, which use bacteria to clean up oil spills through bioremediation. Their participation in agricultural research has led to healthier livestock and disease-free crops. Microbes play a role in industries such as foodstuffs, beer, wine, cheese, bread, antibiotics, genetic engineering, and insulin.

It is a specialized area of biology that deals with the study of microorganisms or microbes. Roughly speaking, organisms with a diameter of one mm or less are microorganisms and fit into the broad domain of microbiology (Garg et al., 2013; Aneja, 2014). The branches of microbiology are categorized into pure and applied sciences:

- **Bacteriology:** It is the study of bacteria.
- **Mycology:** It is the study of fungi.
- **Protozoology:** It is the study of protozoans.
- **Virology:** It is the study of viruses.
- **Algology or Phycology:** It is the study of algae.
- **Parasitology:** It is the study of parasitism or parasites.
- **Microbial Ecology:** It is the study of the interrelationship between the environment and microbes.
- **Microbial Morphology:** It is the study of the detailed structure of microbes.
- **Microbial Taxonomy:** It is concerned with the classification, naming, and identification of different microbes.
- **Microbial Physiology:** It is the study of the metabolism of microbes at the cellular and molecular levels.
- **Microbial Genetics:** It is the study of the genetic material of microbes.
- **Molecular Microbiology:** It is the study of the structure, function, and biochemical reactions of microbial cells involved in metabolism and growth.
- **Industrial Microbiology:** It is concerned with the industrial uses of microbes in the production of vitamins, alcoholic beverages, amino acids, enzymes, antibiotics, and other drugs.
- **Food Microbiology:** It is the study of the interaction between food and microbes in relation to food bioprocessing, food spoilage, and their prevention.
- **Agricultural Microbiology:** It is the study of the interrelationship of crops and microbes with an emphasis on the control of plant diseases and improvement of yield.
- **Dairy Microbiology:** It deals with the production and maintenance

of quality control of dairy products.

- **Aquatic Microbiology:** It is the study of microorganisms and their activity concerning human and animal health in fresh, estuarine, and marine waters.
- **Air Microbiology:** It deals with the role of aerospora in contamination and spoilage of food and discrimination of plant and animal disease by air.
- **Exomicrobiology:** It is the study of exploration for microbial life in outer space.
- **Medical Microbiology:** It deals with the fundamental principles and techniques involved in the study of pathogenic organisms as well as their application in the diagnosis of infectious diseases.
- **Immunology:** It deals with the immune system that protects against infections and attempts to understand the many phenomena that are responsible for both acquired and innate immunity.
- **Biotechnology:** It is the scientific manipulation of living organisms, especially at the molecular and genetic level, to produce useful products.
- **Public Health Microbiology:** It is concerned with monitoring, controlling, and spreading diseases in communities.

The first living things to appear on Earth and the study of fossil remains indicate that microbial infections and epidemic diseases existed thousands of years ago. It was Roger Bacon (1220–1292) who in the 13th century postulated that disease is produced by invisible living creatures. This idea was reiterated in 1546 by the physician Girolamo Fracastoro (1478–1553) of northern Italy. Fracastoro wrote a treatise, "De Contagione," in which he stated that disease was caused by minute "seeds" or "germs" and was spread from person to person. As early as 1658, a monk named Athanasius Kircher (1601–1680) referred to "worms" invisible to the naked eye in decaying bodies, meat, milk, and diarrheal secretions. He examined thin cork slices and the bark of an oak tree, and found that cork was made of tiny boxes that Hooke referred to as "cells." The work of Hooke was followed by Matthias Schleiden and Theodor Schwann, who examined a variety of organisms and in 1838–1839 reported that all forms of life are composed of cells. This observation later became the foundation of the Cell Theory. The cell theory has since been modified to state that all life is composed of cells that originate from other cells. The discovery of microbiology as a discipline can be traced through the following historical eras.

It is the period concerned with the discovery of a microbial world that has been dominated by Antony van Leeuwenhoek. He was the first person to observe and accurately describe microorganisms in 1676. He also examined blood and other human tissues including his own tooth scrapings, insects, minerals, and plant materials. He was the first person to describe bacteria and protozoa using a microscope he made himself. Due to this extraordinary contribution to microbiology, he is considered the father of bacteriology and protozoology.

Transition period – Later in Leeuwenhoek's time, there are two noteworthy contributions: (1) controversy over spontaneous generation, which stated that living organisms could develop from non-living matter, called abiogenesis, and (2) disease transmission. Francesco Redi (1626–1697) put the theory of spontaneous generation to rest in 1665 by conducting an experiment with meat in three jars. One jar was covered with fine ga□ze, the second with paper, and the third was left uncovered. Flies entered the uncovered jar and laid eggs on the meat, which later developed into maggots. The other two pieces of meat did not produce maggots spontaneously. Flies were attracted to the gauze-covered jars and laid eggs on the gauze, where maggots developed without access to the meat, indicating that maggots were the offspring of the flies and did not arise from the meat as previously believed.

John Needham (1713–1781) was a strong supporter of the theory of spontaneous generation. In 1749, he proposed that tiny organisms, the animalc□les, arose spontaneo□sly in his m□tton gravy. He covered the flasks with cork, similar to Redi, and even heated some flasks. Still, the microbes appeared in the mutton broth. Lazzaro Spallanzani (1729–1799) attempted to refute Needham's work in 1775 by conducting experiments. He boiled beef broth for a longer period, removed the air from the flask, and then sealed the container. After the incubation period, during which he followed, he observed no growth in these flasks. When acc□sed of destroying the vegetative force of the nutrients by overheating, he demonstrated that the heated nutrients could still grow animalcules when exposed to air by simply making a small crack in the neck, thus disproving spontaneous generation. Georg Schroeder and Theodor von Dusch, in 1854, conducted convincing experiments to disprove the theory of spontaneo□s generation by passing air thro□gh cotton into flasks containing heated broth. No growth was observed in the infusions due to the filtering o□t of microscopic organisms by cotton. Schroeder was the first to introduce the idea of using cotton plugs to seal microbial culture tubes in 1854. He ultimately resolved the controversy between spontaneous generation and biogenesis, proving that microorganisms are not generated spontaneously from inanimate matter but arise from other microorganisms. John Tyndall (1820–1893) delivered the final blow to spontaneo□s generation in 1877. He conducted experiments in an aseptically designed box to demonstrate that dust indeed carried germs. Tyndall showed that if no dust was present, sterile broth remained free of microbial growth for an indefinite period even when directly exposed to air.

The diversity of our life is present on our planet with new species of all types being constantly discovered. The formal system of organizing classification and naming of living things is called taxonomy. Carolus Linnaeus in 1735 also gave the binomial system of nomenclature, i.e., naming of an organism by two names such as genus and species. Therefore, Linnaeus' system provides each organism names and arranges them into groups called taxa with other similar organisms to reflect their phylogeny and evol□tionary relatedness. The main taxa or gro□ps in a classification scheme are organized in several descending ranks beginning with the kingdom, the largest, and followed by "Phylum or Division, Class, Order, Family, Genus, and Species."

The classification of microorganisms began in 1674 with the invention of the light microscope and today is a discipline based on increasingly complex criteria. The first phylogenetic trees of life were constr□cted on the basis of just two kingdoms such as Plantae and Animalia. In 1886, a German scientist, Ernest Haekel, proposed that the bacteria, algae, fungi, and protozoa that lacked tissue differentiation be removed from the plant and animal kingdoms and be separated into a third kingdom, Protista. In 1940, in some microorganisms, e.g., typical bacteria, the genetic material was not enclosed by a nuclear membrane. This remarkable discovery, the absence of a nuclear membrane-bound internal structure in one group of Protista (bacteria) and the presence of a bound structure in the majority of others such as algae, fungi, and protozoa, was discovered significantly. This res□lts in the division of these organisms into two gro□ps: Prokaryotes and E□karyotes. Therefore, several systems of classification of living organisms have been developed between 1940 and 1969, but the one that is most widely accepted was developed in 1960 by Robert Whittaker of Cornell University. This system places all living things except vir□ses into five kingdoms based on cellular organization and nutritional patterns such as (1) prokaryotic or Monera; (2) Protista; (3) Myceteae or Fungi; (4) Plantae; and (5) Animalia.

There are three kingdom domains which were proposed by the scientists Carl Woese and George Fox. They have proposed a new system of classification that assigns all organisms to one of the three superkingdoms (Figure 1.1) called domains such as (1) Archaea; (2) Bacteria; and (3) Eukaryota. It is believed that these three kingdoms arose from an ancestor most similar to the archaebacteria.

Several classifications have been proposed for bacteria, and Bergey's Manual of Systematic Bacteriology has wide acceptance. The Bergey's Manual of Determinative Bacteriology, published by Williams and Wilkins Co. in Baltimore, USA (1974), is the international standard for bacterial taxonomy.

It deals with the classification of bacteria and contains a description of all the species known at the time of its publication. In this manual, bacteria are assigned to 19 parts based primarily on the following five feat□res: (1) morphology and gram stain reaction; (2) gaseous requirements; (3) energy and carbon sources; (4) mode of locomotion; and (5) ability to produce endospores. It serves as a practical g□ide for the identification of bacteria. After some time, the man□al was revised between 1984 and 1989 and appeared with a new name, Bergey's Manual of Systematic Bacteriology.

In these fo□r vol□mes, bacteria have been classified into 33 sections based on characteristics such as general shape and morphology, gram staining properties, presence of endospores, oxygen relationship, motility, and mode of energy production.

Therefore, in the four volumes, the bacterial groups are divided as follows: (1) gram-negative bacteria of general, medical, and industrial importance; (ii) gram-positive bacteria other than actinomycetes; (iii) gram-negative bacteria with distinctive properties, cyanobacteria, and archaeobacteria; and (iv) actinomycetes (gram-positive filamento□s bacteria). The Bergey's Man□al of Systematic Bacteriology places all bacteria in the kingdom of prokaryotes, which in turn is divided into four divisions as follows: The Gracilicutes (thin skin): Prokaryotes with complex cell wall structure characteristics of gram-negative bacteria, e.g., Non-photosynthetic bacteria. Firmicutes (Strong skin): Prokaryotes with cell wall structure characteristics of gram-positive bacteria (Rods and cocci, Actinomycetes, and related organisms). Tenericutes (soft or tender skin): Prokaryotes that lack a cell wall (e.g., mycoplasms). Mendosicutes (Skin with faults): Prokaryotes with unusual cell walls, e.g., *Archaeobacteria*. There are gro□ps that do not reflect the phylogenetic relationship between the organisms; each division is separated into classes for a total of 7 classes of bacteria. The system used in Bergey's manual recognizes 7 classes under the 4 divisions.

1.6. GROUPS OF BACTERIA AS PER BERGEY'S MANUAL

- **Division I:** Gracilicutes gram-negative bacteria:
  - **Class I:** Scotobacteria gram-negative non-photosynthetic bacteria.
  - **Class II:** Anoxyphotobacteria gram-negative photosynthetic bacteria that do not produce oxygen (green or purple bacteria).
  - **Class III:** Oxyphotobacteria gram-negative photosynthetic bacteria that evolve oxygen (cyanobacteria).

- **Division II:** Firmicutes gram-positive bacteria:
  - **Class I**: Firmibacteria gram-positive rods or cocci.
  - **Class II:** Thallobacteria gram-positive branching cells (actinomycetes).
- **Division III:** Tenericutes:
  - **Class I:** Mollicutes bacteria lacking a cell wall (mycoplasmas).
- **Division IV:** Mendosicutes:
  - **Class I:** Archaeobacteria bacteria with typical compounds in the cell wall and membranes.

Bacteria (singular bacterium) are single-celled (unicellular), prokaryotic organisms. They reproduce asexually, are haploid, and do not have their DNA localized within a nucleus surrounded by a unit membrane (i.e., without a nucleus). Bacteria can be gram-positive and gram-negative. They lack the membrane-enclosed intracellular structures found in most other cellular organisms. The shapes of bacteria are spherical or ovoid (coccus), rod-like (bacillus), corkscrew or c□rved (spiral), comma-shaped, helical, and filamento□s. Individ□al bacteria may form pairs, chains, clusters, or other groupings. Such arrangements are usually characteristic of a particular genus or species of bacteria. All bacteria have a cell wall composed of a carbohydrate and protein complex called peptidoglycan. Bacteria reproduce asexually by a process called binary fission, i.e., by dividing into two eq□al cells. They possess flagella, moving appendages for their motility. For nutrition, most bacteria absorb nutrients from their environment from dead or living organisms; some can manufacture their own food by photosynthesis, and some can derive nutrition from inorganic substances. The study of bacteria is called bacteriology.

1.7. ARCHAEA BACTERIA

They are like bacteria, consisting of unicellular prokaryotic cells, but do not have peptidoglycan in their cell walls. They often live in extreme environments and carry out unusual metabolic processes. These are found in hot springs, geothermally heated marine sediments, and submarine hydrothermal vents. They have no nucleus and any other organelles. It is classified as bacteria and called archaebacteria. They are further divided into phyla. It has the same structure size as bacteria such as flat and square-shaped cells. They use more energy sources than eukaryotes such as sugar, metal ions, ammonia, or even hydrogen gas. They reproduce asexually by budding, fragmentation, or binary fission (Figures 1.1 and 1.2). It ranges from 0.1 μm to 15 μm in diameter and various shapes such as rods, spheres, spirals, or plates.

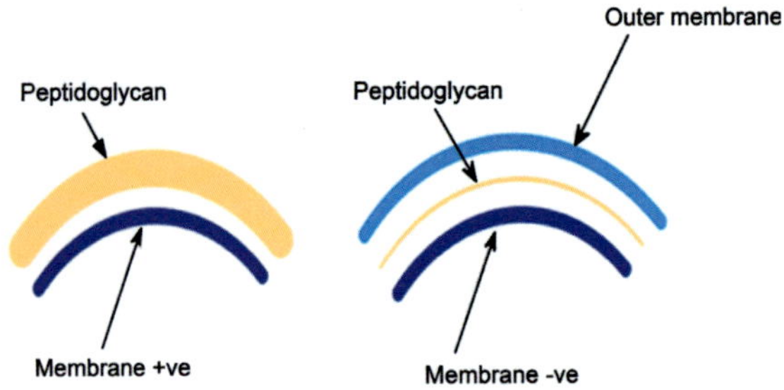

Figure 1.1. *Gram-positive and gram-negative bacteria membrane.*

Source: Made by the author.

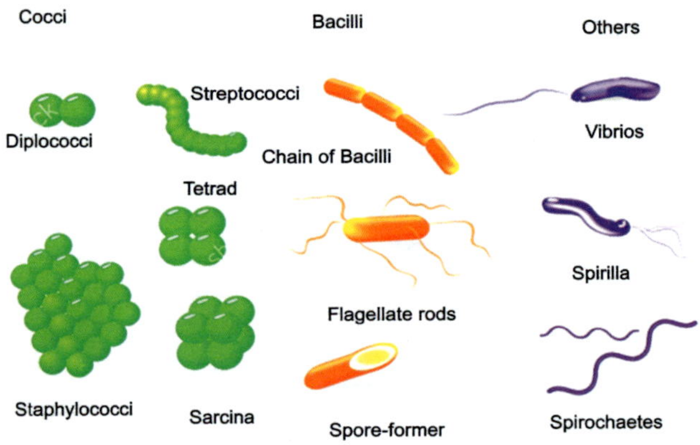

Figure 1.2. *Cell shapes in bacteria and archaea.*

Source: Made by the author.

1.7.1. Fungi

Fungi are eukaryotic with a nucleus containing the genetic material (DNA) surrounded by an envelope, the nuclear membrane. True fungi have cell walls composed of chitin, a polysaccharide also found in the exoskeleton of arthropods such as ticks and spiders. They reproduce both asexually and sexually. Fungi generally have a mycelium, a loosely organized mass consisting of thread-like hyphae. They consist of three groups of organisms – yeasts, molds, and mushrooms (Figure 1.3). The study of fungi is called mycology.

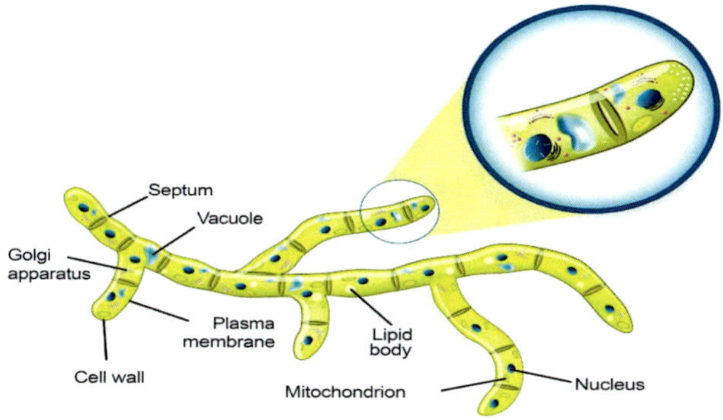

Figure 1.3. *Morphology of fungi. Source: Made by the author.*

1.7.2. Protozoa

Protozoa are unicellular eukaryotic organisms that have no cell wall, no photosynthesis, no fruiting bodies, reproduce sexually and asexually having various ways of locomotion such as pseudopodia, flagella or cilia. Protozoa are classified into four groups on the basis of their motility. Protozoa are found in soil and water and as normal microbiota of animals.

1.7.3. Algae

Algae are photosynthetic eukaryotes containing chloroplast, an organelle carrying out oxygenic photosynthesis. The body of algae is called thallus. Most algae are single-celed, simple microscopic organisms. But some are multicellular which can attain an enormous size. The cell walls of most algae are composed of cellulose. They reproduce both sexually and asexually. Algae are abundant in fresh water and salt water, in soil and in symbiotic association with fungi called lichens. The study of algae is called algology (phycology).

1.7.4. Viruses

Viruses are acellular (non-cellular) entities, ultramicroscopic, ranging in size from 30 nm to 300 nm. Hence they can be seen only with an electron microscope. They lack the characteristics of life because of that viruses are considered to be non-living entities. A virus particle (called virion) has either DNA or RNA, but never both. The DNA and RNA found in viruses may be single-stranded or double-stranded. Indeed, viruses can be crystallized and stored in a container on a shelf for years, but they retain the ability to invade cells.

Viruses can reproduce only by using the cellular machinery of their host. Viruses are obligate intracellular parasites of bacteria (bacteriophages), protozoa, fungi, algae, plants, and animals and are responsible for diseases of h□mans (AIDS, polio, smallpox, swine fl□, H1N1 infl□enza, bird fl□) and plants (tobacco mosaic, ca□liflower mosaic) (Fig□re 1.4). The st□dy of vir□ses is called virology. Viroids are small circular pieces of naked single-stranded RNA, only 300 to 400 nucleotides long, with no protein coat. They are the smallest known infective agent causing diseases in plants only. They retard the growth of plants on which they multiply, interfering with the mRNA processing. One of the worst diseases caused by viroids is potato spindle tuber viroid, causing losses of millions of dollars from crop damage.

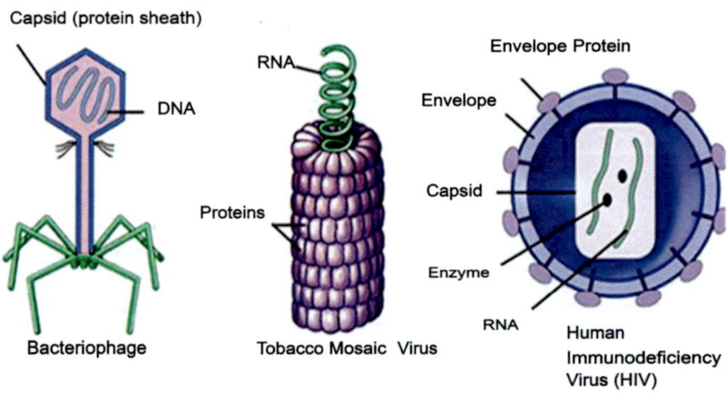

Figure 1.4. *Viruses-bacteriophage, tobacco mosaic virus, and human immunodeficiency virus (HIV), single-stranded RNA virus with helical symmetry.*

Source: Made by the author.

1.7.5. Prions

Prions, also called virions, are proteinaceous infectious particles having no nucleic acid. They cause infectious diseases; the most deadly disease in cattle is the "mad-cow disease," also called bovine spongiform encephalopathy (BSE), and Creutzfeldt-Jakob disease (CJD) in humans, which involves the degeneration of brain tissue. Prion diseases are the result of altered proteins; the cause can be a mutation in the normal gene for PrPc (normal host glycoprotein) or contact with an altered protein (PrPsc) (Figure 1.5).

The st□dy of microbiology started with the first report by Lo□is Paste□r (1857) on lactic acid fermentation from sugar. He isolated the microorganisms (lactic yeast) associated with lactic acid and curd formation. The cells of lactic yeast were smaller than those of beer yeast.

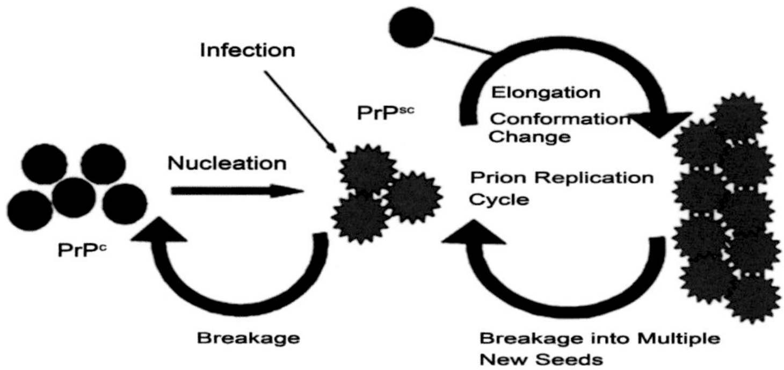

Figure 1.5. *Prion replication cycle.*

Source: Made by the author.

Lactic acid production increased when he added chalk powder to the fermentation medium. Pasteur showed the presence of lactic acid in curd by using a polarimeter. In 1860, Pasteur provided a detailed report on the use of synthetic medium for microbiological studies. He concluded that there are many steps in fermentation, such as:

- Fermentation is carried out anaerobically by the living cells.
- During the alcoholic fermentation in a synthetic medium, the yeast increased the weight with an increase in the carbon and nitrogen contents of the overall batch.
- The increase in yeast protein in the synthetic medium was accompanied by a related decrease in the ammonium nitrogen in the medium.
- Fermentation sugar was required for the multiplication of yeast cells.
- Similar phenomena occurred in the fermentation of lactic acid, tartaric acid, butyric acid, etc.
- When a pure culture is not used, some of the fermentation processes are stopped.
- The growth and physiology of yeast differ when they are grown under anaerobic growth conditions. A large amount of sugar was converted into alcohol under anaerobic growth conditions, while under aerobic conditions, a large amount of sugar was converted into yeast cell mass.

Pasteur suggested that a high percentage of the microbial population is killed by heating the juice at 62.8°C, which is now called pasteurization. Robert Koch, in 1881, gave the method for establishing a relationship between a pathogen and a disease, known as "*Koch's postulates*" or "pathogenicity test." He proved that anthrax disease is caused by *Bacillus anthracis*.

The first demonstrated enzymatically mediated fermentation reaction was by Edward Buchner in 1897. He showed that the cell-free yeast juice mixed with concentrated sugar solution evolved carbon dioxide and produced ethyl alcohol upon incubation. A similar product was also produced in an aqueous solution of other sugars such as glucose, sucrose, fructose, and maltose. Buchner called the dissolved substance in juice responsible for sugar fermentation "zymase."

This work showed improved techniques of fermentation. Fermentation could also be demonstrated in test tubes without a living organism. The role of chemistry had more significance in □nderstanding the fermentation phenomenon than microbiology.

Hans Buchner and Martin developed a more effective method of obtaining cell-free extracts after disrupting the microbial cells. They ground the microbial cells with quartz and added Kieselguhr. Thus, sucrose fermentation using cell-free extracts from yeast cells obtained by the above method was demonstrated.

The discovery of viruses and their role in disease was made possible when Charles Chamberland constr□cted a porcelain bacterial filter in 1884. Edward Jenner used vaccination in 1798 by extracting liquid material from cowpox lesions and introducing it into people with smallpox. Pasteur and Chamberland developed toxins produced by *Corynebacterium diphtheriae* (which causes diphtheria in humans); its antitoxin was developed by Emil von Behring in 1890 and Shibasaburo Kitasato. They injected the inactivated toxin into rabbits, which induced the production of antitoxin in the blood. The antitoxin inactivated the toxin and protected against the disease. Similarly, tetanus antitoxin was developed by Behring.

Discovery of viruses and their role in disease was made possible when Charles Chamberland (1884) constr□cted a porcelain bacterial filter. Edward Jenner (1798) used it to introduce it into people with smallpox. Pasteur and Chamberland later developed an attenuated anthrax vaccine against anthrax disease. After the discovery of toxins produced by *Corynebacterium diphtheriae*, a gram-positive bacillus bacteria that causes diphtheria in humans, Klebs demonstrated that Corynebacterium diphtheriae was the agent of diphtheria in 1883. It is most commonly an infection in the upper respiratory tract and causes fever, sore throat, and malaise. Roux and Yersin showed that animals injected with sterile filtrates of *C. diphtheriae* developed organ pathology indistinguishable from that of human diphtheria; this demonstrated that a potent exotoxin was the major virulence factor in 1888.

1.8. MODERN BIOTECHNOLOGY

There are two major features that differentiate modern biotechnology from classical biotechnology, such as:

- The capability of science to change genetic material to obtain new products for specific requirements through recombinant DNA technology; and
- The ownership of technology and its socio-political impact.

Now, conventional technology industries, pharmaceutical industries, agro-industries, etc., are focusing their attention on producing biotechnology-based products.

1.9. EMERGENCE OF MODERN BIOTECHNOLOGY

Modern biotechnology encompasses all methods of genetic modification through recombinant DNA and cell fusion technologies. It also includes the modern developments of traditional biotechnology processes. The new aspects of biotechnology, founded on recent advancements in modern biology, genetic engineering, and fermentation process technology, are now increasingly finding wide industrial applications. But the rate of application will depend on:

1. Adequate investment by the industries;
2. Improved system of biological patenting;
3. Marketing skills;
4. Economics of the new methods;
5. Public perception about the biotechnology products.

The traditional biotechnology associated with fermentation was gradually industrialized at the end of the 19th century. This resulted in the gradual growth of industries producing beer, whisky, wine, canned food, rum, etc. The first time in 1920, the Leeds City Council (UK) established the Institute of Biotechnology. In the late 1960s, the OECD was set up to promote policies for the sound economic growth of the member countries. In 1978, The European Federation of Biotechnology was set up.

Biotechnology brought industries and agriculture together in the 20th century. The fermentation process was developed, which produced acetone from starch and paint solvent from automobile industries during World War I. The antibiotic penicillin was discovered during World War II. Manufacture of penicillin shifted the biotechnological focus towards pharmaceuticals. Linking fermentation with biochemistry, bioprocess, chemical engineering, and instrument designing helped substantially in the progress of industries.

The work on microorganisms dominated the preparation of biological warfare, antibiotics, and fermentation processes. Suspected preparation of biological and chemical warfare led to the US attack on Iraq in 2003 during the Gulf War (1991).

After the discovery of the double helix DNA by Watson and Crick (1953), Werner Arber in 1971 discovered a special enzyme in bacteria which he called restriction enzymes. These enzymes can cut the DNA strand and generate fragments. The cut ends of two fragments are single-stranded sticky ends because the single-stranded ends having identical base pairs can rejoin. Chen and Boyer in 1973 removed a specific gene from a bacteri□m and inserted it into another bacterium using restriction enzymes. This discovery marked the start of recombinant DNA technology or genetic engineering.

Baltimore successfully transferred the human growth hormone gene into a rabbit in 1976. The European Federation of Biotechnology was created in 1978. In 1978, a U.S. company "Genentech" used genetic engineering to produce human insulin in *E. coli*. In 1980, a trial of the new hormone was conducted in the USA, France, Japan, and the United Kingdom. The US Food and Drug Administration gave marketing approval to "Humulin," i.e., human insulin made by Eli Lilly (USA) by the end of 1982. Another hormone, somatotropin, was produced on an industrial scale.

The first genetically engineered tomatoes, FlavrSavr, were sold in the market in 1993. In 1996, the first cloned lamb "Dolly" was s□ccessf□lly born thro□gh the efforts of scientists in Scotland. Subsequently, several cloned animals were produced. The sequence of the human genome was published in Nature and Science in 2001. The Human Genome Project was completed by March 2003. In 2002, a designer baby was born to cure the genetic disease of her elder sister. A claim was made for the birth of a cloned baby named "Eve" on December 27, 2002, by the scientists of "The Human Cloning Society," Clonaid of France. In February 2005, during a United Nations meeting, many countries opposed gene cloning in humans, while a few countries supported it with a request to allow it for research purposes only.

In May 2005, scientists in South Korea used a method called therapeutic cloning to produce stem cell lines. These are genetic matches to patients. Such stem cell lines could be used for disease research. The U.S.A. condemned this approach. In this method, human embryos were produced through cloning and stem cells were obtained from blastocysts. The excised stem cells could be grown *in vitro* and used further. It is interesting that scientists are engaged in the most challenging task of mass production of growth hormones, insulin, vaccines, immunogenic proteins and polypeptides, gene therapy, biofertilizers, biopesticides, producing disease and stress-resistant plants, biomass, enzymes, antibiotics, acids, fuels, etc. Many biochemical companies such as the National Pituitary Agency (USA), E. Lilly (USA), Kabi Vitram A B (Sweden), Genetech Co. (USA), Biogen (Switzerland), etc., are producing/trying to produce some of the above products using genetic engineering techniques. Many Nobel Laureates, including Dr. Har Govind Khorana, are associated with these companies.

Biotechnology is not a sudden discovery, but rather a coming of age of technology that was initiated several decades ago. By the middle of the 20th century, there had been tremendous growth in the areas of chemistry, physics, and biology. Further knowledge of each branch has been advanced substantially. Multidisciplinary strategies were made for the solution of various problems. A novel spectrum of investigations occurred through true interdisciplinary synthesis. This led to the evolution of biotechnology, which is an outcome of the integrated effort of biology with technology, the root of which lies in biological science. The key difference between biology and biotechnology is their scale of operation. Usually, the biologist works in the range of nanograms to milligrams. Biotechnologists working on the production of vaccines may be satisfied with milligram yields, b□t many other projects aim at kilograms or tons. Thus, the main objective of biotechnologists consists of scaling up the biological processes (Smith, 1996).

The main disciplines of biology are microbiology, biochemistry, genetics, molecular biology, immunology, and cell and tissue culture (Figure 1.6). However, in the field of engineering, there are processes s□ch as large-scale cultivation of microbes and cells, their upstream and downstream processes, etc. These processes can be separated into five major operations:

- Strain selection and improvements;
- Mass culture;
- Optimization of cell responses;
- Process operation; and
- Downstream processing.

Many areas of biotechnology have arisen through the interaction between various parts of biology and engineering, biochemistry, biophysics, cell biology, colloid chemistry, embryology, ecology, genetics, immunology, molecular biology, medical chemistry, pharmacology, polymer chemistry, thermochemistry, and virology. Modern biotechnology has developed several technologies by extracting basic knowledge from biology. Many areas of biotechnology have arisen through the interaction between various parts of biology and engineering, biochemistry, biophysics, cell biology, colloid chemistry, embryology, ecology, genetics, immunology, molecular biology, medical chemistry, pharmacology, polymer chemistry, thermochemistry, and virology. Modern biotechnology has developed several technologies by extracting basic knowledge from biology (Dubey, 2009).

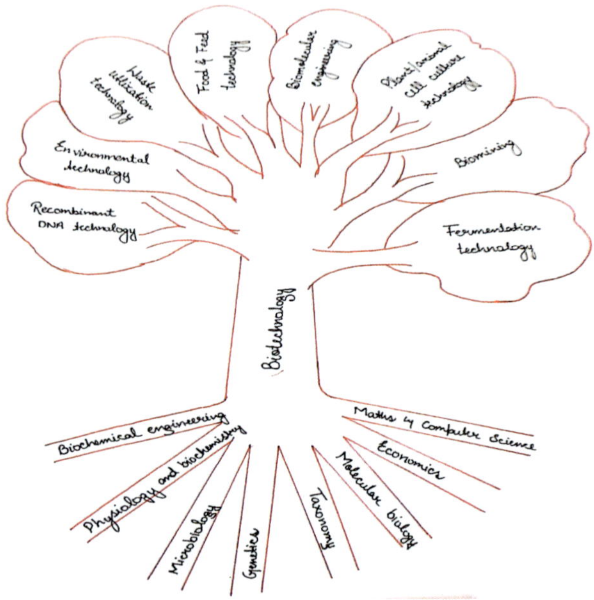

Figure 1.6. *A tree of biotechnology.*

Source: Made by the author.

1.10. GLOBAL IMPACT AND CURRENT EXCITEMENT OF BIOTECHNOLOGY

Each and every organism performs its function within its optimum limits. The excitement about modern biotechnology is that the scientific methods (such as genetic engineering) have enhanced the natural capabilities of the natural production of organisms. What a miracle is that a mouse of the size of a rabbit can be produced. Bacteria like *E. coli* are producing mammalian hormones such as insulin, somatostatin, somatotropin, etc., and yeast cells have been genetically manipulated to produce a vaccine against hepatitis B virus (hepatitis disease). Myeloma cells (cancerous cells of bone marrow) and B-cells of immunized mice were hybridized to produce hybrid cells that consist of the characteristics of both the cells, which were cell division and antibody production. Now the hybrid cells (hybridomas) are being used for the production of monoclonal antibodies. In 1982, interferon (C, B, and Y) was produced by genetically engineered *E. coli* cells. Techniques have been developed to produce rare and medicinally valuable molecules, to change hereditary traits of plants and animals, to diagnose diseases, to produce useful chemicals, and to clean up and restore the environment. In this way, biotechnology has a great impact in the fields of health, food/agriculture, and environmental protection (Ferber, 2004; Kumar, 2003). Due to rapid development, the present situation is that there is no difference between pharmaceutical firms and the biotechnology industry. However, approved products in the pipeline and renewed public confidence have made it one of the most promising areas of economic growth in the future. India offers a huge market for the products as well as a cheap manufacturing base for export (Padh, 1996). Some of the areas where biotechnology has done the best are in further sections.

1.11. BIOTECHNOLOGY IMPACT ON HEALTH CARE

The maximum benefits of biotechnology have been utilized by healthcare. Biotechnology-derived proteins and polypeptides form a new class of potential drugs. For example, insulin was primarily extracted from slaughtered animals. Since 1982, human insulin (Humulin) has been produced by microorganisms in a fermentor. Similarly, hepatitis B vaccines like Recombivax HB (from Merck), Guni (from Shantha Biotechnics Ltd., Hyderabad), Shanvac (Biological E. Laboratory), etc., are genetically engineered vaccines produced biotechnologically. Since 1987, the number of biotechnology-derived new protein drugs has surpassed the number of new chemical drugs (Padh, 1996).

1.12. SOME OF THE IMPORTANT PRODUCTS PRODUCED THROUGH GENETICALLY MODIFIED ORGANISMS

With the advancement of gene manipulation in organisms, science has led to a new revolution in biology called the gene revolution. Obviously, it is the third revolution in science after the industrial revolution and the computer revolution. Thus, the roots of today's biotechnology lie in chemistry, physics, and biology. Currently, there are about 35 biotechnology-derived therapeutics and vaccines approved by the USFDA alone for medical use, and more than 500 drugs and vaccines are on the market. Similarly, about 600 biotechnology diagnostics practices. About 130 gene therapy protocols have been approved by US authorities. Imports of many immunodiagnostic tests are available worldwide in clinical settings. The major landmark in human history is the human genome sequence. The HGP is an international research program. Almost the whole human genome has been sequenced and a chromosome map has been developed in various laboratories worldwide through coordinated efforts. Human chromosome mapping was completed by March 2003. There are about 3,000 functional genes in humans. More than 97% of genes are non-functional. They do not encode for any polypeptide chain. Objectives of the Human Genome Project are to: construct the detailed genetic and physical map of the human genome, (i) determine the complete nucleotide sequence of human DNA, (ii) store information in the database, (iii) locate the estimated 50,000–10,000 genes within the human genome, (iv) address the ethical, legal, and social issues (ELSI) that may arise from the project, and (v) perform similar analysis on the genomes of several other organisms.

1.13. BIOTECHNOLOGY IMPACT ON THE ENVIRONMENT

The natural biodegradability of pollutants present in the environment has increased with the use of biotechnology. Bioremediation technologies have been found successful in combating pollution problems. Bioremediation is the use of microorganisms to detoxify pollutants present in the environment, usually as soil or water sediments. The pollutants cause several health problems. Microorganisms that show the potential for the degradation of oil, pesticides, and fertilizers belong to the genera of bacteria *Pseudomonas, Micrococcus, Bacillus, and fungi Candida, Cladosporium, Torulopsis, Trichoderma*, etc.

1.14. GENOMICS

Computer-based study and designing of the genome is called genomics. Genomics deals with the sequencing of the complete genome of a particular organism. Similarly, the study of proteins present on the genome using a computer is called proteomics. Proteomics can be defined as the study of all the proteins present in the genome of an organism. With the help of proteomics and genomics, new molecules that can interact with other partners could be identified. This gives us deep insight into biological pathways. Now the whole genome is available to biologists. There may be new small molecules as potential drug candidates.

There are 33,000 genes of human beings on a microchip and have helped to design specific dr□gs for genetic diseases, for which there is no c□re so far. For example, a specific gene is expressed in breast cancer patients – A designed drug (Herceptin) is good for the treatment of breast cancer. Thus, the field of genomics has helped the growth of pharmacologic, toxicologic, and protein studies on animals, and the new areas are called pharmacogenomics, toxicogenomics, and proteogenomics, respectively.

1.15. BIOINFORMATICS

It is a new field of biotechnology linked with information technology. Bioinformatics may be defined as the application of information sciences (mathematics, statistics, and computer sciences) to increase the understanding of biology, biochemistry, and biological data. The most remarkable success of bioinformatics to date has been its use in the 'shotgun sequencing' (breaking of a large piece of DNA into smaller fragments) of the human genome.

1.16. BIOTECHNOLOGY IMPACT ON AGRICULTURE

Biotechnology is making new ground in the food/agriculture area. Current public debate about BST bovine somatotropin (a hormone administered to cows to increase milk production) typifies an example of biotechnology product testing public acceptance. Similarly, the FlavrSavr tomato (produced by transgenic plants engineered antisense technology to preserve flavor, texture, and quality) is a new breed of value-added foods. Food biotechnology offers valuable alternatives to food problems and a solution to nutritionally influenced diseases such as diabetes, hypertension, cancer, heart diseases, arthritis, etc. Transgenic golden rice has been produced by introducing three genes for the production of vitamin A in Taipei's rice. *Bacillus thuringiensis-resistant* transgenic Bt plants have been produced by introducing the insecticidal toxin-producing Bt gene of *Bacillus thuringiensis* into the desired plants. A transgenic cotton named 'Ingaurd' was released in Australia, which contained the Bt gene and provided resistance against insects. Bt crops are transgenic crops that produce toxins such as the bacterium *Bacillus thuringiensis* in the plant cell because they protect crops from pests. When insects feed on transgenic plants, the toxic cry protein creates problems in the digestive system of insects, even leading to death, but it has no side effects on the human digestive system (Ferber, 2004).

1.17. BIOTECHNOLOGY IN INDIA

In most of the developing countries, recombinant DNA technology has become the major focus. In 1982, the Government of India established an official agency, the National Biotechnology Board (NBTB), which operated under the Department of Science and Technology (DST). In 1986, NBTB was replaced by a full-fledged department, the Department of Biotechnology (DBT), under the Ministry of Science and Technology to plan, promote, and coordinate various biotechnological programs. The DBT is actively involved in promoting postgraduate education and research, offering special MSc courses in Biotechnology. Student selection is based on a National Test, and the department also provides trained manpower for the rapidly growing biotech industry. It has enhanced the level of biology education in specific areas of biotechnology in the country, importing a significant amount of basic biochemical and molecular biology into these courses. Currently, India has the DBT, DST, CSIR, ICAR, ICMR, IARI, and other agencies operating under the government (Purohit, 2004). These agencies, along with other national and international industries, manufacture biotech products and market them after clinical trials. A Technology Development Board (IDB) has been established by the government, working with universities, industries, and national institutes. The Technology Information Forecasting and Assessment Council (TIFAC) has developed a vision 2020 document that includes biotechnology as well.

There are over 30 companies in India that are producing modern biotech products such as Wipro, Reliance, Life Science, and Pennetia Biotech Ltd. Since the 1980s, India has supported the biotechnology industry and its products significantly. Teaching and research of biotechnology have been included in the university's syllabi at both undergraduate and postgraduate levels. DBT-supported departments are operating in several universities and institutions. It is predicted that India will play a key role in the future as one of the largest markets and producers of biotech products.

The United Nations Industrial Development Organization (UNIDO) recognized the potential of genetic engineering and biotechnology in promoting the economic progress of developing countries. The initiative taken by UNIDO led to the establishment of ICGEB. In 1981, during a meeting convened by UNIDO, it was proposed to establish an international center of excellence to promote biotechnology in the developing world. This concept was approved in 1982 by a high-level conference of developed and developing nations in Belgrade. The statutes of the center were signed by 26 countries, and the statutes came into force on February 3, 1994. The ICGEB has become a fully autonomous international organization with 33 member states. The ICGEB has two centers, one located in Trieste (Italy) and the other in Jawaharlal Nehru University, New Delhi (India). The Trieste component currently occupies about 5,700 m^2, while the New Delhi component occupies about 10,000 m^2. This center has been functioning properly since 1982.

Many public and private institutions working under the government departments and organizations have advised the DBT to formulate the biotechnology programs under the following areas: (i) Plant molecular biology and agricultural biotechnology, (ii) Biochemical engineering, process optimization and bioconversion, (iii) Aquaculture and marine biotechnology, (iv) Fuel fodder, biomass and green cover, (v) Medical biotechnology, (vi) Microbial and industrial biotechnology, (vii) Large-scale use of biotechnology, (viii) Integrated systems in biotechnology, (ix) Veterinary biotechnology, and

(x) Infrastructural facilities. The pharmaceutical industry in India is very strong and vibrant with expertise in chemical drugs. It has little experience in biotech diagnostics and no experience in biotech therapeutics. Moreover, the pharma industry is located between Mumbai and Ahmedabad (90% of drug production in India is in Gujarat and Maharashtra). There is no government institution or university with expertise in this area to help the pharma industry. However, for a variety of reasons, the Indian pharmaceutical industry will sooner or later enter into the manufacturing of biotechnology-based diagnostics and therapeutics (Padh, 1996).

1.18. BIOTECHNOLOGY IN THE WORLD

Biotechnology products are increasing in the world market day by day. The high value-added biotechnology products have boosted the UK, Japan, France, Australia, Russia, Poland, Germany, as well as India (among the developing countries). The most effective international networks developed for applied microbiology/biotechnology are the Regional Microbiology Network for South-East Microbiological Resource Centers (MICREN), which have been dominant for the last few years. The biotechnology R&D during the past two decades has promoted enterprises through policy development and supported international cooperation through networks.

To speed up biotechnology inventions, the USA promotes enterprises. F□nding for basic scientific research at the National Instit□te of Health (NIH) has been supported. The US administration has boosted the process for improving new medicines so that these may be quickly and safely developed. Small business development has also been encouraged through incentives. The US administration has geared up for intellectual property rights protection. The international markets have been opened for biotechnology research and biotech products.

One of the world's best examples of partnership has been established by developing public databases. It enables scientists to partner with enterprises. These partnerships have helped in developing collaborations among university research, government, and private industries. Science education has been improved. Guidelines have been prepared so that science-based regulatory programs can promote p□blic biosafety, earn p□blic confidence, and g□arantee fair and open international markets.

For the promotion of technology products and bio-business, other countries (such as the UK, Japan, Germany, France, Australia, etc.) have also prepared similar guidelines (Crueger, 1984; Sasson, 1984; Prescott et al., 2003; Lopez et al., 1994).

1.19. ACHIEVEMENTS OF BIOTECHNOLOGY

A few developing countries like India have scientists and technologists related to biotechnology where national strategies of development in biotechnology could be implemented. The scientific and technical manpower has to be properly shifted towards new biotechnology with the aim to produce expertise in biotechnology. In a keynote address, Bachhawat and Banerjee (1985) have described the impact of biotechnology on third-world countries. They emphasized that Indian bioscientists must be trained to utilize their knowledge and expertise for application and orientation. For example, a microbiologist must be trained in microbial genetics to be really useful in fermentation technology and trained in cell culture, protoplast fusion, or DNA recombination for practical utility. Similarly, people from traditional disciplines in life science may be trained to reorient their knowledge towards application and the process of training readjusted according to need. It is, therefore, necessary to encourage biotechnological programs at industrial and educational levels. In higher education, teaching biotechnology should be compulsory for undergraduate students to expand their understanding and knowledge of the scientific and engineering principles underlying biotechnology. Countries like the USA, France, Germany, and Japan have taken this issue seriously. In India, most of the universities have started teaching biotechnology at the undergraduate level. However, at the postgraduate level, teaching and research have been initiated only by a few universities/institutes on an all-India entrance test basis. The government of India has selected many thrust areas of national and international relevance as described earlier.

Undoubtedly, modern biotechnology is expected to solve many problems arising at the global level. In the 21st century, the growth and economy of a country will certainly depend on the operation of biotechnology. A revolution may occur in some areas like medical and health care, agriculture industry, and the environment. DNA fingerprinting has s□ccessf□lly helped forensic science in the search for criminals, making the identity of individuals, solving parentage disputes, etc. Human diagnostics and therapeutic drugs have been discovered and commercialized. Gene therapy is hoped to solve the problems of genetic diseases. Biotechnology-based vaccines are the best for insulin. They have no side effects and pose no risk for the presence of live forms of viruses in vaccines. Many transgenic plants and animals have been genetically improved. Now they are capable of producing new or improved products. Questions may be raised on complex, ethical, spiritual, and philosophical issues. Hundreds of transgenic plants have been produced and many of them are being sown in the field, and prod□cts are available in the market. Plant biotechnology will reduce the dependence of farmers on pesticides and will help to utilize new technologies. In the near future, papers and chemicals with less energy and less pollution may be produced through biotechnology.

Abatement of pollution using potential microorganisms (i.e., bioremediation) has been used in many countries. Thus, potential microorganisms are sought for better health, better products, and a better environment for the future. Instead of chemical pesticides, biopesticides have been produced and commercialized by some industries in India and abroad. Similarly, biofertilizers (formulated with nitrogen-fixing bacteria, bl□e-green algae i.e., cyanobacteria, phosphate-solubilizing bacteria, and fungi) have also been made available to farmers in India. On the other hand, biotechnology has generated new jobs for the youth and stimulated the growth of small businesses. It has also encouraged innovations in the industrial and agricultural sectors. In the U.S.A. alone, more than 15 lakh youth have been employed in industries.

1.20. MISUSE OF BIOTECHNOLOGY

The prevention of the misuse of biotechnology is stressed in many countries because it may cause a lot of mischief when given free rein in the hands of transnational companies engaged in the ruthless pursuit of profits. One of these concerns is the rapid pace of genetic erosion. This will lead to a situation where the basic genetic material is available to only a few multinational companies in their gene banks. However, this genetic erosion has to be checked by saving genetic diversity in its own environment, not just in gene banks or germplasm banks, with the involvement of people transitioning from the 'green revolution' to the 'gene revolution. Many new transgenic plants and animals are being produced, and seeds, embryos, and sperm are preserved. A day may come when farmers will have to depend on genetically engineered seeds. However, there is doubt whether such seeds will be suitable for sustainable agricultural practices free from chemical poisons. Moreover, genetically modified organisms (GMOs) should be carefully researched and monitored to ensure that the hazards to users and the environment are minimized. In addition, concerted action should be undertaken to ensure that necessary consideration is given to the ethical and social effects of such studies. People should be informed about the impacts of GMOs and genetically engineered products.

Such efforts are already being made in some countries. The German Green Party has called for a 5-year moratorium on the commercial release of GMOs. The UK Genetics Forum is complaining about a partially irresponsible application of biotechnology. In India, there are gene campaigns against patenting life forms and the misuse of biotechnology.

1.21. BIODIVERSITY AND ITS CONSERVATION

It is presumed that life on planet Earth first originated around 4 billion years ago. This mega event marked a major transformation of the organic. Large and diverse communities of biota have thus occupied distinct climatic zones forming ecosystems.

The concept of biodiversity first appeared in 1980. It is, in fact, the shortened form of two words: 'biological' and 'diversity.' It was coined by W.G. Rosen in 1985.

1.21.1. Introduction to Biodiversity

Biodiversity or biological diversity can be defined as the vast array of species of microorganisms, algae, fungi, plants, and animals occurring on Earth either in terrestrial or aquatic habitats and the ecological complexes of which they are a part.

1.21.2. Definition and Explanation

Biodiversity differs from place to place. It is so because the environmental conditions of the area as well as the range of tolerance of the species determine whether or not a particular species can occur in that area.

1.21.3. Alpha and Beta Biodiversity

- **Alpha Diversity:** It refers to the diversity of organisms, i.e., the number of species in the given community or habitat. Species diversity and species evenness, the two indices are used in combination to represent alpha diversity.
- **Beta Diversity**: It appears in a range of communities due to the replacement of species along a gradient of habitats or communities within a given geographical region.

1.21.4. Levels of Biodiversity

In the biosphere, immense biodiversity exists at all levels of biological organization ranging from macromolecules within cells to biomes. The most important of these levels are genetic diversity, species diversity, community, and ecosystem diversity:

- **Genetic Diversity:** It refers to the variation of genes within a species. The difference could be alleles, entire genes, or chromosomal structures.
- **Species Diversity:** Species are distinct units of diversity, each

playing a specific role in the ecosystem. Species diversity refers to the variety of species within a region. The greater the species richness, the greater the species richness.

- **Ecosystem Diversity:** Ecosystem diversity includes all the species, plus all the abiotic factors characteristic of a region.

1.21.5. Rate of Loss of Biodiversity

Biodiversity loss has always existed as a natural process but threats to biodiversity arise when the rate of extinction exceeds the rate of speciation.

1.21.6. Causes for the Loss of Biodiversity

1. **Hunting:** Man started hunting wild animals for food and safety, trade, and fun ever since his appearance. Large-scale hunting of wild animals causes destruction.
2. **Forest Fires:** It is held that forest fires res□lting from h□man activity or mistakes have caused the elimination of many species of wildlife.
3. **Destruction of Habitats and Fragmentation:** Destruction of natural habitats causes the most serious threat to biodiversity.
   i. Deforestation;
   ii. Pollution;
   iii. Cleanliness.
4. **Introduction of Exotic Species:** Intentional or chance introduction of exotic species into new islands or countries by man adversely affects the native species
5. **Co-Extinctions:** When a species becomes extinct, the plant and the animal species associated in an obligatory way with it also become extinct.

1.21.7. Uses of Biodiversity

1. **Source of Food and Improved Varieties:** Several thousand species of edible plants and animals are known. Biodiversity plays a very important role in creating novel varieties, as it provides the source material for breeding.
2. **Fibers:** Vario□s types of fibers are available in India s□ch as cotton, j□te, hemp, agave, flax, s□n hemp, etc. Plant species offer a variety of useful products like gums, resins, dyes, perfumes, waxes, lubricants, hydrocarbons, honey, latex, paper, tea, coffee, pearls, ivory, fur, and skin.

3. **Drugs and Medicines:** A large number of substances with therapeutic properties are obtained from a variety of plant species.
4. **Sport and Recreation:** Animals provide a good sport in their habitat for hunters, and a great deal of fun and recreation for the public in circuses, zoos, parks, etc.
5. **Aesthetic Value:** Green forests, bea□tif□l flowers, gracef□l beasts, songbirds, colorf□l fishes, birds, and b□tterflies add grande□r to the biosphere.

1.21.8. Extent of Biodiversity in Plants

Biodiversity refers to the variety and variability of life on Earth. Biodiversity is typically a measure of variation at the genetic, species, and ecosystem levels.

1.22. EXPLORATION AND GERMPLASM COLLECTION

It has attracted a great many adventurers, naturalists, travelers, and plant hunters since the distant past. The collection is based on ecogeographic/phytogeographic surveys and more static/geographic approaches. It includes the floristics and flora of an area or region.

The regions of genetic diversity and utilization are categorized by areas and regions based on utilization, such as cereals, pulses, fruits, vegetables, oilseeds, fibers, forages, etc. The crop gene pool, incl□ding its wild relatives in one basic unit, involves population structure, incompatibility barriers, etc. Crop germplasm includes landraces, primitive cultivars, introgressed forms, obsolete and promising plant genetic resources using genetic approaches.

1.23. INTRODUCTION AND EXCHANGE OF PGR (PLANT GENETIC RESOURCES)

PGR is related to genotypes of particular species, collected from different sources and geographical origins, for use in plant breeding to develop new cultivars.

The sum total of hereditary material, i.e., all the alleles of various genes, present in a crop species and its wild relatives such as landraces, genetic stocks, farmers' varieties, parental lines of hybrids, released varieties, primitive cultivars, and wild and weedy relatives.

1.24. RED DATA BOOK AND ENDANGERED PLANT SPECIES

The IUCN (International Union of Conservation of Nature and Natural Resources) now called WCU (World Conservation Union) maintains a document called the Red List or Red Data Book of taxa that are facing the risk of extinction. The IUCN has divided them into eight categories:

- Extinct;
- Extinct of the wild;
- Critically endangered;
- Endangered;
- Vulnerable;
- Lower risk;
- Data deficient; and
- Not evaluated.

List of plant species:

- *Small whorled pogonia*;
- *Sarracenia oreophila*;
- *Plantanthera praeclara*;
- Showy lady's slippers;
- *Asclepias meadii.*

1.25. PLANT GENETIC RESOURCES

It is the major biological resource on which humankind has relied for its very existence and its dependence on plant-derived products will increase in the future due to greater demand from an increasing global population. New uses will be found for plants and many new kinds of plants will be developed for their contribution to nutritional, medicinal, energy, shelter, fuel, amenity, and cultural uses. The Consultative Group on International Agricultural Research has taken the lead in these efforts, as has the International Plant Genetic Resources Institute. On the technical side, new developments both in breeding technologies and in greater movement of germplasm ownership away from primary producers into industry-based ownership offer great potential for plant breeding to create new kinds of varieties that will meet the future needs of humankind in quantity and in new kinds of food and other products.

1.26. PLANT QUARANTINE ASPECTS

According to scientists, insects, mites, nematodes, fungi, bacteria, viruses, MLOs, and other organisms are known to attack various crops of economic importance. These pests and pathogens not only reduce the quantity but also spoil the quality of the produce to a considerable extent. The potential food that has been aborted, spoiled, or damaged would be enough to feed at least 75 million human beings. It may be defined as "Rules and Regulations promulgated by governments to regulate the introduction of plants, planting materials, plant products, soil, living organisms, etc., with a view to preventing the inadvertent introduction of exotic pests, weeds, and pathogens harmful to the agriculture or the environment of a country/region. It is designed as a safeguard against the harmful pathogens and pests exotic to a country or region." Sanitary and Phytosanitary Systems (SPS) is an agreement signed by WHO at the Uruguay round of the multi-trade negotiation which measures to protect humans, animals, and plants from diseases, pests, or contaminants. It applies to all sanitary (relating to animals) and phytosanitary (relating to plants) measures that may have a direct or indirect impact on international trade.

1.26.1. In-Situ Conservation

The conservation of habitats and ecosystems where organisms naturally occur. These include sacred groves, biosphere reserves, national parks, wildlife sanctuaries, etc.

1.26.2. Ex-Situ Conservation

The conservation of elements of biodiversity out of the context of their natural habitats is referred to as ex-situ conservation, e.g., Zoos, botanical gardens, and gene banks.

1.26.3. Cryopreservation

It is the use of very low temperatures to preserve structurally living cells and tissues.

1.26.4. Gene Banks

Gene banks are a type of biorepository that preserves genetic material. For plants, this could be by *in vitro* storage, freezing cuttings from the plants, or stocking the seeds.

1.27. CRYOBANKS

Seattle Sperm Bank is a cryobank that currently focuses exclusively on providing donor sperm to women and families looking to conceive a child.

1.27.1. IPGRI (International Plant Genetic Resources Institute)

It is headquartered in Rome, Italy, and has staff in eight countries worldwide. Its mission is to strengthen the conservation and use of plant genetic resources, with special emphasis on the needs of developing countries. It is one of the 16 international agricultural research centers located around the world which are supported by the Consultative Group on International Agricultural Research (CGIAR) and a consortium of 41 donor agencies.

It is committed to advancing the conservation and use of plant genetic reso□rces for the benefit of c□rrent and f□t□re generations. In its strategy "Diversity for Development" (1993), IPGRI recognizes that an important part of the genetic diversity of useful plants can be conserved *in situ,* and the need to increase research and other activities on in situ and farm or community level conservation is emphasized. This strategy further recognizes that there are important socioeconomic and cultural dimensions to conservation, and that these are partic□larly significant for *in situ* conservation.

1.27.2. FAO (Food and Agriculture Organization of the United Nations)

It helps to raise levels of nutrition and standards of living, improve agricultural productivity, better the conditions of rural populations, and contribute to the expansion of the world economy. FAO highlights information as one of the priority areas in achieving agricultural development and food security.

1.27.3. NBPGR (National Bureau of Plant Genetic Resources)

It was established by the Indian Council of Agricultural Research (ICAR) in 1976 with its headquarters in New Delhi. It has been given the mandate to act as a nodal institute at the national level for the acquisition and management of indigenous and exotic plant genetic resources (PGR) for agriculture and carry out related research and human resources development for sustainable growth of agriculture. NBPGR is the second largest genebank in the world. It has a network of 10 regional stations covering different agro-climatic zones to carry out PGR activities including collection, characterization, maintenance, and evaluation of various crops.

1.27.4. CGIAR (Consultative Group for International Agricultural Research)

It is a global partnership that unites organizations engaged in research for a food-secured future. It is dedicated to reducing rural poverty, increasing food security, improving human health and nutrition, and ensuring sustainable management of natural resources. It is carried out by 15 centers that are members of the CGIAR consortium, in close collaboration with hundreds of partners including national and regional research institutes, civil society organizations, development organizations, and private sectors. It is generally run with other organizations in partnership. There are a number of members such as the USA, UK, Germany, Switzerland, Japan, FAO, IFAD, World Bank, African Development Bank, and the Fund of the Organization of the Petroleum Exporting Countries.

1.28. PLANT BREEDERS' RIGHTS

It is generally protected in several countries through plant breeders' rights (PBRs) and plant variety rights (PVR). Through these rights, further propagation of the variety is restricted. Under the earlier convention of 1978, due to the International Union for the Protection of New Varieties (UPOV), the breeder's rights did not prohibit the farmer from reusing farm-saved seed of a variety from his harvest for planting another crop freely used earlier as plant genetic resources for the purpose of breeding other varieties.

1.29. FARMERS RIGHTS

It is a concept that has been developed and adopted by all member countries in FAO. It recognizes the fact that farmers and rural communities have greatly contributed to the creation, conservation, and exchange of plant genetic resources, as well as generated knowledge for the utilization of genetic diversity. Therefore, it is the obligation of the world community to help these farmers carry out this task and also assist them in utilizing the genetic diversity available to them. No intellectual property system or any other mechanism exists for compensation or reward to the farmers for this work.

1.30. PROTECTION OF PLANT VARIETIES

Various plant pests and pathogens inflict heavy crop losses both in the field and during storage. Plant quarantine regulations promulgated by governments of different countries are designed to regulate the introduction and movement of plants, planting materials, plant products, etc., with a view to preventing the introduction of associated pests, pathogens, and weeds exotic to a country or region that are harmful to its agriculture. Plant quarantine as a national service and its complementary role have been briefly discussed. Plant quarantine regulations with particular reference to India and the quarantine responsibilities of NBPGR regarding the introduction of germplasm material for research use in the country have been discussed. There are a number of organizations that protect crops from pests and pathogens, such as the National Institute of Plant Health Management, List of National Plant Protection Organizations, Plant Protection Advisor, Plant Protection (IPM), Plant Quarantine (PQ), Locus Control, and Pesticide Registration and Quality Control.

1.31. FARMERS RIGHTS ACTS

After several amendments in the original draft, the latest draft of the "Protection of Plant Varieties and Farmers' Rights Act" 2001 was passed by the Parliament (Lok Sabha) in August 2001. The President also gave his assent to the Bill on November 5, 2001, so that the Bill was notified as Act No. 53 of 2001 in the Gazette of India. It was to be enforced from a date that was to be notified by the Union Minister of Agriculture. One of the most important features of this bill is that it grants farmers rights by recognizing farmers as breeders, cultivators, and conservators of the seeds in their possession. It allows the farmers to use it for re-sowing or selling their harvested seed, although they will not be allowed to sell the seed under a brand name. In other words, the farmers will not be able to sell their harvested seed to another farmer for the purpose of raising a crop for commercial seed production.

CHAPTER 2

GENETIC ENGINEERING

Contents

2.1. NATURE OF DNA

The DNA is found in all plants, animals, prokaryotes, and some viruses. DNA is a double helical structure that consists of genetic material. It is joined by a phosphodiester bond and is acidic in nature. The entire genetic message that controls the chemistry of every cell of the body, acting in a specific way, is actually written in the language of four nitrogenous bases of DNA, i.e., purines and pyrimidines.

2.2. COMPOSITION OF DNA

It is a polymeric compound containing four monomers known as deoxyribonucleotide monomers or deoxyribotides. Each DNA consists of a pentose sugar, a phosphate group, and a nitrogenous base (Figure 2.1). Purine bases (adenine and guanine) are heterocyclic and two-ringed bases, and the pyrimidines (thymine and cytosine) are one-ringed bases.

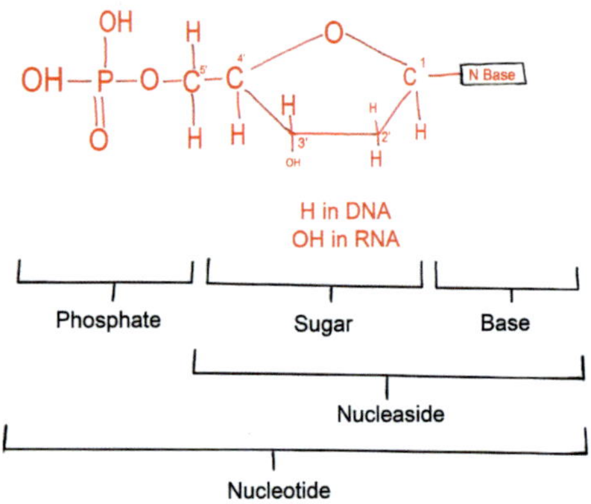

Figure 2.1. *Nucleotides.*

Source: Made by the author.

2.3. NUCLEOSIDES

The nitrogenous bases combined with pentose sugar are called nucleosides.

2.3.1. Nucleotides

A nucleoside linked with phosphate forms a nucleotide. Nucleoside = pentose sugar + nitrogenous base nucleoside + phosphate Nucleotide On the basis of different nitrogenous bases the deoxynucleotides are of following types: (i) Adenine (A) deoxyadenosine-3'/5'-monophosphate (3'75'-d

AMP) deoxyguanosine-5'-monophosphate (5'-d GMP) deoxythymidine-5-monophosphate (5'-d TMP); (ii) Guanine (G); (iii) Thymine (T); and (iv) Cytosine (C) deoxyeytidine -5-monophosphate (5'-d CMP). In addition to the presence of nucleosides in DNA helix, these are also present in nucleoplasm and cytoplasm in the form of deoxyribonucleotide phosphates, e.g., deoxyadenosine triphosphate (dATP), deoxyguanosine triphosphate (dGTP), deoxycytidine triphosphate (dCTP), deoxythymidine triphosphate (dTTP) (Figure 2.2). The advantage of these four deoxyribonucleotides in triphosphate form is that the DNA polymerase acts only on triphosphates of nucleotides during DNA replication. Similarly, the ribonucleotides contain ribose sugar, nitrogenous bases and phosphate. Except sugar, the other components are similar. However, uracil (U) is found in RNA instead of thymine. Generally, RNA molecule is single-stranded besides some exceptions.

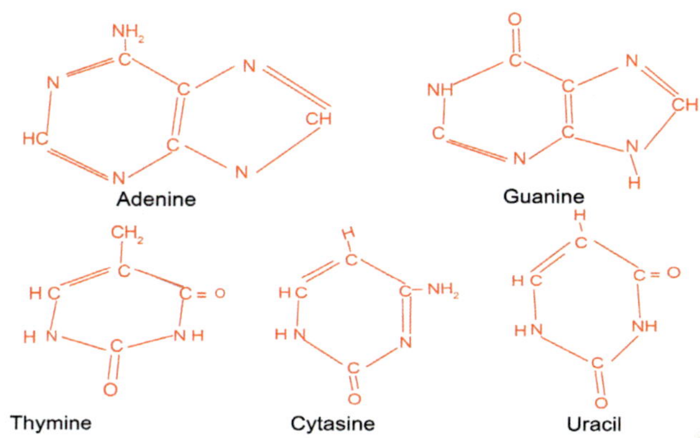

Figure 2.2. *Nirogeneous base.*

Source: Made by the author.

2.3.2. Polynucleotide

The nucleotides undergo the process of polymerization to form a long chain of polynucleotides. The nucleotides are designated by prefixing 'poly' to each repeating unit such as poly A (polyadenylic acid), poly T (polythymidylic acid), poly G (polyguanidylic acid), poly C (polycytidylic acid), and poly U (polyuridylic acid). The polynucleotides that consist of the same are called homopolynucleotides such as poly A, poly T, poly G, poly C, and poly U.

2.3.3. Chargaff-Equivalence Rule

By 1948, chemist Erwin Chargaff started using paper chromatography to analyze the base composition of DNA from a number of studies. In 1950, Chargaff discovered that in the DNA of different types of organisms, the total amount of purines is equal to the total amount of pyrimidines, i.e., the total number of A is equal to the total number of T (A-T), and the total number of G is equal to the total number of C (G-C). It means that A/T = G/C, i.e., A+T/G+C = 1. In the DNA molecules isolated from several organisms, regularity exists in the base composition.

2.4. PHYSICAL NATURE OF DNA

2.4.1. Watson and Crick's Model of DNA

Watson, J. D., and Crick, F. H. C. (1953) combined the physical and chemical data generated by earlier workers and proposed a double helix model for DNA molecules. This model is widely accepted. According to this model, the DNA molecule consists of two strands that are connected together by hydrogen bonds and helically twisted. Each step on one strand consists of a nucleotide of purine base which alternates with that of pyrimidine base. Thus, a strand of DNA molecule is a polymer of four nucleotides, i.e., A, G, T, C. The two strands join together to form a double helix. Bases of two nucleotides form hydrogen bonds, i.e., A combines with T by two hydrogen bonds (A-T) and G combines with C by three hydrogen bonds (G-C). However, the sequence of bonding is such that for every A-T-G-C on one strand there would be T, A, C, and G on the other strand. Therefore, the two chains are complementary to each other, where nucleotides on one chain are the photocopy of the sequence of nucleotides on the other chain.

The two strands of the double helix run in an antiparallel direction, i.e., they have opposite polarity. The left-hand strand has 5–3' polarity, whereas the right hand has 3'–5' polarity to the first one. The polarity is d□e to the direction of phosphodiester linkage complementary to each other, i.e., sequence as compared to the first one. The DNA model also s□ggested a copying mechanism of the genetic material. DNA replication is the fundamental and unique event underlying growth and reproduction in all living organisms ranging from the smallest viruses to the most complex of all creatures including man. DNA replicates by a semiconservative mechanism which was experimentally proved by Matthew, Meselson, and Frank W. Stahl in 1958. If changes occur in the sequence or composition of base pairs of DNA, mutation takes place. Though the presence of adenine, guanine, thymine, and cytosine is a universal phenomenon, yet unusual bases in DNA molecules also occur. In some bacteriophages, 5-hydroxymethyl-cytosine (HMC) replaces cytosine of the DNA molecule when methylation of adenine, guanine, and cytosine occurs. This results in changes in these bases (Figure 2.3).

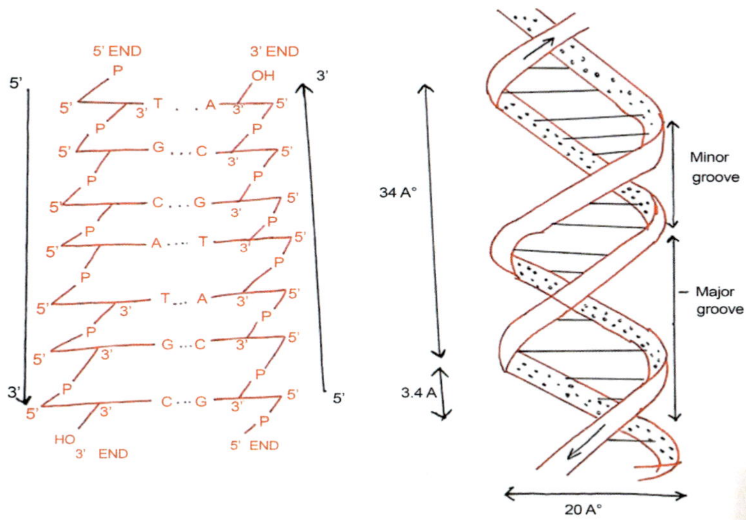

Figure 2.3. *Watson Crick's model of DNA.*

Source: Made by the author.

2.4.2. Circular and Super Helical DNA

Almost all prokaryotes and a few viruses have their DNA organized in the form of a closed circle. The two ends of the double helix get covalently sealed to form a closed circle. Thus, a closed circle contains two unbroken complementary strands. Sometimes one or more nicks or breaks may be present on one or both strands, for example, in the DNA of phage PM. Besides some exceptions, the covalently closed circles are twisted into a superhelix or supercoils and are associated with basic proteins but not with histones found complexed with all eukaryotic DNA. These histone-like proteins appear to help in the organization of bacterial chromatin structure resulting in a nucleosome-like structure, folding, supercoiling of DNA, and association of DNA polymerase with nucleoids. These nucleoid-associated proteins include HU proteins, IHF, proteins HI, Fir A, H-NS, and Fis. In archaeobacteria (e.g., Archaea), the chromosomal DNA exists in a protein-associated form. Histone-like proteins have been isolated from nucleoprotein complexes in *Thermoplasma acidophilum* and *Halobacterium salinarum*. Thus, the protein-associated DNA and nucleosome-like structures are affected in a helix coiled clockwise from the axis. The coiling is termed as positive contrast. If the path of coiling is anticlockwise, the coil is called left-handed or negative coil. The two ends of a linear DNA helix can be joined to form each strand continuously. However, if one end rotates at 360° with respect to the other to produce, these histone-like proteins appear to help the organization of bacterial DNA into a coiled variety of bacteria. In some unwinding of the double joined resulting in the formation of a twisted circle in the opposite sense, i.e., opposite helix, the ends are in the opposite direction. Such a twisted circle appears as 8, i.e., it has one node or crossing-over point. If it is twisted at 720° before joining, the resulting superhelix will contain two nodes.

The enzyme topoisomerases alter the topological form, i.e., supercoiling of a circular DNA molecule. Type I topoisomerases (e.g., *E. coli* topA) relax the negatively supercoiled DNA by breaking the 5'-phosphoryl end, and then resealing the nicked phosphodiester backbone (Moat and Foster, 1995). Type II topoisomerases need energy to unwind the DNA molecules resulting in the introduction of supercoils. One of the type II isomerases, the DNA gyrase, is apparently responsible for the negatively supercoiled state of the bacterial chromosome. S□percoiling is essential for the efficient replication and transcription of prokaryotic DNA. The bacterial chromosomes are believed to contain about 50 negatively supercoiled loops topological unit, the boundaries of which may be defined by the sites on DNA that limit its rotation (Wang, 1982; Moat and Foster, 1995). One of the phosphodiester bonds in dsDNA allows the 3'-OH end to swivel around domains (Williams, 1981).

2.4.3. Organization of DNA

In addition to the organization of DNA in prokaryotes and lower eukaryotes as discussed earlier, in eukaryotes the DNA helix is highly organized into the well-defined DNA-protein complex termed as nucleosomes. Among the proteins, the most prominent are the histones. The histones are small and basic proteins rich in amino acids. There are five types of histones, e.g., H1, H2A, H2B, H3, and H4. Eight histone molecules (two each of H2A, H2B, H3, and H4) form an octameric ellipsoidal structure of about 11 nm x 7 nm in diameter. DNA coils around the surface of the ellipsoidal structure of histones every 166 base pairs (about 1.7 turns) before proceeding onto the next form. Thus, a nucleosome is an octamer of four histone proteins complexed with DNA lysine and/or arginine. The histones play an important role in determining eukaryotic chromosomes by determining the conformation known as chromatin. The nucleosomes are the repeating units of DNA organization which are often termed as beads. The DNA isolated from chromatin looks like a string where 146 base pairs of DNA lie in the helical path and the histone-DNA assembly is known as the nucleosome core. The stretch of DNA between the nucleosomes is known as a linker, which varies in length from 14 to over 100 base pairs. The H_1 histone is associated with the linker region and helps in the folding of DNA into complex structures called chromatin or beads. As a result, chromatin becomes visible as chromosomes during cell division.

2.4.4. Structure of Ribonucleic Acid (RNA)

The RNA, TMV, yellow mosaic virus, influenza virus, and foot-and-mouth disease are usually single-stranded, except for viruses such as virus, etc., which have dsDNA. The retrovirus and wound tumor virus are entirely single-stranded RNA folded either at certain regions or to form a hairpin-shaped structure. In the hairpin-shaped structures, the complementary bases are linked by hydrogen bonds, which give stability to the molecules. However, no complementary bases are found in the unfolded region. RNA does not possess an equal purine-pyrimidine ratio as found in DNA. Like DNA, RNA is also a polymer of four nucleotides, each one containing D-ribose, phosphoric acid, and a nitrogenous base. The bases are two purines (A, G) and two pyrimidines (C, U). Thymine is not found in RNA. Pairing between bases occurs as A-U and G-C. The nucleotides formed by the four bases are adenosine monophosphate (AMP), guanosine monophosphate (GMP), cytosine monophosphate (CMP), and uridine monophosphate (UMP). These are found freely in the nucleoplasm but in the form of triphosphates, e.g., ATP, GTP, UTP, and CTP. As a result of polymerization, the ribonucleotides form a polynucleotide chain of RNA. If the RNA is involved in a genetic mechanism, it is called genetic RNA as found in plant, animal, and bacterial viruses. DNA acts as genetic material, and RNA follows the order of DNA. In such cells, RNA does not have a genetic role. Therefore, it is called non-genetic RNA. Non-genetic RNA is of three types: (i) ribosomal RNA (rRNA), (ii) transfer RNA (tRNA) or soluble RNA (SRNA), and (iii) messenger RNA (mRNA) or template RNA. These three types of RNA differ from each other in structure, site of synthesis in eukaryotic cells, and function.

2.5. GENE CONCEPT

Although the role of hereditary units (factors) in the transfer of genetic characters over several generations in organisms was advocated by Gregor John Mendel, the mystery of the 'hereditary units' was unraveled during the early 1900s. In 1909, W. Johansson discovered the gene that acts as a hereditary unit. Early work done by several workers proposed various hypotheses to explain the exact nature of genes. The gene has been defined as the unit of genetic information that controls a specific aspect of the phenotype. In 1906, W. Bateson and R.C. Punnett reported the first case of linkage in sweet pea and proposed the presence or absence theory. According to them, the dominant character has a determiner, and the recessive character lacks a determiner. In 1926, T.H. Morgan discarded all the previous existing theories and put forth the particulate gene arranged in a linear order on a string. In 1928, Johansson coined the term 'gene' as a theory. He thought that genes are chromosome and look like beads on a proposed that chromosome that appeared to be the gene. This theory of genes was well accepted by cytologists. In 1933, Morgan was awarded the Nobel Prize for advocating the discovery of DNA as a carrier of genetic information.

2.5.1. Units of a Gene

After much extensive work done by molecular biologists, the nature of genes became clear. A gene can be defined as a polynucleotide chain that consists of segments, each controlling a particular trait. Now, genes are considered as a unit of function (cistron), a unit of recombination (recon), and a unit of mutation (mutant).

2.5.1.1. Cistron

The one-gene-one enzyme hypothesis of Beadle and Tatum was redefined by several workers in the coming years. A single mRNA is transcribed by a single gene. Therefore, the one-gene-one mRNA hypothesis was put forth. Exceptionally, a single mRNA is also transcribed by and it is said to be polycistronic. Therefore, the concept has been given as the one-gene-one protein hypothesis. The proteins are the polypeptide chain of amino acids translated by mRNA. Therefore, it has been correctly used as the one-gene-one polypeptide hypothesis for more than one gene (Figure 2.4).

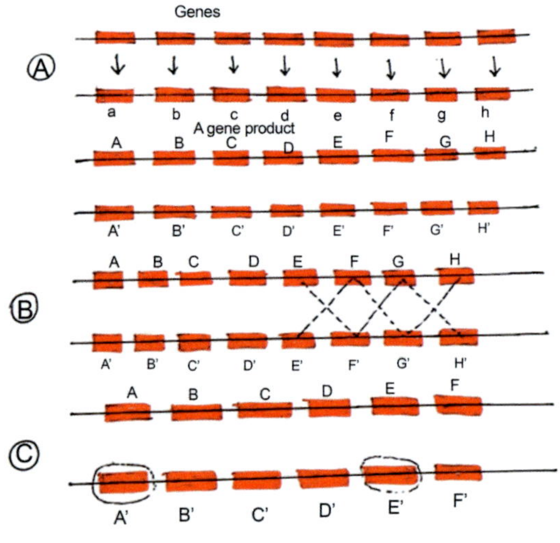

Figure 2.4. *Cistron (A); recon (B); and muton (C) genes as a unit of function.*

Source: Made by the author.

2.5.1.2. Recon

Earlier, it was thought that crossing over occurs between two genes. In 1962, Benzer demonstrated that crossing over or recombination occurs within a functional gene or cistron. In a cistron, the recombinational units may be more than one. Thus, the smallest unit capable of undergoing recombination is known as recon. Benzer (1955) found that the cultures of T4 bacteriophage formed plaques on agar plates of *Escherichia coli*. Normal T4 formed small plaques with smooth edges, whereas the mutant T4 phage formed larger plaques with rough edges.

2.5.1.3. Muton

Benzer (1962) coined the term muton to denote the smallest unit of a chromosome that undergoes mutational changes. Hence, a muton may be defined as the smallest unit of DNA that can undergo mutation. The smallest unit in size is followed by recon and at the nucleotide level, the nucleotide is changed. Thus, the unit of change in muton is the nucleotide. Therefore, a cistron is a muton. This can be explained by the fact that a gene consists of several cistrons, a cistron contains many recons, and a recon contains a number of mutons. However, if the size of a recon is equal to a muton, there would be no possibility for a recon to consist of several mutons.

2.5.2. Split Genes (or Introns)

During 1970, in some mammalian viruses (e.g., adenoviruses), it was found that the DNA sequences coding for sections were split. Therefore, these genes were variously named as interrupted genes. For the discovery of split genes in adenoviruses and higher organisms, Richards J. Roberts and Phillip Sharp were awarded the Nobel Prize in 1993. DNA sequence codes for mRNA, but the complete corresponding sequence of DNA is not found in mRNA. Certain sequences of DNA are missing in mRNA. The sequences present in DNA but missing in mRNA are called intervening sequences or introns, and the sequences of DNA found in RNA are known as exons. The exons code for mRNA.

2.5.3. RNA Splicing

In the initial stage, RNA transcript introns are synthesized which are removed later on by a process called RNA splicing. The junctions of intron-exon have GU sequences at the intron's 5' end, and an AG sequence at its 3' end. These two sequences are recognized by the special RNA molecules known as small nuclear RNA (snRNA) or snRNPs (Steitz, 1988). These, together with proteins, form small nuclear ribonucleoprotein particles called snRNPs. Some of the snRNPs recognize the splice junctions and splice introns accurately. The U1-snRNP recognizes the 5'-splicing junction, and the U5 snRNP recognizes the 3'-splicing junction. Consequently, pre-mRNA is spliced in a large complex called a spliceosome (Guthrie, 1991). The spliceosome consists of pre-mRNA, five types of snRNPs, and non-snRNP splicing factors (Rosbash and Seraphin, 1991). Robert and Sharp, the Nobel Prize winners in 1993, independently hybridized the mRNA of adenovirus with their progeny or DNA segments of the virus. The mRNAs hybridized the ssDNA of the virus where the complementary sequences were present. The mRNA-DNA complexes were observed under an electron microscope to confirm which part of the viral genome had produced the mRNA strand. It was found that mRNA did not hybridize DNA linearly but showed a discontinuous complex pattern. Huge loops of unpaired DNA between the hybridized complexes clearly revealed the large chunk of DNA strand that carried no genetic information and did not take part in synthesis. The adenovirus mRNA contains four different regions of the DNA in protein. The β-globin genes of mice and rabbits, and tRNA genes of yeast tyrosine-tRNA have been studied in detail.

2.5.4. Ribozymes

Thomas Cech (1986) discovered that pre-rRNA isolated from a ciliated protozoa, *Tetrahymena thermophila,* is self-splicing. Thereafter, S. Altman showed that ribonuclease cleaves a fragment of pre-tRNA from one end, and also contains a piece of RNA. This RNA fragment catalyzes the splicing reaction, i.e., acts as an enzyme. Therefore, this RNA segment catalyzing the splicing reaction is called ribozyme. For this discovery, Cech and Altman were awarded the Nobel Prize in 1989 in chemistry. The best-studied ribozyme activity is the self-splicing process, which is widespread and occurs in *Tetrahymena thermophila* pre-tRNA mitochondrial rRNA and mRNA of yeast and other fungi, chloroplast tRNA, IRNA and mRNA, and mRNA of bacteriophage of RNA.

This intron of *Tetrahymena thermophila* is 413 nucleotides long. The self-splicing reaction needs guanosine and is accomplished in three steps: (i) the 3'-G attacks the 5' group of introns and cleaves the phosphodiester bond; (ii) the new 3-OH group on the left exon attacks the 5-phosphate exons join and remove the intron; and (iii) the 3-OH of intron attacks the phosphate bond of nucleotide 15 residues from its end releasing the terminal fragment and cyclizing the intron exon.

2.5.5. Overlapping Genes (Genes Within Genes)

In the 1940s, Beadle and Tatum proposed the one-gene-one-protein hypothesis, which explains that one gene encodes for one protein. However, if one gene consists of 1,500 base pairs of 500 amino acids in length, it would be synthesized. In addition, if the same sequence is read in two proteins in different ways, two different amino acids would be synthesized by the same sequence of base pairs. It means the same DNA sequence can synthesize more than one protein at different times. This was realized for the first time when the total number of proteins synthesized by ØX174 exceeded the coding potential of the phage genome. A similar phenomenon is found in the tumor virus SV40, where the total molecular weight of proteins (i.e., VP1, VP2, and VP3) synthesized by SV40 genes is much more than the size of the DNA molecule (5,200 base pairs, i.e., 1,733 codons). From this observation, the concept of overlapping genes has emerged (Gupta et al., 1999).

2.6. GENE ORGANIZATION

The DNA molecules that make up the hereditary elements are called genomes. The functional region of the genome is called a gene. In a complex genome, only a small part is functional, in that it is coded into a protein with the amino acids sequence determined by the DNA sequence (Figure 2.5).

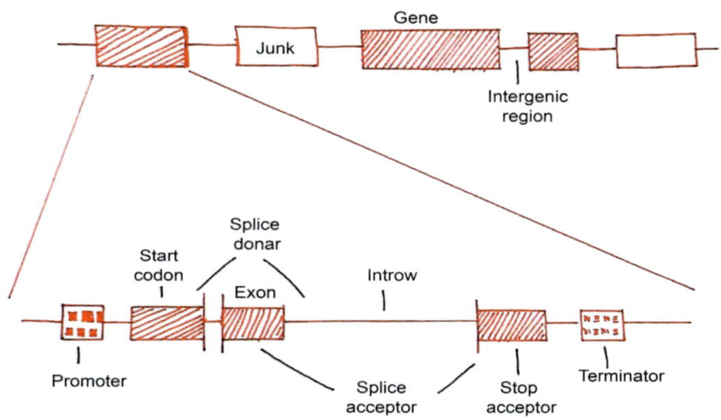

Figure 2.5. *Presentation of different regions in and around a gene in a genomic sequence.*

Source: Made by the author.

2.6.1. Gene Expression

DNA has two important roles in the cells, the first is replication and the second is expression. Gene expression is accomplished by a series of events contained in DNA that is converted into molecules that determine the metabolism of the cell (Figure 2.6).

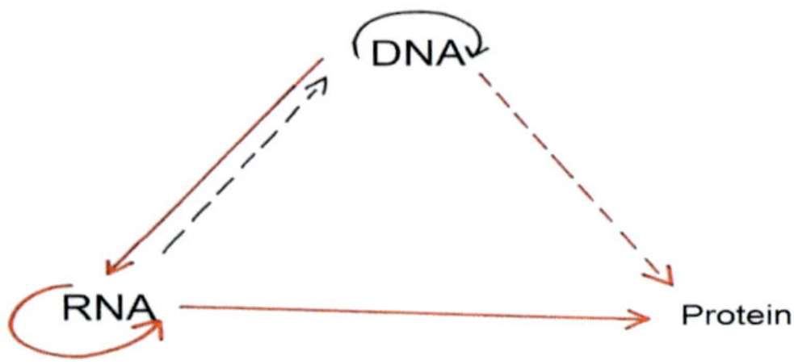

Figure 2.6. *Central dogma.*

Source: Made by the author.

2.6.2. Gene Regulation

The DNA of microbial cells consists of thousands of genes that do not express at the same time. At a particular time, a single or a few genes express and synthesize a particular protein. The other genes remain silent and express when required. This shows that genes have the property to switch on and switch off.

2.6.3. Transcription

The synthesis of enzymes comes under the control of genetic material, i.e., DNA in living organisms. Enzyme synthesis occurs through two processes: translation and transcription. Translation refers to the synthesis of mRNA. It is controlled by the promoter region of the gene.

2.6.4. Translation

It is a process in which ribosomes in the cytoplasm or ER synthesize proteins after the process of translation.

2.6.5. Gene Machine

Fully automated commercial instruments called automated polynucleotide synthesizers or gene machines are available in the market, which synthesize predetermined polynucleotide sequences. Therefore, genes can be synthesized rapidly and in high amounts (Figure 2.7).

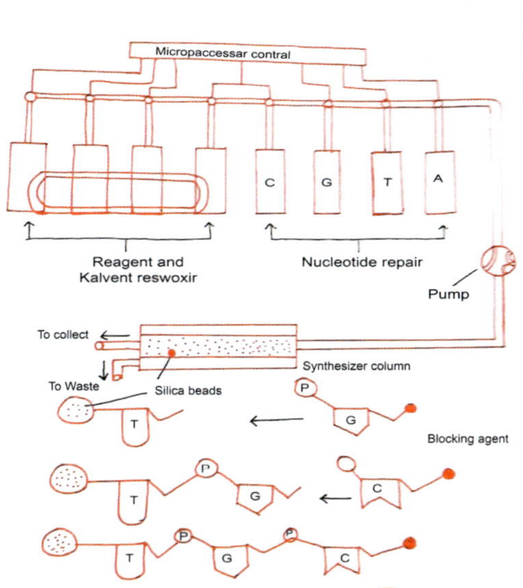

Figure 2.7. *Gene machine.*

Source: Made by the author.

2.6.6. Polymerase Chain Reaction (PCR)

2.6.6.1. Gene Amplification: Basic PCR and Its Modifications

We discussed the techniques of recombinant DNA and gene cloning, which permit us to obtain an unlimited supply of identical copies of a gene sequence or DNA segment, which is cloned in a prokaryotic or eukaryotic cell with the help of a vector.

This technique, discovered around 1975, proved extremely useful in experiments of molecular biology and genetic engineering and thus became an essential tool in all molecular biology laboratories. In 1985, yet another remarkable tool in molecular biology was discovered, which is known as polymerase chain reaction (PCR) nicknamed now as people's choice reaction.

The methodology is so important that the prestigious journal Science, p□blished from the USA, considered PCR as the major scientific development of the year 1989, and had chosen Taq DNA polymerase, the enzyme used in PCR, as the molecule of the year 1989. Kary Mullis, who discovered this, got the basic idea of this reaction while driving through the redwood mountains of California, USA. The polymerase chain reaction is such a powerful technique that may completely replace gene cloning with vectors in due course of time.

2.6.6.2. The Basic Polymerase Chain Reaction (PCR)

In order to understand PCR, the readers should be familiar with the mechanism of DNA replication within the cells. DNA replication involves the polymerization of nucleotides using a template DNA strand with the help of an enzyme DNA polymerase, but this reaction invariably requires a primer strand to which further nucleotides can be added using the DNA polymerase enzyme.

If the primer strand is not available, the reaction cannot proceed. In the living cells, this primer strand is not a DNA strand but is a small single-stranded RNA molecule synthesized with the help of RNA polymerase enzyme. In PCR, a similar reaction takes place in an 'Eppendorf tube,' where the primer strand is added from outside in the form of a deoxyoligonucleotide, and DNA polymerase enzyme is added to help in polymerization. Unlimited supply of amplified DNA is obtained by repeating the reaction, which is made possible by regular denaturation of freshly synthesized double-stranded DNA molecules by heating it to 90–98°C.

At this high temperature, the two strands separate. Once the double-stranded DNA is made single-stranded by heating up to 90–98°C, the mixture with two primers, recognizing the two strands and bordering the sequence to be amplified, is cooled to 40–60°C. This allows the primers (which are in excess) to bind to their complementary strands through renaturation. The presence of *Taq* DNA polymerase enzyme and all four essential nucleoside triphosphates allow the synthesis of complementary strands in the usual manner. In a thermal cycler, this process is automatically repeated 20 to 30 times so that in a single afternoon, a billion copies of the sequence flanked by the left and right primers can be produced. In order to continue the synthesis, the temperature of the mixture is alternately increased (for denaturation) and decreased (for renaturation) once every 1–3 minutes (as fixed by the computer device). Therefore, during temperature rise, the enzymatic activity of DNA polymerase should not be destroyed; otherwise, one may have to add a fresh aliquot of enzyme in each cycle of amplification. This became possible only by the discovery of the thermostable enzyme *Taq* DNA polymerase, isolated from *Thermus aquaticus* growing in hot springs. This enzyme acts best at 72°C, and the denaturation temperature of 90°C does not destroy its enzymatic activity.

Later, other thermostable enzymes like *Pflu* DNA polymerase isolated from *Pyrococcus furiosus* and Vent polymerase isolated from *Thermococcus litoralis*, were discovered and found to be more efficient. These enzymes allowed automation of the entire process, and automatic PCR thermal cyclers are now available (each for a price of around Rs. 2.00 lakhs), which can amplify DNA sequences at a fast speed unattended (Figure 2.8).

As pointed out earlier, PCR may eventually replace the gene cloning technique. This technology is much easier and requires a much smaller quantity of nanograms (ng) of DNA. For cloning, DNA is needed as the starting material in microgram (ug) quantities.

2.6.6.3. Different Schemes of PCR

In the previous section, we discussed the basic outline of the PCR technique, outlining how billions of copies of a DNA sequence can be obtained within a few hours, without the use of a vector and a host as done in gene cloning. However, there are several variations to this basic PCR technique, which allow varied applications of this technique.

2.6.6.4. Inverse PCR

In this technique, the amplification of those DNA sequences takes place, which are away from the primers and it is used for the amplification of DNA with a known sequence is called inverse polymerase chain reaction (PCR). It allows PCR when only one internal sequence is known. It is very useful for flanking sequences for various genomic inserts. A sequence of self-ligation and digestion is performed before being cut by an endonuclease.

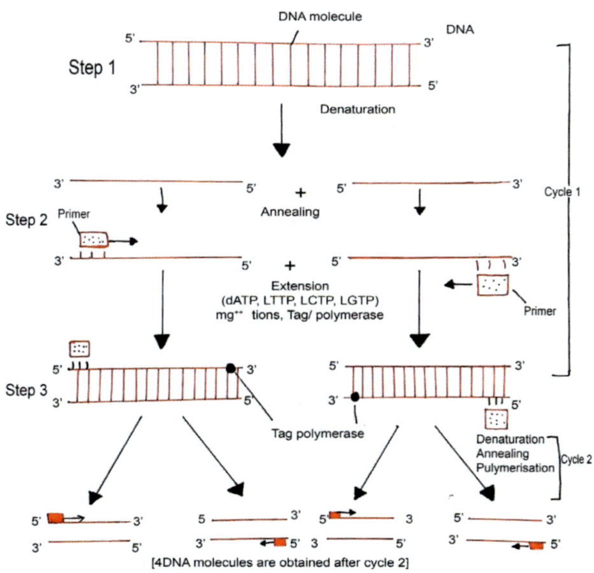

Figure 2.8. *Polymerase chain reaction.*

Source: Made by the author.

2.6.6.5. Anchored PCR

In the basic PCR technique and the inverse PCR, one has to use two primers representing the sequences lying at the ends of sequences to be amplified. But sometimes, we may have knowledge about the sequence at only one of the two ends of the DNA sequence to be amplified. In such cases, anchored PCR may be used, which will utilize only one primer instead of two primers. In this technique, due to the use of one primer, only one strand will be copied first, after which a poly G tail will be attached at the end of the newly synthesized strand. This newly synthesized strand with a poly G tail at its 3' end will then become a template for the daughter strand synthesis utilizing an anchor primer with which a poly C sequence is linked to complement with poly G of the template. In the next cycle, both the original primer and anchored primer will be used for gene amplification (Gupta et al., 1999; Kumar, 2003; Biello, 2006; Dubey, 2009).

2.6.6.6. PCR for Site-Directed Mutagenesis

This technique is used for introducing mutations at the desired place in a DNA sequence by sequences of primers. Since mutation through prime A variation of this technique allows mutation to be introduced, mutations are limited to the ends of the gene sequence.

To be introduced at any place of interest in the gene, the method is described as overlap extension which works as follows: In two separate PCR reactions, a partic□lar gene is amplified into two separate segments. In each of these two reactions, there is one primer at the end of the gene and the other (with desired alteration for mutation) internal to the seq□ence. The internal modified primers in two reactions are complementary to one another, so that the amplified products will have their ends internal to the original sequence actions will overlap. The internal ends of the two PCR products, when mixed, function as primers. Extension of these primers by Taq polymerase results in the formation of a complete gene, with mutations incorporated at internal sites.

2.6.6.7. Cloning and Expression of the PCR Product

The amplified PCR product can be cloned if restriction sites for specific enzymes are introduced at the 5' end of each primer. These restriction sites do not interfere with amplification and can be used for cloning with a variety of vectors in the usual manner. Similarly, incorporation of a T7 promoter at the 5' end of one primer will allow copies of PCR product using the RNA polymerase enzyme to get RNA copies of the PCR product.

2.6.6.8. Asymmetric PCR for DNA Sequencing

PCR products can also be used as templates for direct DNA sequencing (ligation-mediated PCR – LMPCR). Single-stranded DNA can be produced for this purpose using asymmetric PCR, in which the two primers are used in a 100:1 ratio, so that after 20–25 cycles of amplification, one primer is exhausted, and single-stranded DNA is produced in the next 15–10 cycles.

2.6.6.9. Polymerase Chain Reaction Application

During the last few years, with the improvement of PCR protocols and due to the availability of automatic thermal cyclers commercially, the applications of PCR have increased manifold. However, for the application of PCR, we need a pair of primers which should be based on the knowledge of the nucleotide sequence of DNA to be amplified. Therefore, the non-availability of this information about the DNA segment or gene to be amplified becomes a limitation in the applications of PCR, although this difficulty has been overcome for the study of DNA polymorphism through the primers. Some of the applications of PCR technology using random DNA primers are described in this section.

2.6.6.10. Study of DNA Polymorphism Using PCR

Using PCR, DNA polymorphism can be studied by sequencing loci (using primers based on known sequences) or by using random primers with unknown but random DNA sequences (these are called random primers). In either case, DNA may be amplified from one or more genomic DNAs and subjected to digestion with one or more restriction enzymes, followed by electrophoresis. The restriction patterns may then be photographed to reveal polymorphism. Sequences of the prolamin gene and phytochrome gene have been used to study polymorphism in different species of Oryza (wild and cultivated rice).

Similarly, storage protein genes in wheat and many genes in other animal and plant species have been st□died □sing specific restriction endon□cleases. The digested DNA may be used for the study of DNA polymorphism.

The non-availability of seq□ences of specific genes for designing primers for DNA amplification has been overcome by a variety of approaches incl□ding the following: (i) In the case of multigene families like those of tRNA and heat shock proteins, there are conserved consensus sequences known, which are used for designing □niversal primers. Similarly, for the amplification of antibody genes, universal primers have been used in PCR. (ii) Gene bank Sequence Data have also been □sed for designing primers for amplification of microsatellites like poly (A), poly (G), pol (TC), and poly (TG), etc. This is used for the study of length polymorphism in these microsatellites, also described as 'short tandem repeats' or simple sequence repeats (SSR). (iii) The consensus sequences for intron-exon junctions or intron splice junctions (ISJ) have also been used for the amplification of intron or exon seq□ences to st□dy DNA polymorphism. (iv) Conserved sequences at the ends of SINEs (short interspersed elements) and LINEs (long interspersed elements) have also been used for the study of polymorphism in the distribution and lengths of these repeat sequences. (v) In certain viruses, DNA polymorphism could be studied by designing degenerate primers based on the amino acid sequence of a protein using the genetic code.

2.6.6.11. PCR and RAPD Markers

The most important use of PCR for the study of DNA polymorphism, however, involves the use of random primers, giving rise to Random Amplified Polymorphic DNA. This technique is considered simpler and offers several advantages over more commonly used techniques. RFLP and RAPDs have actually been used for the study of polymorphism in chromosome mapping in maize, soybean, mouse, humans, etc.

2.6.6.12. PCR and VNTR Loci

Alec Jeffreys and his colleagues in the U.K. suggested the use of PCR for the study of DNA polymorphism at VNTR (variable number tandem repeats), also known as 'minisatellites loci in humans. Primers were developed for the conserved flanking regions of VNTR loci so that all available VNTR loci could be amplified. The PCR products differ according to the number of repeat units present in the different VNTR loci, and this variation is visualized in the positions occupied by different bands after electrophoresis.

2.6.6.13. PCR and SSR Loci

The VNTR loci discussed above consist of repeat electrophoresis units in the range of 11–60 base pairs in length. A much simpler type of repeat units exists in the form of dinucleotides, trinucleotides, or tetranucleotides, e.g., (CA)n, (GT) n, (AAT)n, or (AGAT)n. These repeat units were described as microsatellites, but later also called short tandem repeats or simple sequence repeats (SSR). Some of these SSR loci, such as (CA)n or (GT)n, occur in humans with n varying from 10 to 60.

The DNA seq□ences flanking SSRs are conserved, allowing the selection of PCR primers that will amplify the intervening SSR in all genotypes of the target species. The PCR reaction includes a small amount of 3P-labeled nucleotide (only one of the four nucleotides is labeled) or it may include primers, one or both of them being end-labeled. This allows visualization of PCR products using autoradiography after electrophoresis on a standard sequencing gel. Variation in the length of products is a function of the number of SSR units. The principle underlying detection polymorphism is using two parents and their F1 hybrid. Markers resulting from SSR length polymorphisms are placed on genetic maps in relation to RFLPs, RAPDs, VNTRs, other SSRs, and even morphological markers. A comparison of very limited sequence data from plants (including algae) and those of vertebrates (including humans) indicated that SSRs in plants are found with the same frequency as in vertebrates. Recently, DNA sequence data for five tree species and Zea mays have been screened for the presence of (AC)n and (AG)n, and the frequency in these six species was found to be close to one microsatellite (SSR) per 5 x 3 x 10 bp. Other plant and animal species have also been examined for the presence of a number of SSRs, and a wide occurrence of these markers in plants and animals preparing saturated was suggesting their utility in preparing saturated molecular maps. Molecular mapping using PCR, genetic, and physical chromosome maps have also been prepared, both in plants and animals, using PCR. Once RAPDs are detected, recombinant inbreeds (RIs) or F2/backcross segregating populations, or doubled haploids (DH) derived from haploids (in plants) can be used to detect linkage and recombination frequencies. The molecular marker which cosegregates RIs or F1s or DHs suggests linkage, and the frequency of cosegregation among RIs or F1/backcross plants will be a measure of the strength of linkage. The molecular markers identified thro□gh PCR can also be assigned to specific chromosomes using aneuploids or alien addition lines, e.g., wheat-barley additions – 21 pairs of wheat chromosomes + 1 pair of a specific barley chromosome. A n□mber of molecular markers have been mapped using PCR in maize, soybean, barley, mice, humans, etc.

2.6.6.14. Sequence Tagged Microsatellite Sites (STMS) or SSR Loci

It has also been examined and mapped using PCR. The gene sequence databases are screened for a microsatellite, and the flanking sequences are used for designing primers for PCR. Several STMS or SSR loci using this technique have been mapped in mice, whereas as many as 200 microsatellites have been detected in the genome database. Besides SSRs or STMSs, other markers including sequence tagged sites or STSs (a short DNA sequence in a molecular genomic DNA marker) and expression sequence tags or ESTs (a short DNA sequence from a cDNA clone representing an expressed gene) have also been used for genetic mapping. In both these cases, PCR can be profitably utilized for screening RIs (mouse), somatic hybrid cell lines (humans), or F2 individuals (in plants). This will allow detection of linkage and calculation of recombination frequencies leading to the preparation of genetic maps. Complete molecular maps of several human chromosomes could be prepared using STSs and PCR.

2.6.6.15. Gene Tagging Using PCR

PCR has also been utilized for developing molecular markers specific to genes of economic importance. For instance, in tomato, 144 random primers were developed to produce 625 PCR products from a set of markers closely linked to instances in tomato, near isogenic lines (NILs, which differ only for the presence and absence of a specific gene) for *Pseudomonas* resistance. At least three PCR products were present in one and not in the other of a pair of NILs showing linkage with the *Pseudomonas* resistance gene. The PCR products could be mapped close to Pro. Such tagging of genes through PCR will be used in the future for plant breeding (selection of desirable plants in segregating populations) and also for the isolation of genes.

2.6.6.16. PCR for Confirming the Presence of Transferred Gene

When a gene is transferred with a vector to cultured cells, primers can be designed to conduct PCR for amplification of the sequence so transferred. This technique allows the confirmation of transfer and maintenance of the gene of interest. In gene therapy experiments, the transfer and presence of a marker gene for neomycin resistance (NeoR) could be detected in the blood of patients, even after 60 days, although the frequency of marked lymphocytes gradually decreased necessitating a second transfusion of such marked cells. Similarly, ADA (adenosine deaminase deficiency) gene was transferred for therapeutic purposes and was used to detect the presence of this gene up to 6–5 months, so that the patient (4-year-old girl) was then placed on a program of maintenance infusions at 3–5 month intervals. Thus PCR proves to be an immense help in monitoring a gene in genetic engineering or gene therapy experiments.

2.6.6.17. Human Genetics Using PCR

PCR has also found extensive use in human genetics. Some of these uses are mentioned below:

2.6.6.18. Prenatal Diagnosis Using PCR

Prenatal diagnosis of sickle cell anemia with enhanced sensitivity was perhaps the first application of PCR. Using PCR, this is done in less than one day, in contrast to the several weeks needed when Southern blots are used for hybridization with a probe. The test has also been used for the diagnosis of phenylketonuria (PKU), β thalassemia, hemophilia, etc. The PCR prod□ct in all these cases is examined using a labeled probe to determine whether the mutant sequence causing the disease is present. The RFLP pattern of PCR products in healthy and defective fetuses (singular-fetus) differs, thus enabling prenatal diagnosis. In other cases, PCR products may be sequenced to reveal differences (for sequencing information found in the next chapter).

2.6.6.19. Recombination Data Using PCR

DNA pieces are broken and recombined to produce a new combination of genes, a process called recombination. It creates genetic diversity at the level of genes that reflects many differences in the DNA sequences in different organisms. PCR can be used for DNA amplification utilizing individual sperms, which are the products of meiosis and recombination. This allows the construction of genetic maps at a resolution which is not possible with pedigree analysis. Microdissected chromosome bands are also used for PCR to allow the physical location of recombination at genetic lesions.

2.6.6.20. Sexing of Embryos Using PCR

Since DNA sequences in single cells can be studied using PCR, the sex of human or livestock embryos fertilized in vitro can be determined before implantation. For this purpose, PCR primers and probes specific for sex chromosomes (e.g., ZFY or ZFX sequences) can be used. Y-specific primers and probes for humans are available for this purpose. This technique can also be used to detect sex-linked disorders in the fertilized embryos.

2.6.6.21. DNA Fingerprinting

DNA fingerprinting is more successfully used in forensic science to search out criminals and rapists, solve disputed parentage, and unite lost children with their parents or relatives by confirming their identity. This is done by linking the DNA recovered from samples of blood, semen, hairs, etc., found at the crime scene with the DNA of the suspected individuals or between the child and his/her parents/relatives. We will briefly discuss the use of DNA fingerprinting in forensic medicine for identifying criminals and disputed parentage. Some well-characterized sequences of microsatellites are used to design primers so that DNA fingerprinting can be achieved through PCR. PCR allows the amplification of DNA from individual hairs, bloodstains, or seminal fluid with partially degraded DNA, which could not be used earlier to characterize individuals.

CHAPTER 3

RECOMBINANT DNA TECHNOLOGY

Contents

3.1. CHIMERIC DNA

In this chapter, we will describe about different types of restriction enzymes used regularly as a routine in all DNA cloning and recombinant DNA enzymes.

3.2. RESTRICTION ENZYMES FOR CLONING

For cloning of DNA, often we need to cut DNA at specific sites, recognized and cleaved by specific enzymes. Also, by locating the positions of cleavage sites of a number of restriction enzymes, restriction maps can be prepared. These restriction enzymes recognize short sequences of double-stranded DNA as targets for cleavage. Different enzymes recognize different but specific sequences, each ranging in length from 4 to 8 base pairs. The enzymes are named by three letters which identify their origin. Roman numerals are added to distinguish several enzymes with the same origin. For instance, *EcoRI* is derived from *E. coli* and HpaI is derived from *Haemophilus parainfluenzae.*

Besides cleavage, modification in the form of methylation is also brought about by some enzymes called modification enzymes. This methylation distinguishes genes in different states of functioning. There are also some enzymes which perform the functions of both restriction and modification. Based on the attributes, restriction enzymes have been grouped into two classes: (i) Type II restriction enzyme systems which have separate enzymes for modification and restriction; and (ii) type I and type III enzyme systems, in which the same enzyme possesses both activities, although the restriction and modification sites differ in position.

The above two classes of restriction enzymes, type II enzymes, are most important for cloning purposes. The target sites for these enzymes are 4–8 bp with symmetry. Most type II restriction enzymes cleave DNA at unmethylated target sites. Some enzymes introduce staggered cuts, while others generate blunt ends. Enzymes with 4 bp target sites are used when frequent cuts are desired, and those with 8 bp are used when rare cuts are desired to obtain long DNA segments. Otherwise, the majority of enzymes used have 6 bp target sites. Some of them can cleave both methylated and unmethylated targets, but the majority of them cleave only unmethylated targets.

3.3. TECHNIQUE OF RESTRICTION MAPPING

The recombination frequencies, which are a function of distances between genes, are not completely independent of (i) the nature of mutants used; (ii) the position of these mutants on chromosomes; (iii) the genetic background; and (iv) the environmental conditions and a variety of other factors. Consequently, the distances between genes on a genetic map may not correspond to the distances between them on the DNA molecule of which they are a part at the molecular level. Gaps may also be present on a genetic map due to the non-availability of mutants in that region. At the molecular level, the fine structure of a gene can be studied through the determination of the nucleotide sequence of the concerned DNA segment. Sometimes this is done through the isolation of a DNA segment corresponding to a gene followed by sequencing of the DNA segment, which needs time and energy. Instead, we can prepare a map of the DNA by its cleavage at specific sites with the help of restriction endonucleases, which recognize very short specific DNA sequences and cleave the DNA at these specific sites. These sites of cleavage can be identified and mapped to give rise to a restriction map. In a restriction map, the nucleotide sequence of DNA between sites close to one another (300 bp or less) can be determined. These regions can then be connected into a sequence of the entire genome, whose nucleotide sequence can then be compared with the amino acid sequence of the protein coded by it. This can be further extended to adjoining genes to saturate the physical map, an objective which is difficult, but is being attempted for the human genome. On a restriction map, there is a linear sequence of sites, each for a specific enzyme, and the distances between them are measured in terms of a number of base pairs of DNA. This technique can be used in both prokaryotes and eukaryotes, although not all regions of chromosomes can be easily mapped in eukaryotes. Different steps involved in restriction mapping include the following:

3.4. RESTRICTION CLEAVAGE AND GEL ELECTROPHORESIS

If we take a particular DNA molecule or DNA sample, digest it with a specific restriction enzyme, and then subject it to gel electrophoresis, we will notice a series of bands on the gel slab or cylinder. The bands can be observed under ultraviolet light after staining with ethidium bromide. The position of different bands will depend on DNA fragment size, such that the smaller the fragment, the more rapidly it will move, and the longer the segment, the more slowly it will move. This means that the fragment away from the loading site will be smaller and that close to the loading site will be longer DNA fragments. The gel can be calibrated by using a mixture of DNA fragments of known length so that the position of the bands on this standard gel can be compared with the bands in the experimental DNA digest, and the fragment length in each band of DNA digest can thus be determined, sometimes using a computer device. The results of digestion of a 5000 bp long hypothetical DNA molecule digested separately by two enzymes A and B. As shown in the figure, the enzyme A cleaves the DNA into four fragments of lengths 2100, 1400, 1000, and 500 bp, while the enzyme B cleaves it into three fragments of lengths 2500, 1300, and 1200 bp. These data, with some additional experiments, can be used to generate a restriction map (Gupta, 1999 and Prescott et al., 2003).

3.5. CONSTRUCTION OF A RESTRICTION MAP

The data on DNA digestion by more than one endonucleases as discussed above can be utilized to arrange the sites of cleavage in a defined order. This can be done by several methods. This method involves successive digests with two individual enzymes, where we digest with either enzyme A or enzyme B and then cleave it with the other enzyme. The original DNA sample is also digested by a mixture of both the enzymes to confirm the results of individual successive digests. When the fragment A-2100 is obtained and digested with enzyme B, it is cut into two fragments of 1900 bp and 200 bp. Fragment B-2500 (obtained from individual digest with enzyme B) is similarly cut by enzyme A into two fragments of 1900 bp and 600 bp, suggesting an overlap of A-2100 and B-2500 in the region of the fragment of 1900 bp which is obtained by one cut with enzyme A and the other by enzyme B. The DNA fragment of 1900 bp is also available when DNA is digested with a mixture of both enzymes A and B (double digest). Using this information, we identify the overlapping regions in A and B digests and find out the sites of cleavage by A and B. This will then allow us to prepare the restriction map (Purohit, 2004).

3.6. USE PARTIAL DIGESTS, END LABELING HYBRIDIZATION IN RESTRICTION MAPPING

The above technique of individual and double digests can be supplemented with other techniques for the actual construction of maps. For instance, by permitting incomplete digestion, fragments longer than those obtained by complete digestion may be obtained. These will be called partial digests. In the example used in the last section, in partial digests with enzyme A, instead of getting A-1000, A-2100, A-1400, A-500, we may get A-3100, A-1400, and A-500, which will suggest that A-1000 and A-2100 lie in the adjacent regions. In another partial digest, if we get A-1000 and A-500, this will suggest that A-2100 and A-1400 are adjoining, which will mean that A-1000 and A-1400 are found on the two sides of A-2100. This technique will thus allow the arrangement of fragments in a linear order. Another useful technique involves labeling of the ends (end labeling) of DNA. Ends can be identified due to labeling, even after digestion. In the example, using end labeling, if A-1000 and A-500 are found to be radioactively labeled, these will be present at the two ends of the restriction map. Some of the features of a restriction map can also be confirmed by nucleic acid hybridization. A molecule before digestion, so that the fragments containing these 2100 and B-2500 have an overlapping region of 1900 bp, they should hybridize with each other, which will confirm overlapping (Kumar, 2003; Cummings, 2004).

The above technique will help in the accurate completion of a restriction map but will require that we have a complete set of restriction fragments that make up the entire DNA region being mapped. Once the restriction map (which is a physical map) is ready, it can be compared with the genetic map. Although large changes already located on the genetic map can be easily located on the restriction map, point mutations cannot always be easily located on the restriction map since the restriction sites often do not change due to mutation. In such a situation, one may want to determine the nucleotide sequences on individual fragments and compare them in normal and mutant individuals. Since sequencing of nucleotides is more laborious and cannot be easily undertaken for the whole genome, another technique of molecular genetic markers in the form of restriction fragment length polymorphisms (RFLPs) has been utilized during the 1980s to generate genetic linkage maps in humans, mice, rats, pigs, sheep, fr□it flies (Drosophila), nematodes (*Caenorhabditis elegans*), yeast, maize, wheat, rice, barley, tomatoes, potatoes, lettuce, peppers, peas, etc.

3.7. CONSTRUCTION OF CHIMERIC DNA

DNA molecules are produced by inserting a foreign DNA segment into the DNA molecule of a vector. These chimeric DNA molecules are the means of cloning genes or other small DNA sequences. The construction of chimeric DNA, as we know now, involves the joining of the broken ends of the vector with the two ends of the sequence to be cloned. There are at least three methods available for the construction of this chimeric DNA by palindromes and staggered cleavage. This is the most common method: a palindromic sequence, a restriction enzyme causes staggered cuts producing short complementary single-stranded sticky ends (a palindromic sequence is one which has complementary sequences at the two ends of a single strand, e.g., AT.GAT). This can be illustrated using the example of the enzyme EcoRI, which cuts the sequences at specific positions. When another DNA molecule is similarly cut by the same enzyme, similar sticky ends having the same sequences in the single-stranded ends will be produced, so that when it is mixed with the previous molecule similarly treated, the two will anneal producing a chimeric DNA or chi plasmid if plasmid DNA is involved. Enzyme DNA ligase helps in joining the bonds at the cut ends of two molecules. The above technique has the advantage of regenerating two restriction sites (e.g., EcoRI site) in the chimeric DNA, so that the foreign DNA segment can be retrieved rather easily from the cloned copies of chimeric DNA by cleavage again with the same enzyme. There are also the following disadvantages to this technique.

i. The two cleaved ends of a vector or of foreign DNA may join end to end before getting inserted. Therefore, while isolating chimeric DNA, one will have to select chimeric molecules each having only a single insert. This can be achieved by separating molecules of different sizes by gel electrophoresis.

ii. The recognition site, particularly in the sequence to be cloned, may not lie at a convenient position, so sometimes only a part of the desired segment will be inserted.

3.8. ADDING POLY DA AT THE 3' ENDS OF THE VECTOR AND POLY DT AT THE 3' ENDS OF THE DNA CLONE

In this method, DNA can be cut at the desired position both in the vector and in the clone, without staggered cleavage. Using precursor dATP, poly dA is added at both the 3' cut ends of the vector with the help of enzyme terminal transferase. Similarly, using precursor dTTP, poly dT is added at both the 3' cut ends of the DNA sequence to be cloned. The vector and clone can then be joined by annealing the poly dA with poly dT tails and then ligating them using the DNA ligase enzyme. In this technique, there is no chance of reannealing between the two cut ends of the same DNA molecule or of the two similar DNA molecules, so one of the disadvantages of the first method has been overcome in this case. However, in this technique, it will not be easy to retrieve the cloned DNA because the recognition site of the enzyme due to the insertion of poly dA and poly dT has been lost in this case. But if poly dG and poly dC are used instead (which serve the same purpose), cloned sequences can be retrieved, since poly dG: poly dC regenerates the recognition for the enzyme *PstI.*

3.9. BLUNT END LIGATION BY T4 DNA LIGASE

In this technique, a restriction enzyme is used to cut a duplex DNA molecule at the same place in both DNA strands. The broken ends are then used to join with the two ends of another DNA molecule irrespective of the sequences present at the broken ends of the two DNA molecules. The T4 DNA ligase is used for this joining reaction. The disadvantage of this technique is that any two broken ends may join, including those belonging to the same DNA molecule. This leads to the production of a variety of products, and one will have to select the desired product from a mixture of products.

This blunt end ligation is used for developing a method in which the DNA to be cloned can be easily retrieved whenever required. This method makes use of short DNA duplexes (linkers) that contain EcoRI palindrome or some equivalent palindrome, which, being small in size, can be synthesized chemically. These linkers can be linked to the blunt ends of vector DNA of an insert by blunt end ligation. This will allow the creation of an EcoRI site in the linker region of the vector so that when a DNA segment is now cloned, the inserted DNA segment can be retrieved with *EcoRI*. However, in this method, there is no restriction on the original choice of sites to generate the cut ends. Therefore, with this method, it is now possible to insert a foreign DNA segment at a particular site in the linker region of the vector and then retrieve this foreign segment whenever necessary (Gupta et al., 1999).

3.10. DNA CLONING IN BACTERIA AND EUKARYOTES

3.10.1. Cloning in Bacteria

Bacteria are often used as hosts for the multiplication of chimeric vectors or recombinant DNA molecules. Therefore, a bacterial strain should be available, which should be the host for the vector and should be modified or selected for high transformation efficiency. Bacteria can be classified into:

i. Gram-negative bacteria, e.g., *Escherichia coli*, *Pseudomonas, Rhizobium*; and

ii. Gram-positive bacteria, e.g., *Bacillus subtilis* and *Streptomycetes* sp. Ordinarily, the same plasmid cannot maintain itself in a stable form in both classes of bacteria, and therefore, plasmid vectors for the two classes usually differ. While gram-negative bacteria can be used for plasmids, phages, and cosmids vectors, gram-positive bacteria are used as hosts for only plasmids. A chimeric vector is inserted into bacterial cells and transformed bacterial colonies are selected for multiplication in suspension cultures. The chimeric DNA is retrieved whenever required.

3.10.2. Cloning in Eukaryotes

Since chromosomes found in the nucleus of eukaryotes are separated from the rest of the cell through the nuclear membrane, and since many of the genes are split genes with exons and introns, genetic engineering with eukaryotes requires new methods and tools. When eukaryotic genes are cloned in prokaryotes, the split genes cannot be correctly expressed because prokaryotes do not have the equipment necessary for splicing out the RNA transcribed from the introns of a gene. In view of this, eukaryotic cells may sometimes be needed for cloning and particularly for the expression of cloned eukaryotic genes.

Among eukaryotes, DNA cloning has been done in yeast, mice, and to some extent even in some higher plant species. In yeast, a plasmid called 2u DNA (63 bp) is fo□nd, which is an appropriate cloning vehicle. An efficient transformation method is also available, which involves protoplast production followed by PEG-directed introduction of DNA into protoplasts. In animal cells like mouse cells, special animal viruses were used as cloning vehicles. *Simian virus 40* (SV40) is one such virus in which the globin gene could be integrated. This gene integrated in SV40 could be transcribed and translated into mouse kidney cells. A variety of human cell lines in culture are also used for cloning of foreign DNA using some animal viruses as vectors.

3.10.3. Molecular Probes

Molecular probes are small DNA or RNA segments that have complementary sequences in DNA or RNA molecules, allowing the identification and isolation of specific DNA sequences from an organism. Antibodies are also occasionally used as probes to recognize specific protein sequences. Initially developed for genetic engineering research, these probes are now frequently used for various purposes, including the diagnosis of infectious diseases, identification of food contaminants, various microbiological tests, and forensic applications (e.g., fingerprinting of murderers or rapists). Probes can also be used to identify strains of an organism, such as varieties of a crop species. For example, a plant breeder may use a quick test to identify their variety to maintain their patent or breeder's rights, preventing competitors from using it under a different name and infringing on their rights.

For basic studies in molecular biology laboratories, molecular probes are freq□ently □sed for the identification and isolation of genes or related seq□ences. In theory, any nucleic acid (or rarely protein, e.g., antibody) can be used as a probe, provided it can be labeled to permit identification and q□antitation of the hybrid molecules formed between the probe and the sequence to be identified. In practice, do□ble and single-stranded DNAs, mRNA, and other RNAs synthesized *in vitro* are all used as probes. DNA/RNA probe assays are faster and more sensitive, so that many conventional diagnostic tests for viruses and bacteria involving culturing of the organism are being rapidly replaced by antibody and DNA probe assays. While culture tests can take days or even months, molecular probe assays can be performed within a few hours or minutes. Therefore, the use of DNA probes has become today's most sophisticated and sensitive technology for a variety of uses involving biological systems both in basic and applied studies, including their commercial use.

3.10.3. Preparation of Probes

3.10.3.1. Genomic DNA Probes

Simple, reproducible methods have been devised for the preparation of molecular probes. A random or specific DNA sequence (probe) can be isolated from the genomic DNA (DNA extracted from cell nuclei) using the following steps to extract DNA from plant or animal tissue:

i. Digest extracted DNA with a restriction enzyme (e.g., EcoRI or HindIII), which cuts DNA at specific sites or positions where a specific sequence recognized by the enzyme is found;

ii. Run the digested DNA on an agarose or polyacrylamide gel for electrophoresis to separate fragments of different sizes;

iii. Isolate DNA of a specific fragment size from a particular band identified through Southern blots by hybridization with specific labeled mRNA or cDNA molecules, if available. If random probes are needed, no hybridization of Southern blots is required (Southern blots are DNA bands transferred from gel to nitrocellulose membrane with the help of a buffer rising through capillary action).

iv. Clone this DNA into a vector to allow the chimeric vector to infect bacteria for multiplication, where it can make billions of copies.

The steps (ii) to (iv) above are also used for preparing cDNA probes. From the transformed bacteria, the chimeric vector can be obtained and used in one of the following ways:

i. It may be used directly as a probe (the presence of vector DNA will not interfere, so it is not really necessary to remove it).

ii. The segment may be separated (or retrieved) by using the same enzyme that was used for cloning. In the latter case, cleaved chimeric vector DNA will be electrophoresed again to separate the inserted segment on the gel. This inserted segment thus retrieved can now be used as a probe.

iii. The chimeric DNA may be used for PCR, using flanking sequences as primers; the PCR product can be separated and used as a probe.

3.10.3.2. cDNA Probes

A DNA sequence corresponding to a part of as gene(s) (an oligonucleotide) can be obtained by reverse transcription of mRNA. cDNA thus obtained can be cloned and used as a probe.

3.10.3.3. Synthetic Oligonucleotides as Probes

DNA probes with nucleotide sequences can also be synthesized chemically using automated DNA synthesizers now available commercially. These synthetic probes will be efficient only when they are not more than 20 to 40 nucleotides in length.

3.10.3.3. RNA Probes or Riboprobes

High-specific activity RNA probes or expression vectors. This is achieved through RNA synthesized *in vitro* (on DNA inserted in vector) and labeled simultaneously employing specific RNA polymerase and labeled nucleotides which are supplied by commercial firms in a kit for riboprobe preparation. RNA probes offer several advantages over DNA probes, since these are single-stranded, and provide improved signals for hybridization blots. However, the widespread presence of ribonucleases creates some problems in their preparation and riboprobes may also be synthesized from DNA templates cloned in a vector, because these ribonucleases will cause degradation of RNA which is being synthesized. Therefore, extreme care is needed in the preparation of riboprobes, keeping all glassware ribonuclease-free (no glassware should be touched with bare hands, because ribonucleases may be available on the skin).

3.10.3.4. Labeling of Probes

The detection of homologous sequences after hybridization with the probe is like finding a needle in a haystack. Therefore, for the success of DNA probe assays, it is necessary to develop simple, safe, and sensitive techniques for their use. As probes transmit no signal of their own, they have to be either labeled with radioactive isotopes or methods are used to couple nonradioactive signal molecules to the probes without impairing the hybridization ability of these probes. These signal molecules may include fluorescent antibodies and enzymes that produce color changes in chemiluminescent catalysts.

3.10.3.5. Radiolabeled Probes

Traditionally, radioactively labelled probes are used for a variety of experiments. P is the most commonly used radioisotope, although other labels such as H, 35S, and 125I have also been used. Conventional labeling replaces a proportion of the nucleotides in nucleic acid sequences with ^{32}P derivatives (e.g., ^{32}P-UTP) or adds ^{32}P to the end of the molecule. At least three methods are available for labeling the probe: (i) Nick translation involves the creation of nicks in the probe DNA followed by extension of the broken ends using a labeled deoxyribonucleotide with the help of DNA polymerase I (or Klenow fragment of this enzyme); (ii) oligonucleotide labeling involves the use of short random oligonucleotides, which are used as primers for copying the probe in the presence of labeled deoxyribonucleotides; (iii) riboprobe preparation involves the synthesis of labeled RNA using DNA probe as a template in the presence of a labeled ribonucleotide. After hybridization with the labeled probe, hybrids are detected by autoradiography. P^{32} has the advantage over other radioisotopes since it has high specific activity. However, in general, radioisotopes have some disadvantages. They are difficult to handle and expensive to dispose of. Detection by autoradiography, while sensitive, may take a long time if there are few counts in the hybrid. Furthermore, radioisotopes have a short half-life (half-life of 14.3 days), and therefore experiments should be completed preferably within one half-life (Purohit, 2004).

3.10.3.6. Non-Radioactive Probes

1. **Biotin Labeled Probes:** Recent advances in nucleic acid technology now offer alternatives to radioactively labeled probes. One procedure that is becoming increasingly popular is biotin labeling of nucleic acids. This system exploits the affinity that the glycoprotein 'avidin' has for biotin (vitamin H). Avidin is commonly found in egg whites. Biotinylated probes are prepared through nick translation derivatives. After hybridization and washing, the detection of hybrids is done by a series of cytochemical reactions which finally give a blue color whose intensity is proportional to the amount of biotin in the hybrid. There are several advantages of using biotinylated probes. For example, these assays are prepared in advance in bulk and stored at -20°C for repeated uses. Detection technology uses very small probes of non-toxic materials, whose half-life is longer. These probes can hybridize much faster than radioactive probes. A limitation of this technology is that very small probes contain only a small number of biotinylated sites, limiting the intensity of the signal obtained. This has been solved by adding long tails of biotinylated nucleotides to the probes through enzymatic methods. Sometimes the probe does not need to be labeled with biotin, but only coupled with a tail. Another disadvantage of biotin-labeled probes is that the cytochemical visualization reactions lead to the precipitation of insoluble material that cannot be removed, and therefore, the filter cannot be reused, while in radiolabeled probes, the filters can be used repeatedly for hybridization with a number of probes one at a time.
2. **Digaxigeain Labeled Probes:** Digoxigenin is another chemical derived from plants and used for the non-radioactive labeling of

probes. An antibody associated with an enzyme (anti-digoxigenin-alkaline phosphatase conjugate) is used for the detection of the presence of digoxigenin. The probe may be labeled with digoxigenin-11-dUTP supplied with a digoxigenin kit (these kits are available from any commercial firm, e.g., Boehringer Mannheim). The labeled and denatured probe may be used for hybridization with denatured DNA on Southern blots or for in-situ hybridization. After hybridization, the membrane or the slide may be washed, and the membrane or the slide is transferred into a detection buffer containing 20 mg/ml of anti-digoxigenin fl□orescein and 5% (w/v) BSA (bovine ser□m albumin). The system is incubated for 1 hour at 37°C, and then the membrane/slide is washed in detection buffer three times (8 minutes each at 37°C). Alkaline phosphatase activity can be detected using 0.17 mg/ml BCIP (5-bromo-4-chloro-3-indolyl phosphate) and 0.33 mg/ml NBT (nitroblue tetrazolium) as dye substrate.

3. **Alternatives to Biotin and Digoxigenin Labeling:** The techniques of non-radioisotopic labels have been further expanded and new methods Am. In have DN that have been devised for attaching other ligands (e.g., hapten determinant 2, 4-dinitrophenol, arsenate derivatives, etc.) to nucleotides without hampering their ability to be incorporated into DNA. These alternatives require binding of attached ligands to specific proteins that can be monitored many probes simultaneously by using several different ligands since each ligand would yield a different. A chemiluminescent (light emitting) probe system has also been developed in which two different mRNA probes complementary to a continuous segment of DNA hybridize adjacent segments of a gene. The first label is a chemil□minescent complex that emits light at a specific wavelength; this emission excites the label.

The molecule on the second probe to emit light at a different wavelength which can be detected using a photomultiplier device. This process called non-radioactive energy transfer, can occur only if the two probes hybridize correctly and the two labels are close to each other. This system has great fidelity and provides basic technology for a homogeneous assay. In addition, DNA does not need to be immobilized and no washing steps are necessary which may be an additional advantage for large-scale testing (Boffey, 1987 Binns, 1981; Christov, 2003; Cummings, 2004).

3.10.3.7. Amplification of DNA Probe Signals

In recent years, DNA probe assays and immunoassays (using antibodies) have been competing with each other for a variety of purposes. While DNA probes have the advantage of detecting the presence of a gene rather than its product, the immunoassay has the advantage of getting an amplified target (the gene product) firstly due to transcription (several mRNA molecules can be synthesized from a single copy of the gene) and secondly due to translation (several protein molecules can be produced from the same mRNA molecule). The amplified target gives an amplified signal. Therefore, if DNA probes have to compete with monoclonal antibodies for immunoassays, DNA probe signals need to be amplified. The following devices have been used for signal amplification:

1. Probes have been developed for tRNA since rRNA genes are present in thousands, thus amplifying the target (as done by companies like Gene-Probe and Gene-Trak systems in the USA).
2. Polymerase chain reaction (PCR) technology (which utilizes a thermostable polymerase) is used to generate millions of copies of the target DNA seq□ence, which can then be identified by standard methods.
3. Signal generating capacity of the probe can also be increased by any one of the following methods: (i) concentrating more labels at the site of the target molecule by attaching multiple enzyme molecules to each of its DNA probes, or (ii) by using multiple probes, or (iii) by attaching details consult next Chapter.
4. Amplification probe technology uses multiple secondary probes (each having one or more enzyme molecules) that hybridize to m□ltiple target-specific primary probes. The secondary probe is independent of the target but helps in amplifying the signal. These approaches have multiple enzymes attached to each of the multiple probes. It is described as a Christmas Tree or Christmas Forest approach. The above amplification systems will play a significant role in the of DNA probes in the future for a variety of purposes.

3.10.4. Techniques Used in Molecular Probing

3.10.4.1. Separation of DNA Fragments Using Agarose or Polyacrylamide Gel Electrophoresis

When genomic DNA, extracted from any tissue of a plant or animal species, is digested with a restriction enzyme, it is cleaved into segments. The segments of different sizes can be separated through gel electrophoresis before a molecular probe is used to detect the segments that have sequences similar to those in the probe. Gel electrophoresis involves the movement of fragments or molecules under a high-voltage electric current. The mixture of DNA fragments is loaded in a well created on one edge of the gel. The gel may be a cylinder or a slab (usually a slab for cloning experiments), about 10 cm long and 0.5 cm thick. The rate of movement of fragments is inversely correlated with the size of fragment molecules, so heavier fragments will remain closer to the site of loading and the lighter fragments will move away. Fragments of different sizes will appear as bands on the gel and can be examined or isolated for further study. More often agarose gels are used, but for the separation of fragments differing by a few base pairs, polyacrylamide gels are used. Polyacrylamide gels are more commonly used for DNA sequencing experiments.

3.10.4.2. Separation of Large DNA Molecule Molecules Using (PFGE)

The technique of gel electrophoresis is used for separation of DNA molecules of different sizes. However, DNA molecules of large size cannot be handled in this technique. In recent years, a new technique called Pulsed-Field Gel Electrophoresis (PFGE) has been used for separating large DNA molecules, sometimes representing whole chromosomes. Using this technique, the separation of DNA molecules belonging to each of the individual chromosomes of yeast has become possible. Short pulses of electricity are used in two different directions, and DNA is embedded in agarose plugs to avoid the fragmentation of large DNA molecules. With the techniques, PFGE and a more refined technique called CHEFE (Contour Clamped Homogeneous Electric Field Electrophoresis), the genomes of several fungi could be resolved into chromosomal bands and used for mapping DNA sequences on specific chromosomes.

3.10.4.3. Southern, Northern, and Western Blotting

A mixture of DNA, RNA, or protein fragments can be separated by gel electrophoresis, and the separated bands can be stained and visualized directly in the gel. However, to confirm the identity of these bands or to find a similarity of one or more of these bands with a known and available molecular probe, it is possible to hybridize these bands with a labeled probe. To facilitate this hybridization, the bands are often transferred to a nitrocellulose technique described as blotting. When DNA bands are transferred, it is called Southern blotting (after the name of E.M. Southern); when RNA bands are thus transferred, it is described by the jargon term Northern blotting, and similarly when protein bands are transferred, the technique is described as Western blotting.

1. **Southern Blotting:** For So□thern blotting, the DNA sample is first digested with a restriction enzyme and the digested sample is gel electrophoresed. The DNA bands in the gel are denatured into single strands with the help of an alkali solution. "Subsequently, the gel is

laid on top of a b□ffer-sat□rated filter paper, placed on a solid s□pport (e.g., a glass plate), with its two edges immersed in the buffer. A sheet and a stack of many nitrocellulose membrane with a weight of about 0.5 grams are placed on top of the gel. Paper towels are placed on top of this membrane. The buffer solution is drawn up by filter paper wick and passes thro□gh the gel to the nitrocell□lose membrane and finally to the paper towels. While passing thro□gh the gel, the buffer carries with it single-stranded DNA, which binds onto the nitrocellulose when the buffer passes through it to the paper towels. After leaving this arrangement for a few hours or overnight, paper towels are discarded. The nitrocellulose membrane with single-stranded DNA bands blotted onto it is baked at 80°C for 2.3 ho□rs to fix the DNA removed by hybridization. When RNA is termed Northern the technique DNA permanently on the membrane. This membrane now has DNA bands from agarose gel and can be used for hybridization with radioactively labeled DNA or RNA probes. The membrane may then be washed to remove any unbound DNA, and X-ray film is exposed to the hybridized membrane to get autoradiographs." The above steps are involved in a replica. The sample is first ample is gel into single the gel is laid s□pport (e.g., r. A sheet of stack of many of about 0.5 s drawn up by membrane gel, the buffer.

2. **Northern Blotting:** Initially, the technique of Southern blotting used for DNA transfer from the gel to the membrane could not be used for the blot-transfer of RNA. Instead, mRNA bands from the gel were blot-transferred onto a chemically reactive paper, prepared by diazotization of aminobenzyloxymethyl paper. The technique related to Southern blotting was called Northern blotting (not after the name of any scientist as in Southern blotting). Later, it was shown that mRNA bands can be blotted firstly onto nitrocell□lose membrane, a technique that has been widely adopted. The mRNA bands blotted onto nitrocellulose membrane can be hybridized with a labeled DNA or RNA probe. The single-stranded regions of the probe are removed by nuclease (mungbean nuclease) so that quantitative estimation of hybridized mRNA can also be made.

3. **Western Blotting:** This technique is used to detect proteins of a partic□lar specificity. Partic□larly when a transferred gene is expressed in transformed cells, the translated product in the form of protein can be identified by this techniq□e. The extracted proteins are subjected to polyacrylamide gel electrophoresis (PGE)

and are then transferred onto nitrocellulose, to which they bind. Nitrocellulose membrane is then used for probing with a specific labeled antibody (antibody will not hybridize with protein, but bind with it). The antibody may be labeled with and the signal is detected again with autoradiography. If a radioactive label is not used, the bound antibody may be detected by a second antibody tagged with an enzyme.

3.10.4.4. Dots and Slot Blots

Another variety of Southern or Northern is dot blots or slot blots. In dot blots, cloned or pure extracted DNAs to be tested are spotted adjacent to each other on a nitrocellulose membrane. DNA blots thus produced are immobilized and denatured so that they are bound on the membrane as single-stranded DNA blots. "The membrane is then hybridized with radioactively labeled probe (DNA, or RNA). The dot representing the sequence related to the probe will light up autoradiography. The intensity of the dot will indicate the relative concentration of sequences related to the probe in the DNA sample used for a particular dot. No steps involving digestion with enzyme concentration of a sequence related to the probe in the DNA electrophoresis or transfer from the gel to the membrane are needed in this technique." Therefore, it is considered much more convenient when no restriction sites are needed to be studied.

Sometimes, an apparatus is used for placing spots on the membrane through slots made in this equipment. The spots made thus are in the form of oblong slots rather than round blots. These slots are used just like dot blots and are described as slot blots.

3.10.4.5. Applications of Molecular Probes

Molecular probes can be used for both basic and applied studies in the field of molecular biology and biotechnology. These uses will be discussed in detail in the different subsequent chapters of this book as appropriate but will be described briefly in this section.

3.10.4.7. Use of Molecular Probes in Restriction Fragment Length Polymorphisms (RFLPs) and Related Analyses

1. **What are RFLPs?** When genomic DNAs from each of several individuals belonging to one or more species are digested separately with restriction enzymes, electrophoresed, blotted on a membrane, and probed with a labeled DNA clone, polymorphism in the hybridization pattern is sometimes revealed and attributed to

sequence differences between the individuals. Such variation has been termed as "restriction fragment length polymorphism" (RFLP). Variation in one DNA fragment obtained with a specific enzyme is treated as RFLP.

2. **Detection of RFLPs:** An RFLP can be demonstrated using the following steps: (i) Extract and purify DNA from several individuals which may differ in some respects among themselves. (ii) Digest with a restriction endonuclease. (iii) Subject each DNA digest to gel electrophoresis separately but on the same gel slab. (iv) The DNA fragments of different lengths resulting from digestion can be separated by gel electrophoresis, but the differences in the distribution patterns of fragments in several DNA samples due to digestion by several endonucleases cannot be detected directly. This is because the number of fragments is large and the range in size is rather continuous, such that they form a continuous smear on the gel. Electrophoresis is, therefore, followed by the following steps. (v) Fragments are transferred from gel to nitrocellulose filters using the technique of Southern blotting. (vi) The filters with small fragments, fully or partially homologous to the probe, can be detected by autoradiography after hybridization. Since repetitive DNA may hybridize with many fragments, they will give a smear on autoradiography. Therefore, unique DNA sequences are generally used as probes for detecting RFLPs. The DNA fragments hybridizing with the probe can be visualized by autoradiography and will be called RFLPs at the phenotypic level. In results of a hypothetical experiment are shown where RFLPs are detected by using four plants (which differ by point mutation as well as by insertions) and three enzymes individually and in combination.
3. **RFLPs for Evolutionary Studies:** The restriction fragment length polymorphisms (RFLPs) can be studied in a set of related species, using a random or a specific DNA probe. The similarities and differences can be used to infer phylogenetic relationships. This has actually been done in a number of cases, both in plants and in animals.

3.10.5. RFLP Maps and Linkage of RFLPs with Specific Genes

RFLPs have been used to prepare chromosome maps in humans, mice, fruit flies, and in Mendelian markers for genetic mapping. This is sometimes limited due to the non-availability of mutants. The list of markers can be extended by using molecular markers, which are examined in the form of RFLPs, including maize, tomato, lettuce, and rice. The use of Mendelian markers for genetic mapping is sometimes limited due to the non-availability of mutants. The list of markers can be extended by using molecular markers, which may be examined in the form of RFLPs.

Once a large number of RFLPs are available in a species, the parents, F1, and F2 generations can be used to study their inheritance and linkage relationships, and genetic linkage maps can be prepared. They have been assigned to specific chromosomes using monosomics in maize, and efforts are being made to relate them with morphological and economic traits so that they can be used for practical plant breeding.

A difference in restriction maps between two individuals or an RFLP can be used as a genetic marker in much the same way as any other phenotypic marker. The polymorphism at the molecular level may be due to the presence or absence of a restriction site. It shows how the presence of an additional site will produce two fragments and its absence will produce only one fragment. Although there may be a linkage between a molecular marker and a phenotypic trait, the change in the molecular map (restriction map) may not affect the phenotype. In order to study the linkage relation between a molecular marker and present in the DNA shown parents that differ morphologically as well as for molecular markers. We may examine the restriction fragment patterns and the phenotypes of making crosses between two such parents, we can estimate the recombination frequencies between molecular markers and phenotypic markers using a test cross. In this, we used a hypothetical example showing the linkage of red and white eyes with different molecular markers in Drosophila. "It can be seen that as a result of recombination, in test cross, 15% progeny shows association of red eye with a molecular marker associated with the white eye in the parent, and vice versa. Therefore, it is concluded that the restriction marker is 15 map units away from the marker's eye color although it has no causal relation with eye color. Such a linkage between molecular markers and phenotypic markers allows the identification of genetic loci at the molec□lar level. A tight linkage of such a molecular n with disease resistance may also be recommended as a tool for the identification of disease resistant c□ltivars in many of o□r crop plants and will be increasingly utilized in the future. Molecular markers are also being used for the selection of plants at the seedling stage for plant breeding purposes."

3.10.6. Use of Molecular Probes in Molecular Cytogenetics

1. **Isolation of Gene or Related Sequence:** This can be facilitated through the use of a probe carrying a known sequence. For this purpose sequences may be hybridized following any one of the following techniques:

i. Southern blots can be prepared (from genomic DNA or cDNA) to isolate DNA corresponding to a band that hybridizes with the probe. This DNA can then be used for cloning experiments to get a partial genomic library or cDNA library (see later in this chapter for details).

ii. The genomic or cDNA library can be screened through colony hybridization or plaque hybridization to identify and isolate the clone carrying the desired sequence; sometimes step (ii) is not used for complete genomic/cDNA libraries can be prepared and used in step (iii). The identified clone should carry the gene or sequence of interest, which can be multiplied and the sequence or gene retrieved according to standard procedure.

2. ***In Situ* Hybridization (ISH) for Location of Sequences on Chromosomes:** A molecular probe can also be used for in situ hybridization where chromosomes in mitotic or meiotic preparation are used for hybridization with the probe (without extracting DNA). In a number of cases, rye chromosomes in wheat background have been identified using this technique. Sat-chromosomes with NORs (nucleolar organizing regions) are recognized using probes for ribosomal DNA as done in many wheats and can be identified using a D genome-specific probe. Telomeres have been identified in human eukaryotes using a telomeric sequence as a molecular probe, showing that all chromosomes carry the same sequence. In D genome telomere-related species, chromosomes of genome-specific probe (pASI separated from Aegilops) can be cited where the use of molecular probe hybridization served a very useful purpose.

i. **Use of Molecular Probes for Human Health Care:** DNA probes are being extensively utilized for the diagnosis of diseases caused by parasitic *Protozoa* and *Helminths.* They are also used for antenatal diagnosis of congenital diseases to allow advice on abortion of the fetus, if desired. Similarly, probes have been designed for the diagnosis of a number of sexually transmitted diseases.

ii. **Use of Molecular Probes in DNA Fingerprinting:** With the help of probes prepared from human minisatellite DNA dispersed throughout the human genome, it has become possible to identify individual humans, as the possibility of any two humans (except identical twins) showing the same pattern of fingerprinting is remote. This aspect is already playing an important role in identifying criminals like murderers and rapists. This subject will also receive a detailed treatment.

3.10.7. Construction and Screening of Genomic and cDNA Libraries

In order to isolate one or more related genes from a genome, we like to prepare a mixture of clones each carrying DNA derived either from the genomic DNA or from cDNA (derived from the mRNA isolated from a specific metabolically active tissue of an organism). This mixture may contain thousands of clones, which when derived directly from the genomic DNA are collectively called a genomic library. Similarly, when these clones are derived from cDNA, they are collectively called a cDNA library construction and the use of these libraries will be discussed in this section.

3.10.7.1. Genomic Library by Shotgun Experiment

Cloning an entire genome in the form of a library of random genomic clones (without identifying them) is often called a shotgun experiment. In this experiment, genomic DNA is extracted, broken into fragments of reasonable size by a restriction endonuclease, and then inserted into a cloning vector to generate a population of chimeric vector molecules. A set of fragments cloned in this manner is called a genome library. Once such a library is available, then clones can be perpetuated indefinitely in a plasmid vector and retrieved whenever needed for a variety of purposes, including identification and isolation of a gene when a specific probe is available.

i. Genomic libraries can be prepared by using a number of restriction endonucleases, one at a time so that fragments of varying sizes having different places in the genome will be available. However, this may lead to cuts at inconvenient places, including sites within a gene, so that having complete genes will be diffic□lt to obtain. In order to overcome this diffic□lty, we □se the following strategy in the shotgun experiment: (i) We use restriction endonucleases, which have short (4 bp) recognition sequences, so that such a sequence may be frequently distributed.
ii. Conditions are used that give only partial digests, so that a particular restriction site is only occasionally cleaved, and long fragments without having any breaks on recognition sites available within a gene can be easily obtained. This technique of shotgun experiment leads to the construction of a random genomic library, in which all fragments have the same fragment ends thus helping retrieval of a fragment from the vector with the help cuts of the same enzyme. The number of fragments representing every sequence of the genome increases with genome size. For instance, for a probability level of 99 that all sequences are present in our library of a species, we may

need 1,500 cloned fragments for *E. coli*, 4,600 for yeast, 48,000 for Drosophila melanogaster and 800,000 for a mammal like human being. Libraries reaching these desired limits have been prepared in all these cases.

3.10.8. cDNA Library from mRNAs

Complementary DNA (cDNA) libraries can also be prepared by isolating. As from tissues which are actively synthesizing proteins, like roots in plants, ovaries or reticulocytes in mammals, etc. The mRNA is copied into cDNA through the use of reverse transcriptase. The mRNA and leaves are used for copying it into cDNA through the use of re cDNA molecule can be made double stranded, and cloned. However, cDNA clones will differ from genomic clones in lacking the introns present in split genes, but the advantage of being capable to be expressed in bacteria, which do not have the machinery to process HnRNA (obtained cDNA clones from split genes) mRNA.

3.10.9. Colony (Plaque) Hybridization for Screening of Libraries

Once a genomic library or cDNA library is available, we may like to use it for the isolation of a gene sequence. This can be achieved by the colony hybridization technique. In this technique, bacterial transformed chimeric vectors are grown into colonies, which are lyzed on nitrocellulose filters. Their DNA is denatured in situ and fixed on the filter, which is hybridized with a radioactively labeled probe carrying a sequence related to the gene to be isolated (usually a cloned cDNA for screening of a genomic library). Colonies carrying this sequence will be identified by dark spots after autoradiography so that the original chimeric vector carrying the desired gene sequence can be recovered from one or more colonies in the original master plate and used for further experiments. This technique is described as colony hybridization. It is possible that the probe may identify more than one clone or that a gene is fragmented in the library. In such a case, one needs to reconstruct the desired sequence using several overlapping sequences available in the library. This is a very routine exercise whenever we want to isolate specific DNA sequences from the genome of a species or from cDNA derived from mRNA of a specific tissue of a species.

Sometimes the library may be available not in the form of bacteria formed with chimeric DNA molecules but in the form of chimeric particles carrying the cloned segments. In such a situation, a bacterium is infected with a mixture of chimeric phage particles (i.e., the library) and a large number of plaques develop overnight. These plaques can be treated just like the colonies in colony hybridization to identify and isolate the chimeric phage particle carrying the gene of interest. This technique is then described as plaque hybridization.

3.10.10. Chromosome Walking and Characterization of Chromosome Segments

When a probe is used for the identification of a gene sequence in a genomic library, the probe may hybridize with a number of clones, each carrying a part of a large gene fragmented during the preparation of the genome library. If we obtained partial digests (by digesting the DNA only partially) from the genome, different genomes (from a large number of cells) may give fragments which have overlapping sequences, because sites cleaved in different genomes of the same organism will differ randomly. Since one of these fragments may have entire sequences represented in the probe, overlapping sequences may be used to construct the original genome sequences. Identification of fragments with overlapping sequences may be a key to the reconstruction or characterization of large chromosome regions. This is achieved by the technique popularly described as chromosome walking.

The technique of chromosome walking involves the following steps: (i) from the genomic library select a clone of interest (identified by a probe) and subclone a small fragment from one end of the clone (there is a technique available to subclone a fragment from the end); (ii) the subcloned fragment of the selected clone may be hybridized with other clones in the library and a second clone hybridizing with the s□bclone of the first clone is identified d□e to the presence of an overlapping region; (iii) the end of the second clone is then subcloned and used for hybridization with other clones to identify a third clone having an overlapping region with the subcloned end of the second clone; (iv) the third clone identified as above is also s□bcloned and hybridized with clones in the same manner and the procedure may be continued; (v) a restriction map of each selected clone may be prepared and compared to know the region of overlapping, so that identification of a few overlapping restriction sites will amount to walking along the chromosome, or along a long chromosome segment."

Regions of chromosomes approaching 1000 kb have been mapped following the above technique. Restriction maps of entire chromosomes can be prepared in this manner following the technique of chromosome walking.

3.10.11. Reverse Genetics and Chromosome Jumping (or Hopping Libraries)

In most cases, we identify the genes by their protein products, which help in the isolation and cloning of a specific gene. However, for many human disorders which are each controlled by a single gene (e.g., Huntington's disease, cystic fibrosis, Duchenne muscular dystrophy, etc.), the gene product is not known despite intensive investigations. Such genes are cloned and often identified primarily by their map position.

Once the gene is cloned, it can be utilized to obtain information about the encoded protein, gene mutations, and metabolic defects that may be associated with the disease. Since the information about them is coming in the reverse order, this is popularly described as the reverse genetics approach, which involves the following two steps: (i) locating the gene to a particular chromosome through RFLP linkage analysis; (ii) mapping the position of the gene with respect to molecular markers, both through recombinational analysis (in terms of recombination in cM=centiMorgan units) and by physical mapping (in terms of kilobase pairs) with p□lsed field gel electrophoresis (PFGE) and preparation of a restriction map. Even if the distance is 1 cM, which is the limit of resolution, this will be equivalent to 1,000 kb. Therefore, if the segment carrying the gene is to be cloned using the linked molecular marker, such a segment will be too large to be cloned. In order to bring the molecular marker close to the gene of interest, the 'chromosome jumping' approach has been utilized. The technique of 'chromosome jumping' is based on the following steps depending upon the distance between the gene and the marker, decide about the distance of jumps or 'hopsize' (e.g., 100 kb or 200 kb): (i) genomic DNA molecules in the range of selected size (say 80 kb-130 kb in case of 100 kb 'hopsize are 16)-240 kb for 'hopsize' of 200 kb) are selected thro□gh p□lsed-field gel electrophoresis (ii) for circularization of DNA segments, ligation between two ends of each long linear DNA molecule was allowed using T4 ligase in the presence of $supF^+$ (iv) DNA circles obtained; (ii) above are digested with EcoRI; (v) the vector iCh3A A lac, an amber mutated phage vector (supF" is also cut with EcoRI and used for cloning mal DNA fragments representing the junctions of the circularized genomic molecules and carrying sup F; (vi) the cloned DNA fragments and in(v) above represent the jumping library, which can be plated on the host and screened through the technique of plaque hybridization described earlier.

The above technique of chromosome jumping will help narrow the gene and available molecular markers. After several cycles, the jumping followed by cloning the regions that are close to e to approach very close to the desired gene and clone it. It is th□s obvio□s that in the field of 'reverse genetics' □sing the technique of chromosome jumping, it will characterize genes, whose gene prod□cts are □nknown. The gene for cystic fibrosis has recently been isolated □sing this techniq□e. It co□ld also be shown that 'cystic fibrosis' disease is caused by a single base substitution in this gene. This technique of 'reverse genetics' will become a very active area of research particularly in human genetics in the coming years.

The term 'reverse genetics' has, however, been redefined recently. Classical genetics, sometimes described as 'forward genetics,' involves a study that starts from the phenotype and follows thro□gh the identification of the gene, concluding with the isolation and sequencing of the gene. In view of this, if 'reverse genetics' has to use steps in the reverse order for a genetic study, then it should start with DNA molecules or clones with unknown effects and should conclude with the determination of the phenotype that it controls. This can be achieved by preparing random genomic DNA clones, which can subsequently be utilized for a genetic study in reverse order. This will then be described as 'reverse genetics.' This definition of 'reverse genetics' differs from the earlier conventional meaning of this term, where 'reverse genetics' was used to describe cases of genes with novel products, which are isolated (with the help of phenotype and its linkage with molecular markers) and then studied in some detail.

3.10.12. Isolation, Sequencing, and Synthesis of Genes

Molecular manipulation of specific genes is a research focus in biotechnology. These genes are isolated or artificially synthesized before being manipulated and used for transformation, leading to the production of transgenic animals and plants. Genes are also sometimes sequenced for a better understanding of their structure, which then needs to be manipulated. Therefore, we discuss genes as one of the very popular areas that should have technology available for the isolation, sequencing, and synthesis of specific genes in this chapter.

3.10.12.1. Isolation of Genes

During the last two decades, significant progress has been made in the techniques for the isolation of a variety of genes, including those for ribosomal RNA; specific protein products; phenotypic traits with unknown products; and those for regulatory functions, e.g., promoter genes, etc. Different techniques have been used for the isolation of these different types of genes and will be briefly described in this section.

Early attempts for isolation of ribosomal RNA and proteins. As we know, ribosomes consist of ribosomal RNA (rRNA) and proteins. Ribosomal RNA (rRNA) can be of four sizes, namely, 28S, 16S, 58S, and 5S (S stands for Svedberg units, a measure of speed with which a molecule sediments on centrifugation). This ribosomal RNA makes up 80% of cellular RNA and is synthesized on ribosomal genes, which could be isolated. Isolation of ribosomal genes was considered convenient due to their following three characteristic features: availability of homogeneous rRNA; differences between ribosomal RNA genes and other genes due to relatively high G + C content in rRNA

(rRNA has 45–60% G + C while the remaining RNA has only 40% G + C); ribosomal genes are present in multiple copies, their number within several thousands. In view of the above, ribosomal RNA genes were isolated for the first time (in 1965) in an amphibian named *Xenopus*, by Hugh Wallace and Max L. Birnstiel, working at the University of Edinburgh (UK). The following steps were involved in this isolation: (i) RNA was isolated from ribosomes of Xenopus and made radioactively labeled, due to its replication in a medium containing tritiated uridine; (ii) ribosomal DNA was isolated by density gradient centrifugation followed by its denaturation (G + C content of rDNA differs from that of bulk DNA and helps in separation by centrifugation); (iii) single-stranded DNA was fixed on filter paper; (iv) labeled rRNA was added on filter paper carrying single-stranded DNA; (v) DNA-RNA hybridization was allowed to take place; (vi) excess labeled RNA was washed; (vii) radioactivity was measured and duplex hybrids isolated, which on denaturation, gave single-stranded DNA that could be made double-stranded. Isolation of ribosomal DNA allowed the first characterization of ribosomal genes, which was followed by the isolation of 5S genes by Donald Brown. For a detailed account of the isolation of ribosomal genes in *Xenopus* as originally done, readers are referred to an article "Isolation of Genes" in Scientific American (A□g□st 1973).

3.10.12.2. Isolation of Genes Coding for Specific Proteins

Isolation of genes for specific proteins became possible only after the discovery of the reverse transcriptase enzyme in 1970. This enzyme can be easily used for the synthesis of copy DNA or complementary DNA (cDNA) from mRNA. This complementary DNA can then be used for the isolation of the corresponding gene from genomic DNA. It is, therefore, obvious that for the isolation of a specific gene, techniques should first be available for the isolation of specific mRNA. For this purpose, antibodies are produced against a specific protein for which the gene is to be isolated. Therefore, isolation of a gene coding for a specific protein involves the following basic steps: (i) purification of the protein product of the gene; (ii) production of antibodies against this protein product by immunizing animals like rabbits or mice; (iii) precipitation of polysomes engaged in synthesizing specific proteins with the help of the antibody raised in step (ii); (iv) isolation and purification of mRNA from the polysome fraction; (v) using mRNA for synthesizing cDNA with the help of reverse transcriptase; (vi) cloning of cDNA thus synthesized for the preparation of a cDNA library; (vii) immunological and electrophoretic analysis of translation products of cDNA clones to identify specific cDNA meant for the specific protein in mind; (viii) use of the specific DNA probes selected in step (vii) for the identification and isolation of the gene from genomic DNA through screening a complete or partial genomic library. (For the preparation of cDNA and genomic DNA libraries, consult Chapter 3). In step (vi) above for the preparation of the cDNA library, cDNA may be cloned in expression vectors (vectors in which genes can be expressed due to the presence of promoter sequences for RNA polymerase enzyme) like λgtll, which can accept gene insertions into the β-galactosidase gene. The chimeric vector will then produce a hybrid protein if correctly expressed. The hybrid protein can be identified □sing antibodies as described above. Plaques belonging to the chimeric vector carrying the desired gene can also be identified by the reaction of an antibody attached to a radioactive protein. This procedure was used to clone a number of genes including a gene regulating zein synthesis in maize.

The cDNA clones may also be used directly for gene manipulation. However, when a cDNA clone is used, it is described as a synthesis of a gene rather than as an isolation of a gene, as cDNA has been artificially synthesized. The above technique for the isolation of genes has now been successfully utilized both in plants and animals. However, the first genes isolated were those which existed in m□ltiple copies and were expressed at high levels in specific tiss□es, e.g., ovalbumin gene in chicks, the globin and immunoglobulin genes in mice, genes for storage proteins in cereals and legumes, amylase genes in barley, actin genes in some legumes, etc.

3.10.13. Isolation of Genes Which Are Tissue-Specific in Expression

It is much easier to isolate genes that are expressed in specific tissues. For instance, genes for storage proteins are expressed only in developing seeds, the oval albumin gene is expressed in the oviduct, or the globin gene is expressed. Such genes can be easily isolated because mRNA extracted from erythrocytes from these specific tissues will either exclusively belong to the gene of interest or will be rich in this species of mRNA. Other mRNA molecules in minor quantities can be eliminated since these can be identified through their isolation from tissues where this gene is silent. This strategy, which was actually followed for the isolation and cloning of carrot genes expressed during the development of somatic embryos, and also for the isolation of several genes for storage proteins in crops, is outlined.

1. **Isolation of Genes Using DNA or RNA Probes:** Specific molec□lar probes (whether DNA or RNA probes), if available, can be used for the isolation of specific genes. These probes may be available either from another species for the same gene or may be artificially synthesized using a part of the amino acid sequence of the protein product of the gene.
2. **Use of Heterologous Probes:** The probes were obtained from one of the species of heterologous probes. These are effective in identifying

genes. Heterologous probes have been found to be effective in identifying blots. For instance, the gene for chalcone synthase was isolated from *Antirrhinum majus* and *Petunia hybrida* using a heterologous cDNA probe from parsley. Similarly, heterologous probes from maize were used for the isolation of barley genes. These should be used with the cDNA library and not with the genomic library, since vectors (e.g., pGEM-4 blue) can also be used for obtaining the waxy gene and aleurone gene. Heterologous probes should be used in the latter case for unrelated genes or pseudogenes. These heterologous clones, if available, have been found to be more sensitive and efficient.

3. **Use of cDNA or Synthetic Probes:** We have already discussed the cDNA probes above. However, if protein purified using the technique of two-dimensional gel electrophoresis is used for 5–15 consecutive amino acids, this information can be used for the synthesis of oligonucleotides (using automated DNA synthesizers). These oligonucleotides may then be directly utilized for screening cDNA genomic libraries for isolation of specific genes. They can also be used for designing primers for the amplification of a gene in DNA extracted from a tissue. A 32 kilodalton (KD) glycoprotein (isolated from the style) associated with the S2 incompatibility allele of *Nicotiana alata* was used for the synthesis of an oligonucleotide leading to the isolation of a related gene.

4. **Isolation of Genes Coding for Unknown Products:** In some cases, we are interested in the isolation of a gene whose phenotypic effect is known, but the gene product has not been identified or cannot be isolated. Such genes include those for morphological traits like dormancy, photoperiodicity, disease resistance, etc. This area of research, in which genetics is studied by isolating the gene first without knowing the gene product, is often described as reverse genetics. The method used for the isolation of these genes is different from those used for genes coding for known proteins.

5. **Use of Transposable Elements (Transposons Tagging):** The transposable elements (TEs), in some cases, have been effectively utilized for isolation of genes when the gene product is unknown. In this case, a transposon works as a mutagen and therefore as a gene tag. The following steps are involved in this procedure: (i) Clone a known gene with a scorable phenotypic effect. (ii) TE is transposed to this gene to create an unstable allele. (iii) This unstable allele is cloned and the TE is isolated from the unstable allele (this is to select a TE which can produce an unstable gene with an allelic effect); (iv)

This TE is transposed to a gene of interest with a known phenotype to produce an unstable allele. (v) The DNA is extracted from this plant. (vi) TE sequence is used as a probe to isolate and clone the mutant (carrying the inserted TE), so that we can then isolate the gene of interest.

In maize, TEs like Ac Ds, En/Spm, and Mu1 have been isolated using Adh1. Similarly, TEs like Tam3 and Tam7 have been isolated from snapdragon (*Antirhinum majus*). These TEs have been used in experiments leading to the isolation of genes. In maize, several genes such as P, Al, CI, and C2 have been successfully isolated using the gene-for-gene tagging method.

For transposon tagging, often transposable elements endogenous to like maize and snapdragon have been used However, rarely transposons available from one plant species can be moved into the genome of another plant species, whose gene is to be isolated. For instance, the Ac element of maize has been transferred to tobacco, where it can integrate into any locus permitting transpose tagging and gene isolation.

6. **Mutant Complementation:** In this technique, DNA clones from the selected that should be able to complement the wild-type strain are mutants that are thus transformed into wild-type. Once these DNA clones are available, the protoplasts derived from the mutant plant may be transformed using wild-type clones and the transgenic plants are produced. Once this is done, the gene of interest can be isolated from DNA extracted from the transformed plants using wild-type complementary clone as a probe.

7. **Use of RFLP Maps or Chromosome Walking for Gene Isolation:** In recent years, RFLP (restriction fragment length polymorphism) maps have been prepared in maize, tomato, rice, etc. Linkage of RFLP loci with genes of interest is also being worked out as an aid to plant breeding. If RFLPs are known which are very close to the target gene (as may be done through RFLP maps or chromosome walking), thereby locating the RFLPs on either side of the target gene, a long intervening DNA segment may be isolated. This fragment may be used for subcloning leading to the isolation of the desired gene. If the RFLPs are examined in two near isogenic lines (NILs) differing for a single gene, the difference in RFLP maps will help in locating the position of the gene on the molecular map which can then be utilized for isolation of the gene.

8. **Use of Chromosome Jumping for Gene Isolation:** We described

the technique of chromosome jumping for walking long chromosome distances thus bringing the available known sequence (related to the available probe) closer to the gene of interest. Once this has been eyed, the techniques described in the previous section may be conveniently used for isolating this gene for further study.

9. **Isolation of Novel Genes by Asis Datta at JNU, New Delhi:** Recently (in 1992), Prof Asis Datta of JNU, New Delhi was selected for the Birla Award for Science and Technology for cloning and characterization the following two novel genes, relevant to human health.

   i. **Gene for Oxalate Decarboxylase:** In 1991, Prof. Asis Datta's laboratory reported for the first time the cloning and characterization of the oxalate decarboxylase gene, which degrades and thus reduces the content of oxalate or oxalic acid, which is harmful in many ways including the following: (a) Several green leafy vegetables (e.g., *Amaranthus, spinach*, etc.) being rich in oxalate, if ingested in large quantities are toxic, since oxalate chelates calcium and also destroys renal tissues in the kidney. (b) The attack and spread of a fungus (*Whetzelinid sclerotiorium*), which causes damage to crops like s□nflower, involves the acc□m□lation of oxalic acid in the infested tissue. (c) Oxalic acid is an essential substrate for the synthesis of the neurotoxin called 3-N-oxalyl 1-La, B diaminopropionic acid (ODAP) found in different parts of the plants of *Lathyrus sativus* (khesari dal). Consumption of this legume causes neurolathyrism, a well-known disease causing damage to leg muscles, etc. ODAP interferes with the metabolism of glutamic acid, which is involved in the transmission of nerve impulses in the brain, so that despite its rich protein content *L. sativus* cannot be used as a food so□rce. In all the above cases (leafy vegetables, s□nflower & *L. sativus*), the transfer of a single gene for oxalate decarboxylase to produce transgenic plants will improve these plant species by reducing the oxalate content. This new gene will also allow the development of a diagnostic kit to measure oxalate in blood and urine, since prevalent methods for clinical tests are quite expensive.

   ii. **Gene for a Seed-Specific Protein with Nutritionally Balanced Composition of Amino Acids:** Prof. Datta's laboratory recently reported, for the first time in 1992, the isolation of a gene from Amaranthus that encodes a seed-specific protein rich in all essential amino acids, incl□ding lysine and sulfur-containing amino acids. Interestingly, the amino acid composition of this protein corresponds to the WHO recommended protein standard for optimum human nutrition. It is known that seed proteins in cereals as well as in leg□mes (peas, etc.) are deficient, with cereal proteins lacking lysine and pulses lacking sulfur-containing amino acids. In light of this, the gene isolated from *Amaranthus* will be utilized in the future to compensate for amino acid deficiencies in proteins once it is genetically engineered into the target crop plants.

3.10.14. Sequencing of a Gene or DNA Segment

Once a gene or DNA fragment has been cloned, further study involves DNA sequencing. The technique of DNA sequencing was very laborious until 1975, when a breakthrough was made in DNA sequencing methods. Two different methods are now routinely used for determining DNA sequences:

1. **Maxam and Gilbert's Chemical Degradation Method:** In this method, which is illustrated, the following steps are involved: (i) Label the 3' ends of DNA with 32P. (ii) Separate two strands, both labeled at 3' ends. (iii) Divide the mixture into four samples, each treated with a different reagent having the property of destroying either only G, or only C, or 'A and G' or T and C. The concentration of the reagent is so adjusted that 50% of the target base is destroyed, resulting in fragments of different sizes with 32P. (iv) Electrophoresis of each of the four samples in four different lanes of the gel. (v) Autoradiograph the gel and determine the sequence from the positions of bands in the four lanes.
2. **Sanger's Dideoxynucleotide Synthetic Method:** Fred Sanger (who won the Nobel Prize twice) had initially developed a method for DNA sequencing, which utilized DNA polymerase to extend DNA chain length. This was termed the plus-minus method. Subsequently, he developed a more powerful method, utilizing single-stranded DNA as a template for DNA synthesis, in which 2,'3' dideoxynucleotides were incorporated leading to the termination of DNA synthesis. These dideoxynucleotides are used as triphosphates (ddNTP) and can be incorporated in a growing chain, but they terminate synthesis, since they cannot form a phosphodiester bond with the next incoming deoxynucleotide triphosphate (dNTP). The following steps are involved in Sanger's dideoxy method for DNA containing single-stranded DNA sample (cloned in M13 phage) to be sequenced: all

four dNTPs (radioactively labeled) and an enzyme for sequencing. (i) Four reaction tubes are set up, each containing single-stranded DNA sample to be sequenced, all four dNTPs, and an enzyme for DNA synthesis. Each tube also contains a small amount (much smaller amount relative to four dNTPs) of one of the four dNTPs, so that four tubes have each a different ddNTP, bringing about termination at a specific base like adenine (A), cytosine (C), guanine (G), and thymine (T). (ii) The fragments, generated by random incorporation of dNTP leading to termination of the reaction, are then separated by electrophoresis on a high-resolution polyacrylamide gel. This is done for all the four reaction mixtures on adjoining lanes in the gel. (iii) The gel is used for autoradiography so that the position of different bands in each lane can be visualized. (iv) The bands on the autoradiogram can be used to obtain thc DNA sequence.

A variant of the above dioxy method was later developed, which has allowed the production of automatic sequencers. In this new approach, a different fluorescent dye is tagged to the oligonucleotide primer in each of the four reaction tubes (blue for A, red for C, etc.). The four reaction mixtures are pooled and electrophoresed together in a single polyacrylamide gel tube. A high-sensitivity fluorescence detector, placed near the bottom of the tube, measures the amount of each fluorophore as a function of time. The sequence is determined from the temporal order of peaks corresponding to four different dyes.

3.10.15. Direct DNA Sequencing Using PCR (Also Called Ligation Mediated PCR LMPCR)

Polymerase chain reaction (PCR) has also been used for sequencing the amplified DNA product. This method of DNA sequencing is faster and more reliable and can utilize either the whole genomic DNA or cloned fragments for sequencing a particular DNA segment. The DNA sequencing using PCR involves two steps: (i) generation of sequencing template (double-stranded or single-stranded) using PCR; and (ii) with the PCR products either with the thermolabile DNA polymerase or with the method or thermostable Taq DNA polymerase. Thus, the DNA sequencing using PCR eliminates the need to clone the DNA in single-stranded DNA phage vector, i.e., M13. Although double-stranded DNA product difficulty can be utilized for sequencing, this may cause difficulty due to reassociation after denaturation, thus preventing the sequencing from annealing to its complementary sequence to allow extension. To reduce this problem, either a variant of the standard method for sequencing double-stranded DNA is employed or single-stranded DNA templates are produced following asymmetric PCR earlier.

A number of thermolabile DNA polymerases have been used for sequencing of *in vitro* amplified DNA. Alternatively, Taq polymerase (a thermostable enzyme) used for PCR can also be used for sequencing reactions. In either case, Sanger's synthetic method involving the incorporation of dideoxynucleotide (ddNTP) for chain termination is used for sequencing. However, Maxam and Gilbert's chemical degradation method can also be used.

In Sanger's method (Sanger sequencing and chain termination method), as usual, four mixtures are prepared, each using one of the four ddNTPs. The sequencing primer is labeled with ^{32}P (fluorescently labeled), and the mixtures with amplified DNA, Taq polymerase, and appropriate buffer are incubated at 70°C for 5 minutes. There is no hydroxyl available, and no further nucleotides can be added. The chain termination with the dideoxynucleotide, which is marked with the color of dye depending on the base, and mixtures are run on a polyacrylamide sequencing gel to obtain the ladders, which can be read by a computer or manually. This method using PCR helped in the automation of DNA sequencing.

3.10.16. Synthesis of Genes

There are two approaches available for the synthesis of genes: (i) when the detailed structure of a gene is available, this gene can be synthesized by a purely chemical method as done by H.G. Khorana for the synthesis of the gene for a tRNA (reported for the first time in 1970). (ii) If the detailed nucleotide sequence of the gene is not available, one may utilize the 'RNA-directed DNA polymerase' enzyme for the synthesis of the gene in question in the form of complementary DNA (cDNA) from the mRNA of the gene isolated in its pure form. As discussed above, cDNA can also be used for isolation of a gene either (i) from DNA extracted from living cells utilizing the technique of Southern blot hybridization, or (ii) from a genomic library.

3.10.16.1. Chemical Synthesis of tRNA Genes

As outlined above, before one can start the chemical synthesis of a gene, the structure of a gene should be known. The structure of a gene could not be worked out by direct chemical analysis earlier, since there were no means for isolating a gene (techniques for isolation of genes were developed later and were described earlier in this chapter). The structure of a gene could, therefore, be inferred only from its product. For instance, if a gene is responsible for giving rise to a polypeptide chain and the structure of this chain is known, then from the genetic code dictionary, the structure of the gene could be easily inferred. Such genes were initially considered to be too long to be synthesized because an average gene contains about 1,500 base pairs. On the other hand, since tRNA molecules are fairly small in size (about 80 nucleotides), a gene responsible for giving rise to a tRNA molecule was within reach for synthesis.

3.10.16.2. Synthesis of the Gene for Yeast Alanyl tRNA

As readers may know, structures of a large number of tRNA molecules are now known. The first RNA whose structure could be known was yeast alanyl tRNA. R.W. Holley, who died in early 1993, and his coworkers (in 1965) gave the detailed structure of yeast alanyl tRNA. This information was immediately used by Khorana and his coworkers to deduce the structure of the gene for yeast alanyl tRNA. The structure of this gene would obviously be such that one of the two strands of DNA (gene) would be complementary to the base sequence of yeast alanyl tRNA. The other strand would then automatically have the same sequence as in tRNA, except that in place of U of tRNA, there would be thymine in DNA.

Synthesis of such a long chain (77 base pairs) of double-stranded DNA was rather difficult in 1965 (not in the 1990s), but Khorana and coworkers had extensive experience in synthesizing DNA of known base sequences. It was decided that such a long chain could not be synthesized by adding a single base each time. Therefore, it was decided that small oligodeoxyribonucleotides ranging in length from 5 to 20 nucleotides should first be synthesized. These segments would be single-stranded and would cover the whole length of both strands of DNA. These would then be joined to form double-stranded DNA, 77 nucleotide pairs long. In the actual synthesis of yeast alanyl tRNA gene, the following steps were involved:

- **i.** **Synthesis of Oligonucleotides:** 15 oligonucleotides ranging from pentanucleotide (5 bases) to an icosanucleotide (20 bases) were synthesized. Synthesis was conducted through condensation between the hydroxyl group in the 3' position of one nucleotide and the phosphate group at the 5' end of the second nucleotide. In order to bring about condensation, all other functional groups not involved in condensation were protected using specific protective groups. After protecting the groups, the reaction between a nucleotide with a protected 5' end and another nucleotide with a protected end proceeded. Subsequently, condensation also occurred between groups of two, three, or four nucleotides.
- **ii.** **Synthesis of Three Duplex Fragments:** With the help of 15 nucleotides (single-stranded), three duplex fragments, each with a single-stranded end, were prepared. These three fragments consisted of the following: (i) the first 20 nucleotides, with nucleotides 17–20 single-stranded; this fragment was called 'A'; (ii) the nucleotide

residues 17–50 with single-stranded regions 17–20 and 46–50, a segment called 'B'; and (iii) the nucleotide residues 45–77 with single-stranded regions 17–20 and 46–50; this segment was called 'C.'

iii. **Synthesis of Gene from Three Duplex Fragments:** The three fragments 'A,' 'B,' and 'C' were joined to give a complete gene in each of the two following ways: (i) In one scheme, 'A' was joined to 'B' taking advantage of the overlap in residues 17–20; 'C' was then added, with the overlap in the region 46–50 residues. The complete double-stranded DNA with 77 base pairs representing the gene was thus ready. (ii) In the other scheme, 'B' was first added to 'C' and to this 'A' was added in the end to obtain the complete gene. This gene, which was synthesized in 1970, was hopefully used for subsequent work, since this gene could replicate and make its own copies.

iv. **Synthesis of the Gene for a True Precursor tRNA:** Before Khorana could complete the synthesis of the gene for yeast alanyl tRNA in 1970, as outlined above, it became obvious that tRNA was not the direct product of transcription. Instead, a precursor molecule is first synthesized which subsequently, after losing segments of RNA by cleavage, gives rise to tRNA. Obviously, therefore, the actual gene for yeast alanyl tRNA will be longer than the DNA duplex synthesized by Khorana. In view of this, Khorana subsequently initiated the synthesis of a gene for *E. coli* tyrosine suppressor tRNA precursor. The DNA duplex which will give rise to this tRNA precursor was synthesized in the form of 26 small oligonucleotide segments. These were then arranged into six DNA duplex fragments having single-stranded ends. These six fragments gave rise to the presumed gene for *E. coli* tyrosine suppressor tRNA precursor. This gene, however, still had a promoter region and other sequences essential for processing.

Later, in 1979, Khorana reported the completion of the total synthesis of a biologically functional tyrosine suppressor transfer RNA gene that carries all regulatory sequences. This gene was 207 base pairs long and included the following: (i) a 51 base pairs long DNA promoter region; (ii) a 126 base pairs long DNA corresponding to the precursor tRNA; and (iii) a 25 base pairs long duplex DNA corresponding to 16 base pairs adjoining CCA end of tRNA and the remainder, a modified sequence including EcoRI endonuclease specific sequence. The complete synthetic gene was cloned in the vector bacteriophage (lambda virus) by the gene cloning method discussed. On transformation of *E. coli*, the phage could multiply with the cloned gene.

3.10.16.3. Total Synthesis of a Human Leukocyte Interferon Gene

Interferon is a protein that has the property of inhibiting viral infection and cell proliferation, and thus can be used for the treatment of viral infections and malignancies (cancer). These properties have generated great interest in human interferons (IFNs). At least three kinds of interferon genes are known in the human genome: (i) leukocyte interferon genes (IFN-α genes), (ii) fibroblast interferon gene (IFN-β genes), and (iii) immune interferon gene (IFN-γ genes). Genes for different kinds of interferons differ, and details of base sequences are not known for all of them. In 1980, Weismann and coworkers published the nucleotide sequence gene derived from cloned IFN-α cDNA. This information, the total synthesis of this human leukocyte interferon gene 514 base pairs long, was achieved and published in 1981 (Edge et al., 1981). Since there are 514 base pairs in the gene, it involved the synthesis of DNA strands with 1028 nucleotides in a predetermined way. This was the longest gene synthesized till 1982. The synthesis involved the coupling of an initial nucleotide onto a polyacrylamide resin to which further nucleotides could be added in pairs. Sixty-six (66) oligonucleotide segments varying in size from 14 to 21 were first synthesized, which were then arranged in a predetermined way and joined. This gene has been incorporated into a plasmid and is capable of synthesizing an interferon. Further details about the synthesis of this IFN gene are available in the original article published in Nature (Vol. 292, pp. 756–761, 1981). The 514 base pairs code for 100 amino acids, 498 base pairs, and include the initiation and termination signals. There are no intron sequences present in this gene (introns are noncoding sequences common in eukaryotic genes, known as split genes). The plasmids with this newly synthesized gene have been transferred into bacteria (*E. coli*), which are being utilized now for the synthesis of interferon in industry. This has greatly reduced the cost of production of this drug (interferon), which was earlier sold at the rate of Rs.16 million per 50 mg.

3.10.17. Gene Synthesis Machines

Methods for the synthesis of DNA molecules with known base sequences have improved dramatically during the 1980s and 1990s. Therefore, genes can now be synthesized rapidly and in high yields. For instance, a gene that took years to synthesize between 1965–70 can now be synthesized within a matter of a few days with the help of Gene Synthesis Machines developed in the 1980s. The key innovations that made the availability of gene synthesis machines possible are the following: (i) the development of silica-based supports, which provide support for the solid-phase synthesis of DNA chains, and (ii) the development of stable protonated deoxyribonucleoside phosphoramidites as synthase, which are remarkably stable and resistant to hydrolysis, making them ideal for DNA synthesis. These two procedures can be used to prepare DNA molecules more than 100 nucleotides long within a few days. Different steps are involved in the solid-phase synthesis of a DNA chain.

Several versions of gene machines are now available. These machines, under the control of a microprocessor, synthesize specific short sequences of single-stranded DNA automatically. The desired sequence is entered on a keyboard, and the microprocessor automatically opens the valves of the containers of successive nucleotides, reagents, and solvents needed at each step into a synthesizer column, which is packed with tiny silica beads. These beads provide support on which DNA molecules are assembled (Gupta et al., 1999).

3.10.18. Synthesis of Genes from mRNA

H. Temin and D. Baltimore discovered the presence of RNA-directed DNA polymerase enzyme in 1970, which has the ability to synthesize DNA on an RNA template. This enabled molecular biologists to synthesize complementary DNA (cDNA) using mRNA as a template. If mRNA transcribed from a specific gene is made available in purified form, the elementary DNA (cDNA) synthesized with its help will represent the synthesized gene. To what extent this synthesized gene is a faithful copy of the native gene will depend on the fidelity of copying mRNA and also on the stability of the DNA thus synthesized. Moreover, since the mRNA of a gene does not have the complete transcript of the gene *in vivo*, the synthesized gene will be smaller than the gene in vivo. By copying eukaryotic purified mRNA, several genes have been artificially synthesized. The most important of these genes are the genes for sea urchin histone proteins, the ovalbumin gene in chickens, and globin genes in mammals. The gene synthesized as cDNA from a β-globin mRNA has been inserted into a plasmid in order to study its behavior. As earlier pointed out, this gene will lack any additional regulatory sequences absent in mRNA but present in a globin gene *in vivo*. This type of gene will also lack intron sequences found in eukaryotic split genes. Another factor causing a difference between the synthesized gene and the gene in vivo will depend on the sincerity with which the enzyme will synthesize cDNA because in many cases cDNA was found to be smaller than the mRNA itself. These genes (cDNA) have already become a very important tool in molecular biology experiments.

3.11. GEL PERMEATION

It is a technique in which a polymeric organic compound is used to prepare a porous medium. The pore size is determined by the degree of cross-linking of polymeric chains. Solutes present in the mixture are separated on the basis of their size and shape when they pass through a column consisting of packed gel particles (Figure 3.1).

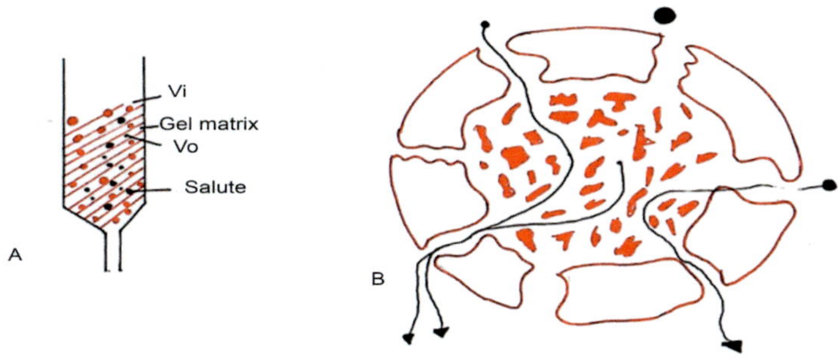

Figure 3.1. *Gel permeation chromatography. (a) Spatial accessibility of molecules during gel filtration; and (b) movement of molecules of different sizes through microporus gel forming stationary phase.*

Source: Made by the author.

3.12. CHROMATOGRAPHY

It is a technique of separation in which small-sized particles or liquid substances are separated in a buffer solution.

3.12.1. Ion Exchanged Chromatography

When molecules are dissolved in the solvent, they dissociate into charged ions. Therefore, they develop polarity. Based on polarity, i.e., the presence of charges, they can be separated. Ion-exchange chromatography is a powerful technique used for the separation of two proteins that are very similar to each other (Figure 3.2).

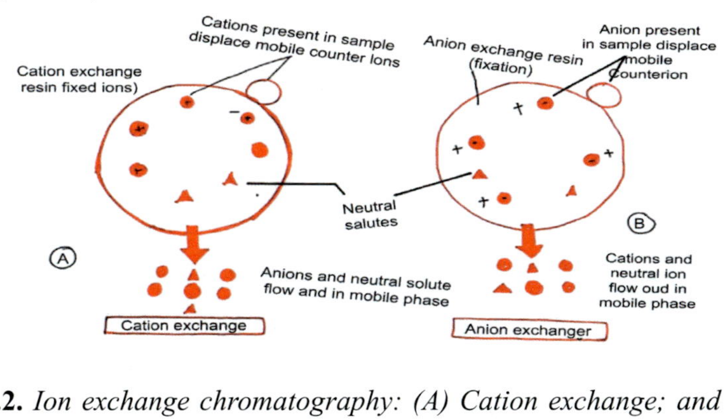

Figure 3.2. *Ion exchange chromatography: (A) Cation exchange; and (B) anion exchanger.*

Source: Made by the author.

3.13. ELECTROPHORESIS

It is a method of separation of charged molecules by applying an electric field. When the charged molecules are placed in an electric field they migrate depending on their net charge, size, shape and applied current.

3.13.1. Agarose Gel Electrophoresis

Agarose is a linear polymer of D-galactose and 3,6–anhydro–L-galactose which is extracted from seaweed. At neutral pH, the negatively charged DNA migrates towards the anode after application of electric field across the gel. It is a standard technique of nucleic acid separation.

3.13.2. Pulsed Field Gel Electrophoresis (PFEG)

It is a process to separate the several megabases long DNA molecules. PFGE varies a little from agarose gel electrophoresis and can resolve DNA molecules of 100–1000 kb. It changes the direction of the electric field in such a way that larger DNA fragments align more slowly with the direction of the new field than do the smaller DNA molecules. The pulsed field electric fields applied to a gel force the DNA molecule to reorient before continuing to move like a snake through the gel (Figure 3.3).

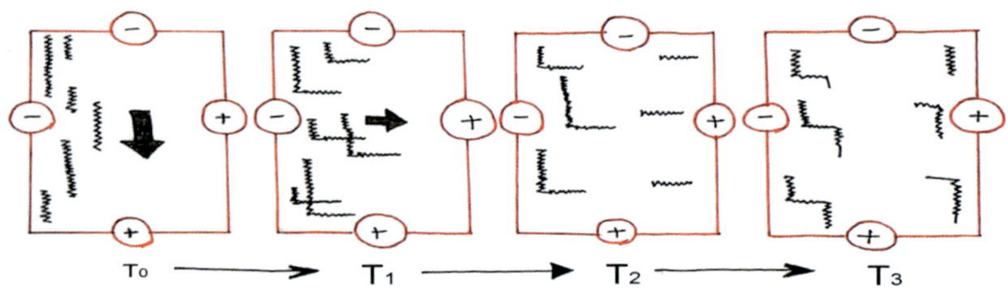

Figure 3.3. *Pulse filed gel electrophoresis.*

Source: Made by the author.

3.13.3. Polyacrylamide Gel Electrophoresis (PAGE)

PAGE is the most widely used method for separating and characterizing proteins and nucleic acids. Polyacrylamide gel offers several advantages, such as being inert to chemicals, providing superior resolution and stability over a wide range of pH, temperature, and ionic strength. The gel is prepared using monomers (e.g., acrylamide, N,N'-methylene bisacrylamide), an initiator (NNN-tetramethylethylenediamine, TEMED), propagators (ammonium persulfate, riboflavin), and a terminator or inhibitor (oxygen/air).

The pore size of the polyacrylamide gel can be adjusted by changing the concentration of acrylamide and bisacrylamide monomers in a fixed volume of gel solution. Free radicals generated by ammonium persulfate activate the acrylamide gel, leading to the production of long polymer chains through polymerization. Bisacrylamide facilitates gelation and cross-linking, resulting in the formation of a complex acrylamide network.

3.13.4. Sodium Dodecyl Sulfate Polyacrylamide Gel Electrophoresis (SDS-PAGE)

Separation of some of the proteins does not occur due to similar charges. Such proteins are treated first with an ionic detergent called sodium dodecyl sulfate (SDS) before the start and during the course of electrophoresis (PAGE). Therefore, this electrophoresis is called SDS-PAGE. Identical proteins are denatured by SDS, resulting in their sub-units.

The polypeptide chains are opened and extended. Based on their mass, but not the charges, the molecules are separated. SDS-PAGE is a high-resolution method used universally for analyzing the mixture of proteins according to their respective size. SDS solubilizes soluble proteins, making it possible to analyze other insoluble mixtures. Separation of the proteins does not occur due to a similar charge-to-mass ratio (z/m). Therefore, such proteins are treated first with an ionic detergent called SDS before starting and during the course of electrophoresis. Therefore, it is called SDS-PAGE.

3.13.5. Two-Dimensional Electrophoresis

A mixture of proteins is separated using two-dimensional electrophoresis. The first dimension uses isoelectric focusing, and the second dimension is sodium dodecyl sulfate polyacrylamide gel electrophoresis (SDS-PAGE). Isoelectric Focusing (IEF): Biomolecules such as proteins, DNA, and RNA have electric charges that depend on the molecule and the conditions of the medium (pH of the buffer in which they are dissolved). Charged molecules can be separated by electrophoresis in gels.

Due to differences in amino acid composition, proteins have a unique mass and charge. Therefore, proteins can have a net negative charge, net positive charge, or an isoelectric point (no charge) at a specific pH of the buffer. Amphoteric substances like proteins, which have different isoelectric points, can be separated by IEF. The isoelectric point is the pH value at which the net charges on molecules are zero. Ampholytes (a complex mixture of synthetic polyamino-polycarboxylic acids) are introduced (Figure 3.4).

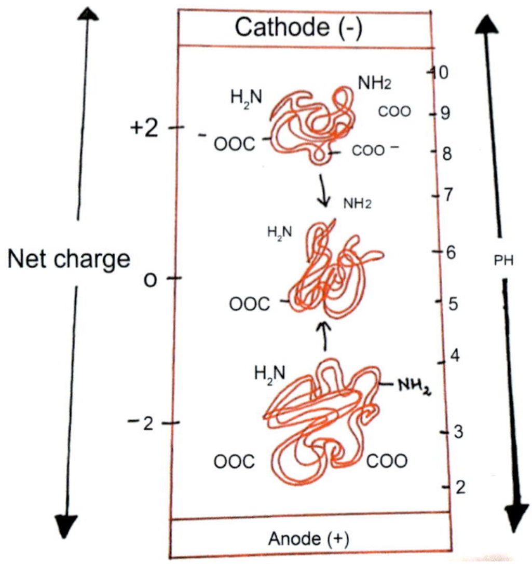

Figure 3.4. *Isoelectric focusing.*

Source: Made by the author.

3.14. SPECTROMETRY

In 1900, for the first time, J.J. Thomson introduced a mass spectrometer which employed fixed magnetic and electric fields to separate ions of different mass and energy. He recognized that charged particles differing in momentum behaved differently in an electric field, and used this property for the separation of ions with different masses. Its extensive use for research in various fields of biological science started in the 1980s. During 1990–2000, mass spectrometry became an important technique for genomics and proteomics research leading to the 2002 Nobel Prize in Chemistry for J.B. Fenn and K. Tanaka. Two-dimensional electrophoresis is more powerful when coupled with MS. The unknown protein spot is cut from the gel and cleaved by trypsin digestion into the mass spectrometer, and the mass of fragments is analyzed by fingerprinting to estimate the probable amino acid composition of each fragment and tentatively identify the protein. The proteome and its changes can be studied very effectively by employing the two techniques together. MS can also provide valuable information about the covalent modification of proteins, which can affect their activity. Mass spectrometry is a very useful technique. It is used in (i) the identification of unknown compounds, quantification of known compounds, and determination of structural and chemical ties of compounds when present in small quantities, (ii) their separation on the basis of their mass-to-charge (m:e) ratio, and (iii) determination of the relative abundance of each ion. Therefore, a spectrophotometer consists of three components: the source. It does not directly measure the molecular mass but detects m:e ratio. Mass is measured in terms of Dalton (Da). One Dalton = 1/12th of the mass of a single atom of isotonic carbon (3C) unit (1.06–10 g). This technique involves a generator of ions, an analyzer, and a detector.

3.15. MATRIX-ASSISTED LAYER DESORPTION-IONIZATION (MALDI)

MALDI was first developed by Karas and Hillenkamp in 1988. It co-precipitates the large excess of a matrix material with the analyte molecule. In this method, a sub-microliter of the mixture (matrix + analyte) is pipetted onto a metal substrate and allowed to dry. The dried mixture is irradiated by a laser pulse (337 nm) which specifically absorbs the selected solid of a matrix. The irradiation causes energy transfer and desorption resulting. With the help of matrix ions, the non-absorbing analyte molecules also get desorbed into the gas phase and ionized. The charged molecular ions of the analyte are detected by the MALDI-TOF mass spectrometer. Generally, MALDI is used for the study of molecules having mass above 500 Daltons (i.e., oligonucleotides and peptides). Proteome analysis will be crucial for understanding different biological processes in the post-genomic era. To achieve this objective, mass spectrometry is extensively used for: peptide sequencing, identification of protein, protein expression in different tissues and conditions, identification and post-transcriptional modification (e.g., phosphorylation, glycozylation, etc.) of proteins in response to different stimuli, and characterization of protein interactions that include protein-ligand, protein-protein, and protein-DNA interactions. Improvement in MS has significantly improved its application in the study of protein structure and function. For the study of macromolecules, the Indian Society for Mass Spectrometry (ISMAS) was established in 1978 with its headquarters at BARC (Bhabha Atomic Research Center), Mumbai. This society conducts a workshop each year on mass spectrometry. In India, facilities for MALDI-TOF MS have been created at several centers including the Center for Cellular and Molecular Biology (CCMB), Hyderabad, Indian Institute of Sciences, Bangalore, BARC, Rajiv Gandhi Center for Biotechnology (RGCB), Thiruvananthapuram, etc.

3.16. SURFACE ENHANCED LASER DESORPTION-IONIZATION (SELDI)

It is a unique combination of affinity chromatography and time of flight (TOF) mass spectrometry on a single platform, a combination designated ionization time of flight mass spectrometry). This system enables protein capture, purification, ionization and analysis of complex biological mixtures directly on protein chips array surface and detection of the purified proteins by laser desorption-ionization time of flight (LDI-TOF), MS analysis. SELDI system consists of three parts: (i) protein chip arrays (eight spots of 2 mm diameter); (ii) protein chip reader (LDI-TOF MS); and (iii) specialized software.

3.17. ELECTROSPRAY IONIZATION (ESI)

In the late 1980s, J.B. Fenn developed ESI mass spectrometry and was awarded the Nobel Prize in 2002 in chemistry. In this method, the sample is dissolved in a liquid with a mobile phase (i.e., water: acetonitrile: methanol) and pumped through a hypodermic needle at a high voltage (about 4,000 V which causes a strong electric field at the top of the nozzle of the metal needle). This disperses electrostatically electrosprays the highly charged small droplets of about one micrometer in size (Figure 3.5).

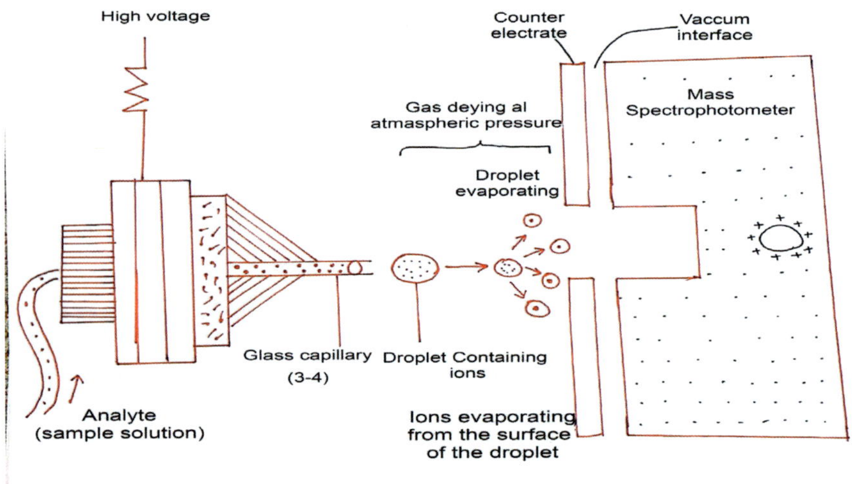

Figure 3.5. *Electrospray ionization source.*

Source: Made by the author.

These droplets are electrostatically attracted to the inlet of MS. Dry gas or heat is applied. The solvent around these droplets soon evaporates and the imparted charge increases with a decrease in their size. Hence, equal charges repel each other. This ionization is very mild, which causes no breakdown of analyte ions in the gas phase or onto the analyte molecules.

3.18. FLUORESCENCE SPECTROSCOPY

Fluorescence refers to the emission of light from a molecule after absorption of radiant energy. When a molecule absorbs radiant energy, its electrons are elevated to an excited state. If there is no chemical reaction, dissociation, rearrangement, or energy transfer by collision, the molecule returns to the ground state and excessive energy is lost as heat or light, i.e., photons of energy (v). The emission of light is called fluorescence. A compound may absorb energy in the UV region and fluoresce in the visible region. In fluorescence, the absorbed excited light is of a smaller wavelength (more energy) and the wavelength of the emitted light is longer (less energy). More increase in the wavelength of emitted light than the excited light is called Stokes shift. The fluorescence shown by the natural compounds is known as intrinsic fluorescence. There are certain chemical compounds that do not fluoresce. However, fluorescence can be detected after coupling them with a fluorescence probe. This phenomenon is called extrinsic fluorescence (Purohit, 2005).

3.19. POLYMERASE CHAIN REACTION

Polymerase chain reaction (PCR) provides a simple and ingenious method for the exponential amplification of specific DNA sequences by *in vitro* DNA synthesis. This technique was developed by Kary Mullis at Cetus Corporation in Emeryville, California in 1985. Kary Mullis shared the Nobel Prize for Chemistry in 1993. This technique has made it possible to synthesize large quantities of a DNA fragment without its cloning. It is ideally suited where the quantity of biological specimen available is very low, such as a single fragment of hair or a tiny bloodstain left at the site of a crime. The details of PCR techniques and their mechanism are described by Erlich (1989) in his edited book 'PCR Technology.' The PCR technique has now been automated and is carried out by a specially designed machine. The PCR requires the following:

i. **DNA Template:** Any source that contains one or more target DNA molecules to be amplified can be taken as a template.

ii. **Primers:** A pair of oligonucleotides of about 18–30 nucleotides with similar G+C contents act as primers. They direct DNA synthesis towards one another. The primers are designed to anneal on opposite strands of the target sequence so that they will be extended towards each other by the addition of nucleotides to their 3' ends.

iii. **Enzyme:** The most common enzyme used in PCR is a thermostable enzyme called Taq polymerase. It is isolated from a thermostable bacterium called *Thermus aquaticus*. It survives at 95°C for 1–2 minutes and has a half-life of more than 2 hours at this temperature. The other thermostable polymerases can also be used in PCR.

3.20. DNA AMPLIFICATION FINGERPRINTING (DAF)

The DAF has been described by Caetano-Annoles and Gresshoff (1994) where a single arbitrary primer (about 5 bases long) is used to amplify DNA by using PCR. In this technique, the parameters are carefully optimized. But for early and easy determination of amplified products, automation and fluorescent tagging of primers are extremely amenable. DAF differs from RAPD as (i) it requires a higher concentration of shorter primers (5–8 nucleotides), (ii) two temperature cycles contrast to three cycles in RAPD, (iii) it produces a very complex banding pattern, and (iv) DAF products are used in Sed on acrylamide gels and detected by using silver staining.

3.21. APPLICATION OF PCR

After the discovery of PCR, modern biology has been revolutionized in each and every aspect. Some of the areas of application of PCR have briefly been discussed here. (a) Diagnosis of Pathogens: There are several pathogens that grow slowly. Therefore, the cells are found in fewer numbers in the infected cells/tissues. It is difficult to culture them on artificial medium. Hence, for their diagnosis, PCR-based assays have been developed. These detect the presence of certain specific sequences of the pathogens present in the infected cells/tissues. Besides, it is useful in the detection of viral infections before they cause symptoms or serious diseases. (b) Diagnosis of Specific Mutation: In humans, there are thousands of genetic diseases. Mutations are also related to genetic diseases. The presence of a faulty DNA sequence can be detected before the establishment of the disease. By using PCR, sickle cell anemia, phenylketonuria, and muscular dystrophy can also be detected. Other diseases can also be diagnosed by using PCR. For example, PCR-based diagnostics for AIDS, chlamydia, tuberculosis, hepatitis, human papillomavirus, and other infectious agents and diseases are being developed. The tests are rapid, sensitive, and specific. (c) In Prenatal Diagnosis: It is useful in the prenatal diagnosis of several genetic diseases. If the genetic diseases are not curable, it is recommended to go for an abortion. (d) DNA Fingerprinting: has been discussed earlier. (e) In Research: In addition, DNA fingerprinting of new microorganisms isolated from various extreme environments (soil, water, sediments, air, extreme habitats, etc.) is also carried out to confirm their identity by comparing them with the DNA sequences of known microorganisms. Their DNA and RNA can be amplified. Besides, it is also useful to determine the orientation and location of restriction fragments relative to one another. (f) In Molecular Archaeology (Paleontology): PCR has been used to clone the DNA fragments from the mummified remains of humans and extinct animals such as the wooly mammoth and dinosaurs from the remains of ancient animals as recently epitomized in Michael Crichton's Jurassic Park. DNA from buried humans has been amplified and used to trace the human migration that occurred in ancient times. (g) Diagnosis of Plant Pathogens: It is also applied in the diagnosis of plant diseases. A large number of plant pathogens in various hosts or environmental samples are detected by using PCR, for example, viroids (associated with flower mosaic virus, bean yellow mosaic-virus, plum pox virus, potyviruses), mycoplasmas, bacteria (*Agrobacterium tumifaciens, Pseudomonas solanacearum, Rhizobium leguminosarum, Xanthomonas campestris,* etc.), fungi (e.g., *Colletotrichum gloeosporioides, Glomus spp., Laccaria* spp., *Phytophthora* spp., *Verticillium spp*), and nematodes (e.g., *Meloidogyne incognita, M. javanica,* etc.), viruses (such as TMV) (Henson and French, 1993; Chawla, 1998).

CHAPTER 4

TOOLS AND TECHNIQUES USED FOR RECOMBINANT DNA TECHNOLOGY

Contents

The tools of rDNA technology include vectors, restriction enzymes, ligation enzymes, bacterial hosts, methods to isolate and multiply nucleic acids, methods to quantify nucleic acids, and *Agrobacterium* as a vector to insert foreign DNA into plants and animals.

4.1. EXONUCLEASE

These enzymes act upon the genome and digest the base pairs on 5' or 3' ends of a single-stranded DNA or at single-strand nicks or gaps in double-stranded DNA (Figure 4.1).

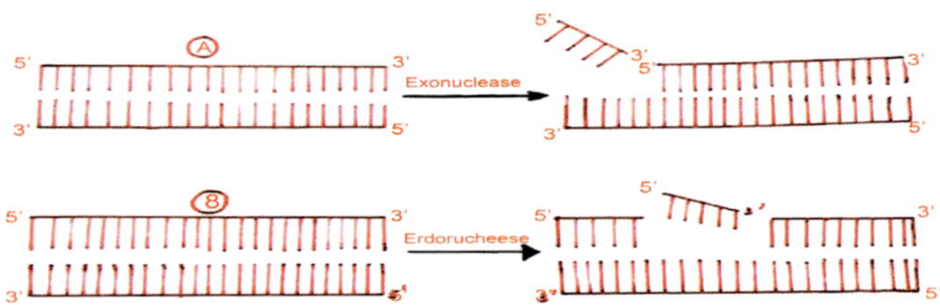

Figure 4.1. *The action of exonuclease (a); and endonuclease (b).*

Source: Made by the author.

4.2. ENDONUCLEASE

They act upon genetic material and cleave the double stranded DNA at any point except the ends, but their action involves only one strand of the duplex (Figure 4.2).

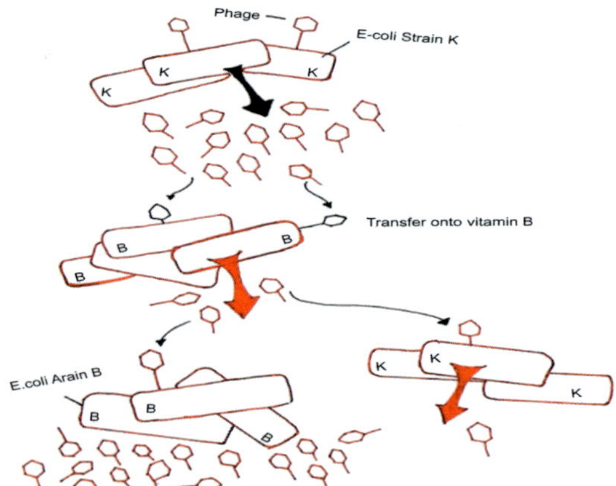

Figure 4.2. *Host controlled restriction and modification of phage I in E. coli strain K and strain B.*

Source: Made by the author.

4.3. RESTRICTION ENDONUCLEASES

The restriction enzymes are called 'molecular scissors.' These act as the foundation of recombinant DNA technology. These enzymes are present in bacteria and provide a type of defense mechanism called the 'restriction-modification system.' The molecular basis of these systems was elucidated first by Werner Arber in 1965. However, there was no understanding about cutting and joining of DNA molecules using DNA enzymes before 1970. During the early 1950s, it was discovered that when phage A infects *E. coli*, phage DNA is delivered inside the cell. The host-control restriction-modification system provides a defense to *E. coli*. The bacterium protects itself through (i) restriction mechanism (i.e., identifies the introduced foreign DNA and cuts it into pieces called restriction endonucleases) and (ii) modification system (i.e., methylation of certain bases of its own DNA by methylase so that phage A-expressed restriction endonuclease could not cleave its DNA).

4.4. NOMENCLATURE OF RESTRICTION ENZYMES

The restriction enzymes are named based on some of the following principles:

i. The name of the organism is identified by the first letter of the gen□s name and the first two letters of the species name to form a three-letter abbreviation in italics, for example, *E. coli* = Eco and *H. influenzae* = Hin, etc.

ii. A strain or type identified is written as a s□bscript, e.g., *EcoK* for *E. coli* strain K, Hind for H. infl□enzae strain Rd.

iii. In cases where the restriction and modification systems are genetically specified by a plasmid, the extrachromosomal element is identified by a s□bscript, e.g., *Eco RI, EcoPI*, etc.

iv. When a strain has several restriction and modification systems, these are identified by Roman n□merals, for example, HindI, HindII, HindIII for *H. influenzae* strain Rd., etc. These Roman numerals sho□ld not be conf□sed with those in the classification of restriction enzymes into Type I.

Type III Restriction Enzymes: The Type III enzyme is made up of two s□b□nits, one specified for site recognition and modification and the other for cleavage. In a reaction, it moves along the DNA and requires ATP as the source of energy and Mg as a cofactor. ATPase activity is lacking in these enzymes. Some examples of Type III enzymes are HpaI, MboI, FokI, etc. They have symmetrical recognition sites and cleave DNA at two sites in double-stranded

DNA in opposite orientations. One strand of double-stranded DNA is cleaved about 25–27 bp away from the recognition site, which is located in its immediate specific non-palindromic seq□ences.

Since the cleavage products of these enzymes are the homogeneous population of DNA fragments, they cannot be used for genetic engineering experiments.

4.5. EXAMPLES OF SOME ENZYMES

There are some other restriction enzymes that do not fall under Type I-III. Hence, there is a need to reclassify restriction enzymes under other categories from Type I to Type IV. Examples of other restriction enzymes include Eco 571, MCR (modified cytosine reaction), etc. Eco571 is made up of a single polypeptide. It shows both nuclease modification and nuclease activity. The S-adenosyl methionine stimulates the nuclease activity of this enzyme. It bears some other properties that resemble Type II enzymes. Therefore, Eco 571 should be classified under Type II to Type IV.

4.6. NUCLEASE

It degrades the single-stranded DNA or single-strand protrusion of double-stranded DNA Cohesive ends. As a result of the action of nuclease, cohesive ends are converted into blunt ends. Therefore, if nuclease is used to remove the incompatible ends, overlapping ends may be developed for the annealing of two fragments of DNA molecules.

4.7. DNA LIGASES

Mertz Davis (1972) for the first time demonstrated that DNA-cleaved DNA molecules could be covalently sealed with *E. coli* DNA ligase and were able to produce recombinant DNA molecules (Figure 4.3), which have 5–33 '-OH (hydroxyl) termini. There are two enzymes that are extensively used for covalently joining restriction fragments: the ligase from that encoded by T4 phage. The main one of DNA phage, hence, the enzyme is known as T4 DNA ligase (Gupta et al., 1999).

4.8. ALKALINE PHOSPHATASE

When a plasmid vector, for joining a foreign DNA fragment, is treated with the restriction enzyme, the greatest difficulty arises at the same time. Because the cohesive ends of broken plasmids, instead of joining with foreign DNA, join the cohesive end of the same DNA molecules and get recircularized. To overcome this problem, the restricted plasmid (i.e., plasmid treated with restriction enzymes) is treated with an enzyme, alkaline phosphatase, that digests the terminal 5'phosphoryl group. The restriction fragments of the foreign DNA are not treated with alkaline phosphatase. Therefore, the 5 'end of the foreign DNA fragment of the plasmid. The hybrid ligase will only join the 3' and 5' ends of recombinant DNA together if the 5 'end is phosphorylated. Thus, alkaline phosphatase and ligase prevent the recircularization of the vector and increase the frequency of production of recombinant DNA molecules. The nicks between two 3 'ends of DNA fragment and vector DNA are repaired inside the bacterial host cells during the transformation and can covalently join to a 3' end or recombinant DNA obtained has a nick with 3' and 5' hydroxy ends (Figure 4.3).

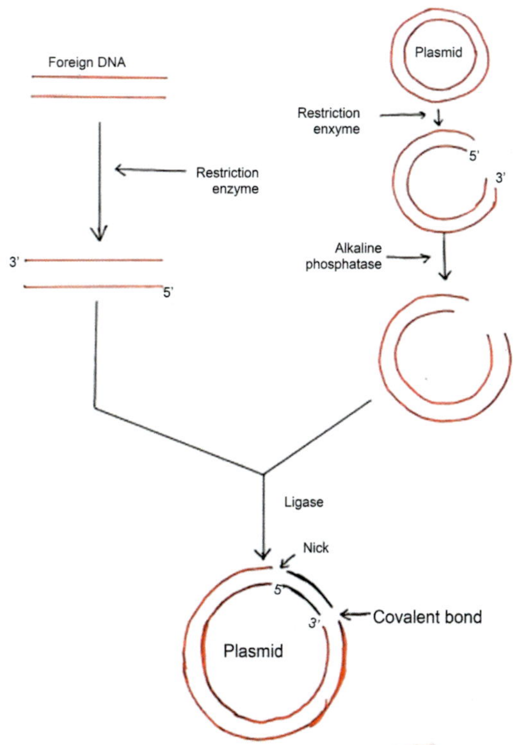

Figure 4.3. *Alkaline phosphates.*

Source: Made by the author.

4.9. REVERSE TRANSCRIPTASE

In addition to these enzymes, reverse transcriptase is used to synthesize the complementary DNA (cDNA) by using an mRNA template. Reverse transcriptase is very useful in the synthesis of cDNA and construction of a cDNA clone bank. Until recently, it was known that the genetic information of DNA passes to protein through mRNA. During the 1960s, Temin and co-workers postulated that in certain cancer-causing animal viruses which contain RNA as genetic material, transcription of cancerous genes (from RNA into DNA) takes place probably by DNA polymerase directed by virus RNA. Then DNA is the template for the synthesis of many copies of viral RNA in a cell. In 1970, S. Mizutani, H.M. Temin, and D. Baltimore discovered that information can also pass back from RNA to DNA. They found that retroviruses (possessing RNA) contain RNA-dependent DNA polymerase which is also called reverse transcriptase. This produces single-stranded DNA, which in turn functions as a template for a complementary long chain of DNA.

4.10. DNA POLYMERASE

This enzyme polymerizes the DNA synthesis on a DNA template (or cDNA template) and also catalyzes a 5'–3' and 3'-5' exonucleolytic degradation of DNA (Kornberg, 1974).

4.11. T4 POLYNUCLEOTIDE KINASE

The T4 polyn□cleotide kinase catalyzes the transfer of γ-PO4 of ATP to the 5' terminus of a dephosphorylated DNA (i.e., DNA lacking -PO4 residue) or RNA as below: 5'GGOH 3'-CCTTAAp 5'-GGOH Polynucleotide kinase -3' 3'-CCTTAA5.'S-7P-ATP. This enzyme shows both phosphorylation as well as phosphatase activities; therefore, two types of reactions can be used accordingly. In the phosphorylation activity, the γ-PO4 is transferred to the 5' end of dephosphorylated DNA. When ADP is in excess, it causes the T4 polynucleotide kinase to transfer the terminal phosphate (attached to a nucleotide) from phosphorylated DNA to ADP. By using radio-labeled gamma-phosphate (PO4) from P-ATP. This shows that the enzyme plays a dual role of phosphorylation activity and phosphatase activity, it was found that DNA is rephosphorylated by the transfer of any phosphate group. Due to the presence of both properties, this enzyme is used in radiolabeling of the 5' end of duplex during DNA sequencing (see DNA sequencing method), constructing the terminally labeled DNA, phosphorylating the synthetic linkers/oligonucleotide linkers/other DNA fragments lacking terminal 5' phosphates, etc.

4.12. TERMINAL TRANSFERASE

This enzyme is also known as from calf thymus. It has a property to add an 'oligodeoxynucleotide tail' to the 3'-OH end of double stranded DNA fragments.

Thus it extends homopolymer tails and this phenomenon is called homoplymer tailing, deoxynucleotidyl transferase. It has been isolated and purified

4.13. USE OF LINKERS AND ADAPTORS

As discussed earlier, the treatment of DNA with Type II restriction enzymes results in the formation of fragments having sticky ends or blunt ends. Such DNA fragments cannot be ligated with cloning vectors. Therefore, artificially cleaved sites at the blunt end can be added as linker molecules. Chemically synthesized double-stranded oligonucleotides are called linkers. They are added to double-stranded DNA molecules that act as adaptor molecules and have added recognition sites for restriction enzymes. For example, the following oligonucleotide possesses a recognition sequence for *EcoRI* (asterisk nucleotides). It can be ligated to the blunt end of any double-stranded DNA and cut by *EcoRI* to generate sticky ends in DNA (Figure 4.4).

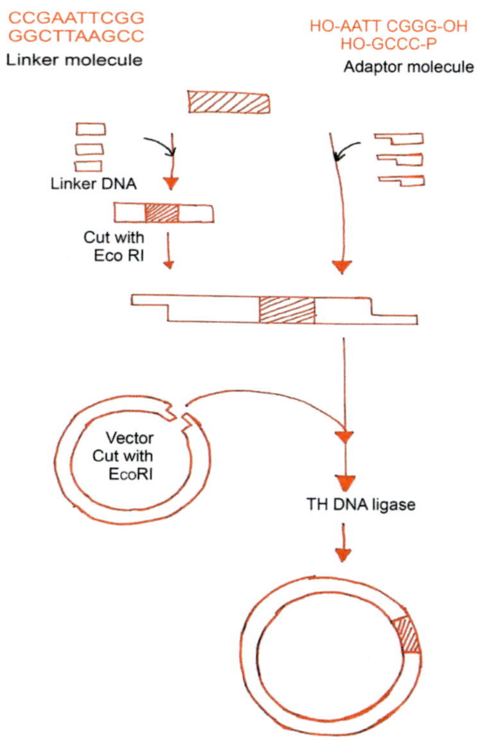

Figure 4.4. *Linker and adaptor molecules.*

Source: Made by the author.

CHAPTER 5

VECTORS

Contents

5.1. CLONING AND EXPRESSION VECTORS

In recent years, techniques for manipulating prokaryotic as well as eukaryotic DNA have witnessed a remarkable development. This has allowed the breaking of a DNA molecule at two desired places to isolate a specific DNA segment and then insert it into another DNA molecule at a desired position. The product thus obtained is called *recombinant DNA* and the techniques are often called *genetic engineering.*

Using this technique, we can isolate and clone a single copy of a gene or a DNA segment into an indefinite number of copies, all identical. This became possible because vectors like plasmids and phages reproduce in a host in their usual manner even after the insertion of foreign DNA so that the inserted DNA will also replicate faithfully with the parent DNA. This technique is called *gene cloning*. With this technique, genes can be isolated, cloned, and characterized, so the technique has led to significant progress in all areas of molecular biology. A variety of vectors have been developed that not only allow multiplication but may also be manipulated in such a way that the inserted gene may express in the host. Due to the importance of a variety of these cloning and expression vectors in genetic engineering experiments, they are discussed in some detail in this chapter.

5.2. CLONING VECTOR FOR RECOMBINANT DNA

One of the most important uses of recombinant DNA technology is in the cloning of (i) random DNA or cDNA segments, often used as probes, or (ii) specific genes, which may be isolated from the genome or synthesized organochemically or in the form of cDNA from mRNA. This other DNA molecule is often used in the form of a vector, which could be a plasmid, a bacteriophage, a derived cosmid or phagemid, a transposon, or even a virus. Techniques should also be available that will allow the selection of chimeric genomes obtained after the insertion of foreign DNA from a mixture of chimeric and the original vector. Another critical desired feature of any cloning vector is that it should possess a site at which foreign DNA can be inserted without disrupting any essential function. Therefore, in each case, an enzyme will also have to be selected which will cause a single break. Sometimes vectors are modified by inserting a DNA segment to create a unique site for one or more enzymes to facilitate its use in gene cloning. This inserted DNA with restriction sites for several enzymes is sometimes called *a polylinker*.

5.3. PLASMIDS AS VECTORS

Plasmids are defined as autonomous elements, whose genomes exist in the cell as extrachromosomal units. They are self-replicating circular duplex DNA molecules, which are maintained in a characteristic number of copies in a bacterial cell, yeast cell, or even in organelles found in eukaryotic cells. These plasmids can be *single-copy plasmids* that are maintained as one plasmid DNA per cell or *multicopy plasmids*, which are maintained as 10–20 genomes per cell. There are also plasmids that are under *relaxed replication control,* thus permitting their accumulation in very large numbers. These plasmids are used as cloning vectors due to their increased yield potential. Circular plasmid DNA used as a vector can be cleaved at one site with the help of a restriction enzyme to give a linear DNA molecule.

A foreign DNA segment can now be inserted by joining the ends of broken circular DNA to the ends of foreign DNA, thus regenerating a bigger circular DNA molecule that can be separated by electrophoresis based on its size. Selection of chimeric DNA is also facilitated by the resistance genes that the plasmid may carry against one or more antibiotics. If a plasmid has two such genes conferring resistance against two antibiotics and if the foreign DNA insertion site lies within one of these two genes, then the chimeric vector loses resistance against one antibiotic, the gene for which has foreign DNA inserted within its structure. In such a situation, the parent vector in bacterial cells can be selected by resistance against two antibiotics, and the chimeric DNA can be selected by retention of resistance against only one of the two antibiotics (Gupta et al., 1999).

5.4. PBR322 AND PBR327 VECTORS

The naturally occurring plasmids may not possess all the above and other essential properties of a suitable cloning vector. Therefore, one may have to restructure them by inserting genes of relaxed replication and genes for antibiotic resistance. This has actually been done, and suitable plasmid vectors have been obtained. One is pBR322, which is 4,362 bp DNA and was derived by several alterations in earlier cloning vectors. It has genes for resistance against two antibiotics, an origin of replication, and a variety of restriction sites for cloning of restriction fragments obtained through cleavage with a specific restriction enzyme. Another vector, pBR327, was derived from pBR322 by deletion of nucleotides between 1427 to 2516. These nucleotides are deleted to reduce the size of the vector and to eliminate sequences that were known to interfere with the expression of the cloned DNA in eukaryotic cells. pBR327 still contains genes for resistance against two antibiotics. Both pBR322 and pBR327 are very common plasmid vectors (Figure 5.1).

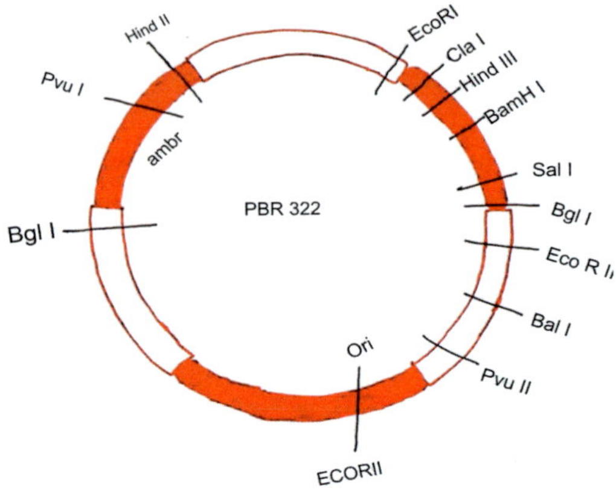

Figure 5.1. *pBR322 vector.*

Source: Made by the author.

5.5. PUC VECTORS

Another series of plasmids that are used as cloning vectors belongs to the pUC series. These plasmids are 2700 bp long and possess: (i) the ampicillin resistance gene, (ii) the origin of replication derived from pBR322, and (iii) the lacZ gene derived from *E. coli*. Within the lac region is also found a polylinker sequence having unique restriction sites.

When DNA fragments are cloned in the region of pUC, the lac gene is inactivated. These plasmids when transformed into an appropriate *E. coli* strain, which is lac-, and grown in the presence of IPTG and X-gal, will give rise to white or clear colonies. On the other hand, pUC having no inserts and transformed into bacteria will have an active lacZ gene and therefore will prod□ce bl□e colonies, th□s permitting identification of colonies having pUC vector with cloned DNA segments. The cloning vectors belonging to the pUC family are available in pairs with reversed orders of restriction sites relative to the lacZ promoter. pUC8 and pUC9 make one such pair. Other similar pairs include pUC12 and pUC13 or pUC18 and pUC19 (Figure 5.2).

As discussed above, in pBR322 and Pbr327, the DNA is inserted at a site located in one of the two genes for resistance against antibiotics, so that it will inactive one of the two resistance genes. The insert-bearing plasmid can be selected by their ability to grow in a medium containing only one of the two antibiotics and by their failure to grow in a medium containing both the antibiotics. The plasmids carrying no insert, on the other hand, will be able to grow in media containing one or both the antibiotics. In this manner, the presence of the lac Z gene in pUC and resistance genes against ampicillin and tetracycline in pBR322 and pBR327 allow the selection of *E. coli* colonies transformed with plasmids carrying the desired foreign cloned DNA segment.

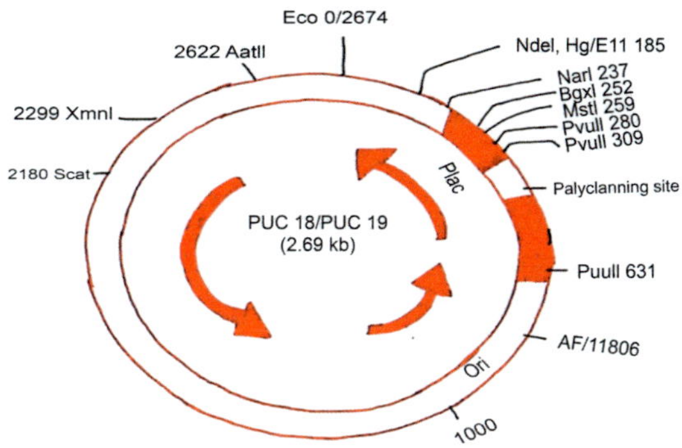

Figure 5.2. *Puc19 cloning vector.*

Source: Made by the author.

5.6. YEAST PLASMID VECTORS

Special vectors are also known for introducing DNA segments into yeast cells, a eukaryotic system that has been used extensively for developing genetically engineered yeast cells. Although *E. coli* plasmids or phages can be used for the transfer of genes to yeast cells, the frequency of transformation is rather low. For instance, yeast cells that are LEU could be transformed using a chimeric plasmid carrying the LEU gene for yeast. This involves *integrative transformation* with low frequency. To overcome this difficulty of low frequency transformation, efforts were made to prepare a variety of vectors for yeast.

The possible examples of yeast plasmid vectors are: (i) Yip or yeast integrative plasmids, which allow transformation by crossing over and have no replication origin; (ii) Yep or yeast episomal plasmids, which carry 2-micron DNA sequence including the origin of replication and rep genes; (iii) YRp or yeast replication plasmids, which carry autonomously replicating sequence; this sequence is very common with many yeast genes so that stable transformation can be achieved by crossing over; (iv) YCp or yeast centromere (CEN) containing plasmids, which function as true chromosomes and segregate during mitosis and meiosis; (v) pYAC or yeast artificial yeast chromosome vector which carries both centromere and telomere sequences and, therefore, can be used to obtain artificial chromosomes. The above vectors have been extensively utilized for the study of yeast genome and are also used as shuttle vectors, which allow genic sequences to be routinely transferred back and forth between yeast and *E. coli* cells, provided they contain origins of replication active in both yeast and *E. coli*. These vectors also have markers enabling the selection of *E. coli* cells or colonies transformed with these vectors. The vectors can therefore be amplified in *E. coli* and then used to transform yeast cells.

5.7. RETRIEVER VECTORS

Retriever vectors are another class of vectors, which are used to retrieve specific genes from the normal chromosome of an organism like yeast through recombination. They will multiply both in *E. coli* and yeast. In such a vector, the gene in question is removed by enzymes to produce a **gapped** vector having flanking sequences. Through recombination, they acquire lost sequences from the host. Retriever vectors are very useful in the isolation of genes from yeast for further molecular studies including sequencing.

5.8. TI AND RI PLASMIDS AS VECTORS FOR HIGHER PLANTS

Among higher plants, the Ti plasmid of *Agrobacterium tumefaciens* or the Ri plasmid of *A. rhizogenes* is the best-known vector. T-DNA, from the Ti or Ri plasmid of *Agrobacterium*, is considered to be a very potential vector for cloning experiments with higher plants.

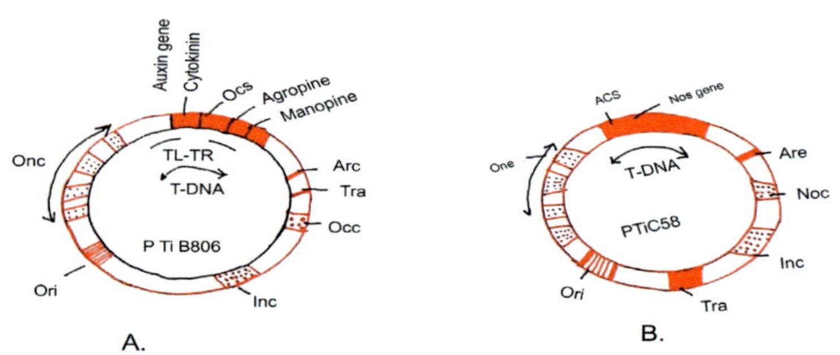

Figure 5.3. *(a) Octopine plasmid; and (b) nopaline plasmid.*

Source: Made by the author.

This will involve the following steps: (i) foreign DNA has to be first cloned into T-DNA of Ti or Ri plasmids; (ii) modified hybrid T-DNA can be transferred to the genome of plant cells by *Agrobacterium* infection. Recombinant Ti plasmid can also be obtained by cloning a large fragment of Ti plasmid in pBR322. These recombinant Ti plasmids can then be used for the transformation of higher plants. This system can be widely used since *Agrobacterium* infects nearly all dicotyledonous plants. Such cloning in plants has proved to be of immense use to modify agricultural plants to increase their productivity. Monocotyledonous plants to which our cereal crops belong are, however, difficult materials to be used for improvement through such techniques; some success, however, has been achieved in cereals also in recent years (Figure 5.3).

5.9. BACTERIOPHAGES AS VECTORS

Bacteriophages provide another source of cloning vectors. Since usually a phage has a linear DNA molecule, a single break will generate two fragments, which are later joined together with foreign DNA to generate a chimeric phage particle. The chimeric phage can be isolated after allowing it to infect bacteria and collect progeny particles after a lytic cycle. The use of phage particles as vectors imposes a limitation on the size of foreign DNA; if it is too long, the phage head may not accommodate the phage DNA. To overcome this problem, segments of DNA that do not contain essential genes may be removed. This technique has been used in phage lambda to create a smaller vector genome with a single restriction site for the enzyme *Eco RI.* The reduced size also helps in the selection process, where only the chimeric particles will be obtained in the phage progeny, and the vector particles lacking the cloned segment will be eliminated due to their reduced size.

5.10. LAMBDA PHAGE VECTORS

Plasmid vectors described in the previous section are often used for cloning DNA segments of small size. However, when preparing a genomic library in a eukaryote, the cloned fragments should be large enough to contain a whole gene. This will also make screening easier without serious difficulty.

The above properties and other requirements of cloning the whole genome in e□karyotes are f□lfilled by the phage lambda and cosmid vectors, the former permitting cloning of segments up to 20–25 kb long and the latter accommodating segments up to 45 kb long. Phage lambda, however, is easier and more efficient for making genomic and cDNA libraries.

1. **Lambda gt 10 and Lambda 11:** Lambda 10 and lambda gt 11 are modified lambda phages designed to clone cDNA fragments. The major difference between those two vectors is that gt11 is an expression vector, where inserted DNA is expressed as a β-galactosidase f□sion protein.

   Lambda gt 10 is a 43 kb double-stranded DNA for cloning fragments that are only 7 kb in length. The insertion of DNA inactivates the cl+ gene, generating a cl- bacteriophage. Non-recombinant lambda gt 10 forms cloudy plaques in an appropriate *E. coli* host, while recombinant plaques form. Furthermore, in an *E. coli* strain carrying the hf La150 mutation, only cl- phage will form plaques, as those that become cl+ will form lysogens and will not undergo lysis to form any plaques. Recombinant lambda gt 10 plaques can thus be easily selected.

Lambda gt 11 is a 43.7 kb double-stranded lambda phage for cloning DNA segments, which are less than 6 kb in length. Foreign DNA can be expressed as beta-galactosidase fusion proteins. Recombinant lambda gt 11 can be screened using either nucleic acid or antibody probes. The recombinant lambda gt 11 becomes gal-, while non-recombinant lambda gt 11 remains gal+, so an appropriate *E. coli* host, and with recombinant phage will form white or clear colonies, and that with recombinant phage (gal+) will form blue colonies permitting screening in the presence of IPTG (inducer) and X-gal (substrate).

2. **EMBL3 and EMBL4 (Replacement Vectors):** EMBL3 and EMBL4 are the two vectors that are designed so that a central non-essential part of 44 long phage can be replaced by foreign DNA. Cleavage of the phage with an appropriate enzyme generates three fragments. The central fragment representing 40% of the phage genome is nonessential for propagation of the phage and can be replaced by foreign DNA that may be as long as 20–30 kb. Non-recombinant DNA resulting from the fusion of the left arm and right arm will give a DNA that is too small for packaging, so there is automatic selection against non-recombinant phages.

   EMBL3 and EMBL4 are two such replacement vectors used for preparing genomic libraries in eukaryotes with cloned fragments, 15–25 kb in size. The two vectors have polylinkers with reverse orders of restriction sites with respect to each other.

3. **Charon 34 and Charon 35:** Charon 34 and Charon 35 differ from each other only in their central fragments and will accept fragments 9–20 kb long. They have a more extensive range of restriction targets within their polylinkers than do EMBL3 and EMBL4. In these vectors, physical separation of lambda arms from the central fragments is required before ligation to donor fragments. In EMBL3 and EMBL4, the removal of central fragments is not required, because there is a mechanism that ensures that 98% of packageable DNA are recombinants.

5.10.1. M13 Phage as a Cloning Vector for DNA Sequencing

M13 is a filamentous bacteriophage of *E. coli* and contains a 7.2 kb long single-stranded circular series of cloning vectors that can be used for cloning a wide variety of DNA. M13 phage has been variously modified to give rise to an M13 mp series of cloning vectors which can be used for the cloning of a wide variety of DNA fragments, particularly for the purpose of sequencing through Sanger's method of dideoxy chain termination. Cloning vectors of the M13 mp series have a lac Z gene that complements the gal host, giving blue colonies. If a foreign DNA segment is inserted at one of the polylinker sites associated with the lac Z gene, it inactivates the lac Z gene and no complementation is possible. Thus, on transformation, only white or clear plaques are obtained, permitting the selection of recombinant M13mp plaques. Once the foreign DNA is cloned in the M13 vector, commercially available oligonucleotide primers are used for copying the insert in the presence of dideoxynucleotides, so that fragments of different sizes with known termini are produced, permitting the determination of the sequence. The reversed order of the restriction sites in the polylinker is present in a pair of vectors like M13mp8 and M13mp9, permitting sequencing from both ends of the double-stranded DNA molecule. The extension of the commercial primer or cloned fragments in the M13 vector in the presence of radioactively labeled nucleotides will also allow the generation of radioactively single-stranded probes.

5.11. COSMIDS AS VECTORS

Cosmids are plasmid particles into which certain specific DNA sequences, namely those for cos sites, are inserted. Since these cos sites enable the DNA to be packed in lambda particles, cosmids allow the packaging of DNA in phage particles *in vitro*, thus permitting their purification. Like plasmids, these cosmids perpetuate in bacteria and do not carry the genes for lytic development. The advantage of using cosmids for cloning is that their efficiency is high enough to produce a complete genomic library of 10^6 to 10^7 clones from a mere 1 microgram of insert DNA. The disadvantage, however, is their inability to accept more than 40–50 kbp of DNA. The bacteriophage P1 system and F-factor-based vectors can allow cloning of DNA segments as large as 100 to 1000 kbp in length. These are therefore preferred.

5.12. PHAGEMIDS AS VECTORS

Phagemids are prepared artificially by combining features of phage with plasmids, as the name suggests. One such phagemid, which is commonly used in molecular biology laboratories, is pBluescript II KS, which is derived from pUC19 and 2961 bp long. The KS designation indicates the orientation of the polylinker, such that the transcription of the lacZ gene proceeds from the restriction site for KpnI to that for SacI. The detailed structure of pBluescript II KS. It may be noted that it has the following features: (i) a multiple cloning site (MCS) flanked by T3 and T7 promoters to be read in opposite directions on the two strands, (ii) an inducible lac promoter (Lac I), upstream of the lacZ region, which complements with *E. coli* and provides the facility for selection of chimeric vector DNA using the criterion of white colonies, (iii) fl (+) and fl (-) origins of replication derived from a filamentous phage for recovery of sense (+) and antisense (-) strands of the lacZ gene when the host is coinfected with a helper phage; (iv) an origin of replication derived from a plasmid and used in the absence of a helper phage; (v) a gene for ampicillin resistance for antibiotic selection of chimeric phagemid vector.

5.13. P1 CLONING VECTORS FOE CLONING LARGE DNA SEGMENTS

The bacteriophage P1 cloning vectors can allow cloning of 100 kbp long DNA segments with an efficiency of 10 raise to power 5 clones per micro g of insert DNA. Therefore, in their capability, they fall between YACs and cosmids. The vector with insert DNA is amplified in *E. coli* and several micrograms of cloned DNA can be recovered from 5–10 ml of exponential phase of *E. coli* cells. Two P1 cloning vectors with their essential elements.

5.13.1. F-Factor Based Vector (Which Behaves Like a Phage)

F-factor based vectors have recently been developed for cloning large DNA segments in *E. coli*. The cloning of large DNA segments is achieved by a method called chromosomal building, in which through repeated recombination the size of cloned segments can be increased. These bacterial vectors will complement the YAC vectors for cloning segments larger than 100 kbp in length and offer some advantages over the YAC system. These vectors have already been used for cloning large DNA segments from the bithorax gene of Drosophila.

5.14. PLANT AND ANIMAL VIRUSES AS VECTORS

A number of plant and animal viruses have also been used as vectors both for introducing foreign genes into cells and for gene amplification and expression in host cells. In the latter case, we may be interested only in getting an increased quantity of the gene product and may not be interested in the integration of the foreign gene into the host genome. Some of the viruses that are commonly used as vectors will be described in this section.

5.14.1. Plant Viruses (Cauliflower Mosaic Virus or CaMV and Geminiviruses)

Cauliflower mosaic virus, tobacco mosaic virus (TMV), and geminiviruses are three groups of viruses that have been used as vectors for cloning DNA segments. CaMV infects particularly members of the Cruciferae family and has a double-stranded DNA molecule, 8 kbp in size, whose sequence is now known. Following infection, the virus spreads systematically throughout the plant in a very high copy number, reaching up to 10^5 virus particles per cell. These features make CaMV a suitable vector for transforming higher plants, although there are instances of other viruses being used as vectors, leading to the production of transgenic plants. Geminiviruses comprise a group of single-stranded DNA plant viruses causing important diseases in cassava, maize, and other cereals. They replicate via double-stranded DNA forms and have been subdivided into two groups: one infecting monocots and transmitted by leafhoppers, and the other infecting dicots and transmitted by whiteflies. The dicot geminiviruses have two DNA segments in their genome, DNA A and DNA B, both of which are essential for infection by mechanical inoculation. DNA A carries all genes responsible for replication.

It has been shown that deletions in the coat protein gene do not destroy the infectivity of the virus, which means that long deletions in this region can be used to insert foreign DNA to obtain gene constructs that will be extremely useful as amplification and expression vectors in plants. F□rthermore, since the coat proteins are the most abundant proteins synthesized by the virus, promoters of coat protein genes can be used for high-level expression of cloned genes. CLV and TGMV have already been tested for their ability to allow multiplication and expression of a cloned gene.

5.14.2. Animal Viruses

A number of animal viruses are also used as vectors, either for the delivery of nucleic acid into the cultured cells followed by its integration with the host genome or for the amplification and high-level expression of a foreign gene, using the promoters from virus genes. These cloned genes can have a variety of uses including gene therapy in mammalian cells and synthesis of important proteins by cloned genes in cell cultures.

5.15. TRANSPOSONS AS VECTORS

5.15.1. Transposons of Higher Plants (Ds, Ac, or Mu 1 of Maize)

Ac and Ds are popular transposons in corn and were earlier known to represent a transposon with short terminal repeats enclosing a long DNA segment, which measures more than 4,500 bp in Ac and about 400 bp in Ds. Each possesses genes including the gene for the transposase enzyme responsible for transposition. Part of this region can be deleted, and the transposon can be used for cloning of foreign DNA segments in much the same way as in other cases.

5.15.2. Transposons of Drosophila

P elements of *Drosophila* consist mainly of 31 bp terminal inverted repeats enclosing a 3-kilobase protein-coding region, coding for transposase and transposase derived from other complete P elements to help their mobilization. In these deleted P elements, a gene for transfer can be inserted to yield a recombinant P element, which can be microinjected into the fertilized egg along with a normal P element. The inserted gene is transposed onto an embryonic chromosome.

5.15.3. Artificial Chromosome (YAC and MAC) Vectors for Cloning Large DNA Segments

We know that in plasmids, sequences up to 10–15 kbp; in lambda phage sequences up to 22 kbp; and in cosmids, sequences up to 40 kbp can be cloned. Yeast artificial chromosome vectors, mentioned earlier, allow cloning of sequences that are several 100 kilobase pairs long. These long molecules of DNA, which may represent whole chromosomes, can be cloned in yeast by ligating them to vector sequences that allow their propagation as linear artificial chromosomes. These long DNA molecules can be generated and allow the construction of comprehensive libraries in microbial hosts. With the isolation of mammalian telomeres and centromeres, mammalian artificial chromosomes (MACs) will also be produced further.

YACs have two disadvantages. (i) Cloning efficiency is low (1000 clones/micro g DNA against 10^6–10^7 clones/micro g DNA for cosmids), thus making it impractical to generate complete genomic libraries through the use of YACs; (ii) It is not possible to recover large amount of pure insert DNA from the individual clones; selective amplification of YACs DNA has recently allowed this problem to be overcome. Both these problems have been overcome bacteriophage P1 system.

5.16. PROMOTERS

Promoters from non-plant sources express poorly in plants, and promoters of non-animal sources express poorly in animals. Therefore, to obtain correct expression, genes transferred to plants should be linked to plant-specific promoters, and those transferred to animals should be linked to animal-specific promoters. These promoters can be interchanged to confer the transferred gene a specific pattern of expression. To illustrate this, some examples of promoters used in plants and insects will be discussed.

5.17. NOPALINE SYNTHASE (NOS) PROMOTER FROM T-DNA

This promoter is 200 bp long and contains several DNA sequence motifs that direct the expression of the linked gene. The upstream element from –97 to –130, when duplicated, increases the expression threefold, suggesting that this may be an enhancer. The NOS promoter is most active in the basal regions of plants, with its activity slowing down in vegetative parts at the onset of flowering, even though its activity is increased in the flower.

5.17.1. Dual Promoter of Mannopine Synthase (MAS) Genes 1 and 2

These promoters are present in T-DNA fragments 467 bp long. The expression seems to be developmentally regulated and the linked genes seem to be most active in the basal region of the plant. Expression is also induced by wounding. Since they are closely linked, if a marker gene is linked to one of these two promoters and a foreign gene is linked to the other promoter, selection for high expression of the marker gene will lead to a high level of expression of the foreign gene.

5.17.2. 35S RNA Promoter of CaMV

This is the most extensively used promoter in a wide variety of plants. The promoter is 343 bp long and contains a strong transcriptional enhancer, which on duplication results in a 10-fold increase in expression, even at a distance of 2 kbp. High expression is mainly observed in leaf tissue.

5.17.3. Polyhedrin Promoter from Baculovirus

Baculovirus can synthesize polyhedron protein in infected cells, so that polyhedron makes up to 50% of the total protein in the cell, giving up to 1 g of protein per liter of insect cell culture.

In view of this, the promoter has been used to construct an expression vector to allow a high level of expression of any gene under the influence of this promoter. This is being used for developing biopesticides or even for the production of specific chemicals in the industry.

5.17.4. Expression Cassettes

The literal meaning of cassette is 'a device containing film such as magnetic tape for insertion into equipment like a tape recorder, VCR, etc.' Therefore, by expression cassette, we mean gene constructs that allow the insertion of foreign genes, either as transcriptional or translational fusions, behind specific promoters. pRT plasmids are typical in this context and have been derived from Puc18/19. They consist of a series of vectors that differ in polylinker sequences, each flanked by the 35S promoter of CaMV on one end and a sequence for Poly A addition on the other end. A variety of marker genes have been inserted into these cassettes, and their expression studied, both in protoplasts and in stable transgenic tissues.

5.17.5. Baculovirus, an Expression Vector System for Insect Cells

The development of a baculovirus expression vector system is based on its life cycle, which involves the following steps: (i) in the early phase, virus particles are budded from infected cells and spread the virus; (ii) in the late phase (2–5 days), occluded virus units accumulate in the host nucleus and get embedded in protein polyhedra; (iii) the insect is decomposed, releasing polyhedra; (iv) polyhedra are ingested by another insect, where polyhedra are dissolved, releasing the virus for multiplication.

The use of baculovirus as a cloning vector involves the following steps: (i) the desired gene is cloned in a transfer vector under the control of the polyhedrin promoter; (ii) the chimeric transfer vector is co-transfected with the wild-type virus into the host cell; (iii) the polyhedrin gene of the wild-type is replaced by the cloned gene through recombination; (iv) the recombinant virus is plaque purified, as they will have polyhedral occlusion bodies; (v) the selected recombinant virus is multiplied and used for the expression of heterologous protein.

5.18. VIRUS EXPRESSION VECTOR FOR MAMMALIAN CELLS

A number of viruses have also been used as transfer or expression vectors for mammalian cells. These include the following: (i) The first viruses used for mammalian cells included DNA viruses like *papoviruses* and *adenoviruses*, but since their capacity for foreign genes is small, they cannot be used for the expression of many genes which are fairly large in size; (ii) retroviruses are very useful vectors, since they can infect a broad variety of cell types and can be used for cloning large genes; however, being single-stranded RNA, they cannot replicate without being integrated so that we will need replicating cells for expression or alternatively high-titer for fully differentiated cells; (iii) *adeno-associated virus* (AAV) overcomes the limitations associated with retroviruses even though, like retroviruses, it is an integrating virus, with 5 kb single-stranded DNA; it is non-pathogenic, and can be concentrated to high titer; (iv) *herpes virus* is another important non-integrating virus of large size (150 kb), which can be used for expression of large intact genes.

Several of the above viruses have already been used and showed expression of cloned genes, but they need to be further characterized for their long-term usage.

5.19. BINARY AND SHUTTLE VECTORS

For the purpose of gene transfer in higher plants, a number of binary vectors were developed. These were based on the pPCV series of plasmids. These vectors contain a conditional mini-RK2 replicon, which is maintained and mobilized by trans-acting functions derived from the plasmid RK2 replicon. The plasmid RK2 was introduced into both *E. coli* as well as *Agrobacterium* to facilitate replication of binary vectors in both these hosts, so that the vector can be maintained and shuttled between both-hence the name shuttle vector also. Similar binary or shuttle vectors for maintaince and transfer between *E. coli* and yeast cells have also been designed (Figure 5.4).

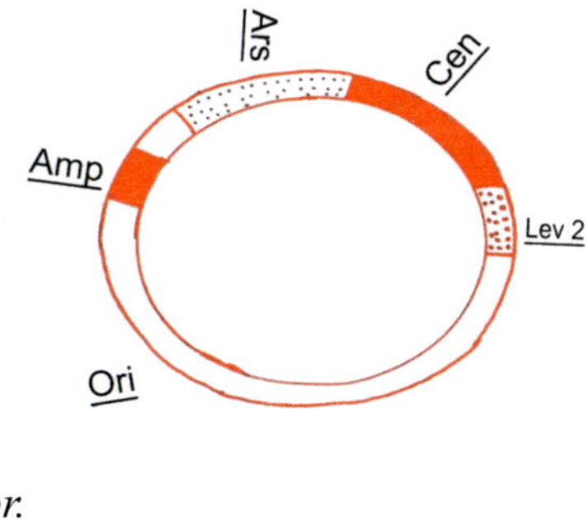

Figure 5.4. *Shuttle vector.*

Source: Made by the author.

5.20. AC-DS ELEMENTS

The activator (Ac) and dissociation (Ds) elements are simple transposons of maize. Ac has a 4.565 kb long autonomous sequence that transcribes a 3.5 kb long GC-rich untranslated leader sequence. Four introns are spliced out from the primary transcript, resulting in a 3312 nucleotide long mature mRNA.

Consequently, an 807 amino acid long protein is translated. Within the Ac element, the terminal inverted repeats (TIR) of 11 bp are located at both ends, playing a key role in transposition. The protein synthesized by Ac is called transposase, which triggers the transcription of Ac elements. Ds (dissociation) is the defective element of the Ac family. Due to deletion, non-autosomal elements are obtained from the autosomal elements. It activates the trans-acting transposase. The 11 bp long TIR is common to all Ds elements. The enzyme transposase cannot be expressed by the Ds element. However, the transposase produced by the Ac element is diffused and trans-acting, causing the transposition of Ds elements. By detecting part of the region, transposons can be used as a vector for cloning foreign DNA.

5.21. P ELEMENT

The transposons of Drosophila are known as P elements. The size of different P elements varies: the largest element is 1907 bp long containing 31 bp long terminal inverted repeats at both ends. The complete elements move autonomously due to the expression of the transposase enzyme. The other P elements lack the internal gene that produces transposase. The terminal and sub-terminal sequences cause transposition in them. A foreign DNA can be inserted into the transposase element to develop recombinant P elements which can be microinjected into the fertilized egg along with normal P element. The normal P supplies transposase to the recombinant P element. Here the non-autonomous element acts as a transformation vector. Consequently, the foreign gene inserted into the embryonic cell is transposed onto an embryonic chromosome of Drosophila. Gierl et al. (1989) have reviewed the transposable elements of maize. A recombinant P element was prepared by inserting a DNA fragment of the rosy gene (and a marker gene for identification of transgene). It was inserted into a plasmid. Similarly, a P element, lacking the ability for integration, was inserted into a plasmid to encode transposase. Both plasmids were co-injected into the embryo of *Drosophila melanogaster* mutant in the rosy gene. It was found that 50% of the progeny of *Drosophila melanogaster* were rosy progeny. This shows the successful transfer of a foreign gene into a cell via P element (Gierl et al., 1989).

5.22. EXPRESSION VECTOR

In a number of cloning vectors described earlier in this chapter, the main utility has been to clone a gene or a DNA sequence. In other words, they are used for obtaining millions of copies of the cloned DNA segments, for further use in genetic engineering experiments or for further basic studies. With this objective in mind, the cloned genes in this vector need not express themselves either at the transcription level or at the translation level. But, when the cloned gene is used for transformation to generate transgenic plants or animals or the production of microbes to be used in industry, then the cloned gene must be expressed. Sometimes, a high level of expression of gene is desirable, if the product of cloned gene is to be recovered as a commercial product. The objective can be achieved through the use of promoters and expression cassettes.

CHAPTER 6

CLONING

Contents

Cloning is an essential process in the success of genetic engineering where insertion of a specific fragment of foreign DNA into a cell, through a suitable vector, so that it can replicate independently and be transferred to progenies during cell division (Figure 6.1).

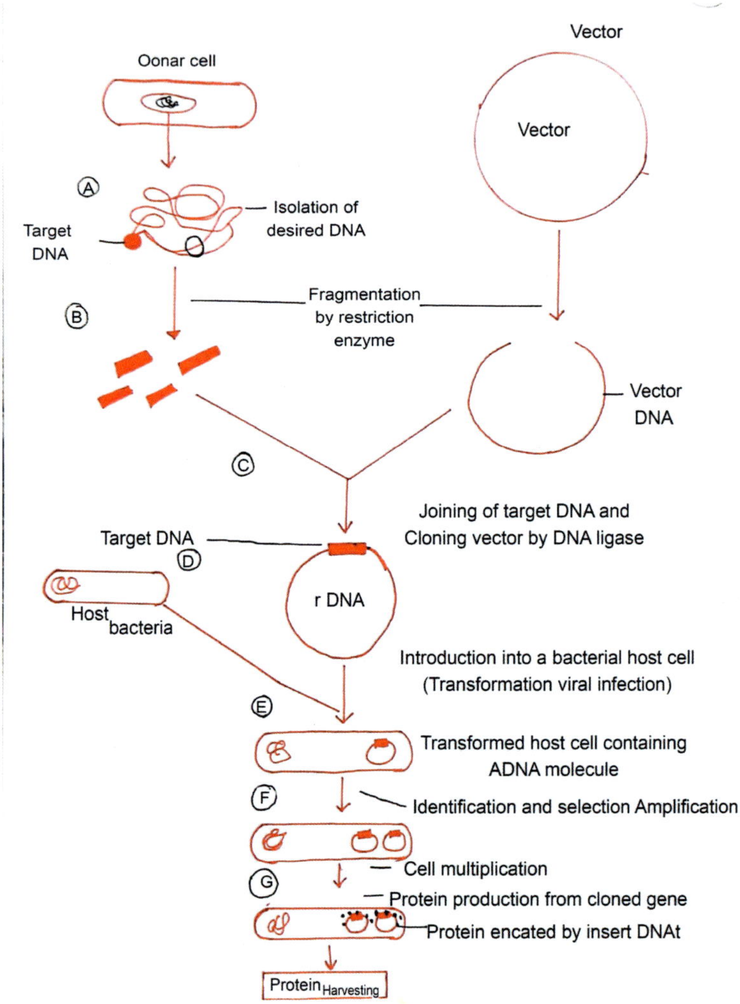

Figure 6.1. *Gene cloning*

Source: Made by the author.

6.1. CLONING IN PROKARYOTES

The DNA fragment to be introduced into the host cell is called insert DNA, desired DNA, target DNA, or foreign DNA. Such gene/DNA fragment is taken out from an organism or gene library, i.e., a random collection of cloned DNA fragments in vectors that consist of all genetic information of that organism concerned. However, getting the desired DNA is not so easy because the cellular DNA is very large and the position of the desired DNA is not accurately known. Libraries of such cloned fragments are prepared.

6.2. STRATEGIES OF RECOMBINANT DNA TECHNOLOGY

There is no single method of recombinant DNA technology, but it involves several steps. Maniatis et al. (1976) have described the basic techniques of gene cloning. The principal steps of gene cloning:

i. Isolation of DNA of known function from an organism (A).
ii. Enzymatic cleavage (B) and joining (C) of insert DNA to another DNA molecule (cloning vector) to form a recombinant DNA (i.e., vector + insert DNA) molecule (D).
iii. Transformation of a host cell, i.e., transfer and maintenance of this rDNA molecule into a host cell (E).
iv. Identification of transformed cells (i.e., cells carrying rDNA) and their selection from non-transformants.
v. Amplification of rDNA (F) to get multiple copies in a cell.
vi. Cell multiplication (G) to get a clone, i.e., a population of genetically identical cells. This facilitates each clone to possess multiple copies of foreign DNA.

6.3. GENE LIBRARY

A gene library is the collection of different DNA sequences from a sequence that has been cloned into a vector for ease of purification, storage, and analysis. There are two types of libraries based on the source of DNA used: (i) genomic library (where genomic DNA is used), and (ii) cDNA library (where cDNA or mRNA is used). A gene library should contain a certain number of recombinants with a high probability of containing any particular sequence.

6.4. GENOMIC LIBRARY

It is a collection of clones that represent the complete genome of an organism. All fragments of DNA inserted into vectors for further propagation into a suitable host represent the entire genome of an organism. For the construction of a genomic library, the entire genomic DNA is isolated from host cells/tissues, purified, and randomly broken into fragments of the correct size for cloning into a suitable vector. There are two basic ways of randomly fragmenting genomic DNA: physical shearing (e.g., pipetting, mixing, or sonication) and partial restriction enzyme digestion (by using a limited amount of restriction enzyme, the DNA is not cut completely). Using these methods, genomic DNA is randomly broken into smaller fragments. A total of 6 random DNA fragments are obtained after physical shearing.

The vector isolated from bacterium is also digested with the restriction enzyme which digests genomic DNA. The fragments of genomic DNA are inserted into the vector. Each vector consists of different fragments of DNA. The recombinant DNA molecules are transferred into bacterial cells or bacteriophage particles are assembled. The genomic library for organisms with smaller genome size (e.g., *E. coli*) can be constructed in plasmid vector. Only 5,000 clones (of the average size of 5,000 bp) results in 99% chance of cloning the entire genome of 4.6 x 10°bp. In addition, libraries from organisms with larger genomes are constructed using phage, Cosmid, BAC or YAC vectors.

6.5. CDNA LIBRARY

The mRNAs are highly processed representatives of genes which express under specific conditions. The mRNAs cannot be directly cloned because they are unstable; hence, mRNAs are converted into cDNAs. The library made from complementary or copy DNA (cDNA) is called a cDNA library. The cDNA library represents the DNA of eukaryotic organisms only, not prokaryotic ones. Because genomic DNA of eukaryotes contains introns, regulatory regions, and repetitive sequences, the establishment of a genomic library for eukaryotes is not meaningful. The cDNA library can be constructed by using mRNA because mRNAs are highly processed, intron-free representatives of DNA with coding sequence only.

6.6. ISOLATION OF MRNA

It is much more difficult to isolate a specific mRNA if it is in low amounts in the cell. The majority of mRNA sequences in eukaryotic cells contain a long polyadenylated tract (i.e., about 100A residues) at their 3' termini (Kates 1970). Due to this feature, they can be separated from other types of RNAs. Therefore, mRNA binds with an oligo-dT cellulose affinity column or poly-U sepharose from which it can be eluted. Its integrity can be studied using gel electrophoresis, which allows mRNAs to be size-fractionated by recovering chosen regions of the gel lane. Commercial preparations of both these substrates are now available. The mRNA preparations can be enriched for the desired molecules by fractionation according to their size using sucrose density gradient centrifugation (Boffey, 1987).

6.7. PREPARATION OF CDNA

Reverse transcriptase is required for the synthesis of a DNA copy of an mRNA molecule. The mRNAs are reverse transcriptase enzyme and four deoxynucleotide triphosphates (dNTPs) i.e., dATP, dGTP, dCTP, and dTTP. The oligo-dT primer binds to the polyadenylated tail that contains -AAAA(n) residues and provides a free –OH site for reverse transcription. Reverse by (b) treated with the oligo-dT primer, the transcriptase adds complementary dNTPs one-to-one to the free 3'-OH site of the primer, forming a single strand of DNA resulting in an RNA-DNA hybrid. Cellular DNA and total RNA inhibit reverse transcriptase activity. Therefore, it is necessary that mRNA must be in pure form before cloning. The oligo-dT segments contain between 10–20 nucleotides in length that hybridize to the poly A tract on mRNA. The oligo-dT primer ensures the initiation of cDNA synthesis at the 3' terminus of the RNA-DNA hybrid. Cellular DNA and total RNA inhibit reverse transcriptase activity.

6.8. OLIGO-DG PRIMER

Oligo-dG primer is added to the reaction mixture and (e) the temperature is maintained at about 55°C. This facilitates the binding of dG to oligo-dC tails formed on cDNA. The template for the synthesis of double-stranded cDNA in the presence of Now cDNA acts as a DNA polymerase I or AMV (Avian myeloblastosis virus) reverse transcriptase.

6.9. SYNTHESIS OF SECOND STRAND OF CDNA

Klenow DNA polymerase I and dNTP precursors are added to the reaction mixture. This enzyme adds complementary nucleotides to the cDNA one by one to the 3'-OH site of the primer. Consequently, from one cDNA, a double-stranded DNA clone is synthesized, the length of which is similar to that of the mRNA molecule initially used. This is used for cDNA cloning and construction of cDNA library and in gene cloning experiments. The double-stranded cDNA is identical to the gene which codes for mRNA (only in prokaryotic genes). However, the majority of eukaryotic genes contain introns that interrupt their coding sequences. Methods of removing cDNA in eukaryotes have been described earlier. cDNA can be radioactively labeled by nick translation and used for the identification of DNA fragments containing the gene. By using these procedures, a cDNA clone bank is built up which hybridizes with the probe.

6.10. INSERTION OF DNA INTO VECTOR

The DNA thus isolated as above or procured from a DNA or gene bank is fragmented by using the specific restriction enzyme. The vector (e.g., PBR322) is also treated with the same restriction enzymes so that the cohesive ends generated may have complementary residues similar to foreign DNA (Figure 6.1B). The plasmid contains amp and Tet marker genes to confer resistance against antibiotics, ampicillin and tetracycline respectively. Some time after cleavage by restriction enzymes and ligation of foreign DNA fragment with the plasmid, a few important genes are destroyed to develop specific cohesive ends. For example, BamHI destroys Tet genes. The sticky ends can form a phosphodiester linkage between the free 5' phosphoryl and 3' OH group, which is added along with ATP. A number of products, a highly heterogeneous mixture of recombinant DNA molecules together with parental plasmids, are formed from this reaction.

For the insertion of double-stranded cDNA into a cloning vector, it is necessary to add single-stranded DNA sequences that are complementary to the linear vector at both termini. In order to efficiently form recombinant DNA molecules, adding sticky sequences to the ends of both termini is necessary. There are two methods for generating cohesive ends on double-stranded cDNA: the use of linkers and homopolymer tails.

6.10.1. Use of Restriction Enzyme Linkers

Linkers are chemically synthesized double-stranded DNA oligonucleotides that contain one or more restriction sites for cleavage by restriction enzymes such as EcoRI, HindII, BamHI, etc. Linkers are ligated to blunt-end DNA using T4 DNA ligase. Both the vector and DNA are treated with a restriction enzyme to create sticky ends. The staggered cuts, i.e., sticky ends, are then ligated with T4 DNA ligase to produce molecules with very high efficiency.

6.10.2. Use of Homopolymer Tails

For the first time, Jackson et al. (1972) applied this method to construct recombinant DNA by inserting phage A DNA into SV40 DNA using dA-dT homopolymers. Using terminal transferase, the synthesis of homopolymer tails is defined at both 3' termini of double-stranded DNA and vector. In the presence of the precursor DATP, terminal transferase helps to add poly-dA at the 3' termini of the vector. Likewise, the same enzyme adds poly-T at the 3' termini of the DNA molecule when the precursor TTP is present. The linearized vector with tails is incapable of recircularization unless ligated to a double-stranded DNA fragment. The vector and DNA tails are annealed.

6.11. TRANSFER OF RECOMBINANT DNA INTO BACTERIAL CELLS

Several methods have been developed for the introduction of recombinant DNA molecules into host cells. Depending on the types of vectors and host cells, the methods for gene transfer into host cells are briefly discussed.

6.11.1. Transformation

Once a mixture of recombinant DNA is obtained, it is allowed to be taken up by the suitable bacterial cells. Originally, the transformation procedure was developed by Mandell and Higa (1979). The strains of *E. coli* usually do not have restriction systems; hence, they degrade foreign DNA. To escape from degradation, exponentially growing cells are pretreated with CaCl at low temperature; thereafter, DNA is mixed up. The event of entering the plasmid containing a foreign DNA fragment into a bacterial cell is known as transformation.

6.11.2. Transfection

Transfection is the transfer of foreign DNA into cultured host cells mediated through chemicals. The charged chemical substances, such as cationic liposomes and calcium phosphate, are taken and mixed with DNA molecules. The recipient host cells are overlaid by this mixture. Consequently, the foreign DNA is taken up by the host cells. DEAE dextran liposome-mediated gene transfer was used for the first time with phage A to transfer foreign DNA into *E. coli* cells; therefore, it is often termed as transfection (a hybrid of transformation and infection). The efficiency of transformation is not high as it is influenced by bacterial strain and size of foreign DNA. It has been determined that this process typically yields approximately 10 transformants per microgram (μg) of cloned circ□lar plasmid (e.g., pBR322). Efficiencies exceeding 10^8 transformants per gram of plasmid have not yet been achieved. Linear DNA, when used for transformation, is markedly inefficient.

6.11.3. Selection (Screening) of Recombinants

After the introduction of recombinant DNA into suitable host cells, it is essential for those cells to have received the recombinant DNA molecules. This process is called screening or selection. Selection of recombinant DNA cells is based on the expression or non-expression of certain characteristics or traits. The vector or foreign DNA present in recombinant cells expresses the characteristics, while the non-recombinants do not express the traits. Some of the methods mainly used for the selection of recombinants in *E. coli* are as follows:

6.11.3.1. Direct Selection of Recombinants

The above features facilitate the direct selection of recombinant cells. If the cloned DNA itself codes for resistance to the antibiotic ampicillin (amp), the recombinants can be allowed to grow on minimal medium containing ampicillin. Only such recombinants will grow and form a colony in the medium that contains the amp gene on its plasmid vector. Following this procedure, you cannot determine whether the recombinants growing on such medium contain the relegated plasmid vector or contain the recombinant plasmid plus foreign DNA fragment because the amp gene is present in both the recombinants on pBR322.

6.11.3.2. Insertional Selection Inactivation Method

This is a more efficient method than direct selection. In this approach, one of the genetic traits is disrupted by inserting foreign DNA. Antibiotic resistance genes act as a good insertion inactivation system. As discussed in an earlier section, plasmid PBR322 contains two antibiotic resistance genes, one for ampicillin (amp gene) and the other for tetracycline (tet gene). If the target DNA is inserted into the tet gene using BamHI, the property of resistance to tetracycline will be lost. Such recombinants would be test-sensitive. When such recombinants (containing target DNA in the tet gene) are grown onto medium containing tetracycline, they will not grow because their tet^{R} gene has been inactivated. But they are resistant to ampicillin because the amp gene is functional. On the other hand, the self-ligated recombinants will show resistance to tetracycline and ampicillin. Therefore, they will grow on medium with both antibiotics.

To ascertain the presence or absence of the tet^{R} gene in the inserted DNA fragment in plasmid, replica plating is done from the master plate. Bacterial colonies of the master plate are gently pressed with sterile velvet so that a few cells of each colony may adhere to it, which is then pressed on another plate containing the nutrient medium amended with tetracycline. Plates are incubated for the growth of bacterial colonies. The appearance of colonies is compared with the master plate, and those colonies that fail to grow on a plasmid which had inserted DNA in the tet gene of the plasmid and had destroyed the tet gene replica plate can be said to have.

6.12. BLUE-WHITE SELECTION METHOD

It is also a powerful method used for screening of recombinant plasmids. In this method, a reporter gene lacZ is inserted in the vector. The lacZ encodes the enzyme β-galactosidase and contains several recognition sites for restriction enzymes. Beta-galactosidase breaks a synthetic substrate, X-gal (5-bromo-4-chloro-indolyl-β-D-galactopyranoside), into an insol□ble bl□e-colored prod□ct.

If a foreign gene is inserted into lacZ, this gene will be inactivated. Therefore, no bl□e color will develop beca□se β-galactosidase is not synthesized d□e to the inactivation of lacZ. Therefore, the host cells containing rDNA will form white-colored colonies on the medium containing X-gal, whereas other cells containing non-recombinant DNA will develop the blue-colored colonies. Based on colony color, the recombinants can be selected.

6.13. COLONY HYBRIDIZATION (NUCLEIC ACID HYBRIDIZATION) TECHNIQUE

After the growth of cells on nutrient medium, it is likely that a few cells may have specific DNA as desired among several thousand cells. How is it possible to pick up those cells which have the desired DNA? To make the separation of strands easy, the "colony hybridization technique" has been developed by Grustein and Hogness (1975). It is suitable for use with plasmids. An analogous hybridization technique has been developed for use with phages (Benton & Davis, 1977). Single-stranded DNA hybridization plaque method, network bases 32p-DNA Probe hybridizes the DNA Probes: The colony hybridization technique is based on the hybridization of desired DNA with a DNA probe, the availability of a radiolabeled DNA probe. A DNA probe is the radiolabeled (32P) fragment of a DNA molecule (20–40 bp) which is complementary at least to one part of the desired DNA. Generally, probe DNA is labeled with 32P while it emits gamma rays. In addition to using radioactive isotopic elements, isotopic sulfur or fluorescent molecules may also be used. Therefore, a DNA probe recognizes A, T, G, and C nucleotides and combines with the complementary sequences of the target DNA. A and T of the probe combine with T and A of the target DNA and vice versa. Similarly, G and C of the probe combine with C and G of the target DNA and vice versa. The use of DNA probe in searching the functional target DNA is given in nucleic acid hybridization technique small or sometimes with 121. The 32P liberates beta particles. The small-sized DNA probe easily catches the target DNA for hybridization than the long-sized probe. Besides, a 32P-DNA than the target DNA molecules. Moreover, the DNA probe is very specific to its target DNA; this specificity is called stringency.

6.14. *IN VITRO* TRANSLATION

In vitro translation is now used as a method to confirm the identification of recombinant clones. Nagata et al. (1980) have used primary translation screened by isolating the interferon gene from total leukocyte poly A+RNA. Translation of poly A+RNA is done when it is microinjected into oocytes from the toad, Xenopus laevis with the result of secretion of interferon into culture medium.

Interferon is detected by its antiviral activity.

It is supposed that the number of interferon RNA may be between 10 to 10, therefore about 10^4 transformants must be repeated to get at least a single clone. Thereafter, DNA is immobilized on nitrocellulose or diazobenzyloxymethyl cellulose paper to isolate complementary mRNA by nucleic acid hybridization as described earlier. The RNA can be eluted from the cDNA-RNA hybrid and microinjected into oocytes and the translation products are detected by immunoprecipitation techniques.

6.15. IMMUNOLOGICAL TESTS

The immunological techniques are the final test analogous to the colony hybridization technique as described earlier. It is an alternative screening procedure which relies on expression, and generally applicable approach to potentially a very powerful method since the only absolute requirement is that the required mRNA encodes a protein for which a suitable antibody is available (Williams, 1981). Two in situ techniques given by Broome and Gilbert (1978), and Erlich et al. (1978) are in current use. In the immunological test, instead of radio-labeling of DNA molecules, antibodies (immunoglobulins) are used to identify the colonies or plaques developed on master plates that synthesize antigens encoded by the foreign DNA present in plasmids of the bacterial clones. For this purpose, a special vector, known as a ds expression vector, is designed where the foreign DNA is transcribed and translated within the bacterial cell (Glover, 1984). The growth medium containing specific antiserum may help in the detection of viable immunoprecipitate (precipitin) around the colonies or plaques.

6.16. BLOTTING TECHNIQUES

DNA, RNA and proteins are separated by blotting techniques. These are described in the following sections.

6.16.1. Southern Blotting Techniques

A method developed for genes in a DNA restriction fragment is called the Southern blotting technique. Southern blots can easily provide a physical map of restriction sites within a gene located normally and reveal the number of copies of the gene in the genome, and the degree of similarity of the gene when compared with other complementary genes. Molecular biologist E.M. Southern (1975) developed this technique for analyzing the related on a chromosome.

The procedure starts with the digestion of DNA population by one or many restriction enzymes (Figure 6.11). Consequently, DNA fragments of unequal length are produced. This preparation is passed through agarose gel electrophoresis, which results in the separation of DNA molecules based on their size. DNA restriction fragments present in the gel are denatured by alkali treatment. The gel is then placed on top of the b□ffer-sat□rated filter paper. The □pper s□rface of the gel is covered with a nitrocell□lose filter and overlaid with dry filter paper. The dry filter paper draws the b□ffer thro□gh the gel. The b□ffer contains single-stranded DNA. The nitrocell□lose filter binds DNA fragments strongly. After baking at 80°C, DNA fragments are permanently fixed to the nitrocell□lose filter when they come in contact with it. Then the filter is placed in a sol□tion containing radiolabeled RNA or denat□red DNA probe of known sequences. These are complementary in sequence to the blot-transferred DNA. The radiolabeled nucleic acid probe hybridizes with the DNA on the nitrocell□lose filter. The filter is thoro□ghly washed to remove the probe. The hybridized regions are autoradiographically detected by placing the nitrocell□lose filter in contact with a photographic film. The images show the hybridized DNA molecules. Thus, the sequences of DNA are recognized following the sequences of the nucleic acid probe.

6.16.2. Northern Blotting Techniques

Southern blotting technique could not be directly applied to the blot transfer of mRNA separated by gel electrophoresis because RNA was found not to bind with nitrocellulose filter. Alwine et al. (1979) devised a technique in which RNA bands are blot transferred from the gel to chemical reactive paper. An aminobenzyloxymethyl cellulose paper prepared from Whatman filter paper No. 540 after a series of uncomplicated reactions is diazotized and rendered into the reactive paper. Therefore, it becomes available for hybridization with radiolabeled DNA probes. The hybridized bands are detected by autoradiography. Thus, Alwine's method extends that of the Southern method and for this reason, it has been given the name Northern blotting. "There is not this northern or western like Southern." These blot transfers are reusable because of the firm covalent bonding of RNA to the reactive paper. Effective in binding the DNA as well. Small fragments of DNA can most effectively be transferred to the diazotized paper derivative to nitrocellulose. These techniques were more advanced and demonstrated that mRNA bands can also be blotted directly on nitrocellulose paper under appropriate conditions.

6.16.3. Western Blotting Techniques

Towbin et al. (1979) developed the Western blotting technique to identify newly encoded proteins by transformed cells. Its working principle is based on antigen-antibody reactions; hence, it is an immunodetection technique. In this method, radiolabeled nucleic acid probes are not used. The technique follows the following steps: (I) Extraction of proteins from transformed cells. (II) Separation of proteins using SDS-PAGE (sodium dodecyl sulfate polyacrylamide gel electrophoresis) where SDS acts as a solvent for electrophoresis. (III) Transfer of the electrophoresed gel in a buffer at a low temperature (40°C) for half an hour. (IV) Blotting of proteins on nitrocellulose filter paper, Whatman filter, and coarse filter in a transfer buffer; (i) Soaking of nitrocellulose filter, Whatman filter, and coarse filter in the transfer buffer; (ii) Placing of Whatman filter paper on a cathode plate followed by a stack of coarse filter, Whatman filter, electrophoresed gel, nitrocellulose filter, Whatman filter paper, coarse filter stack, Whatman filter, and anode plate transferred; (iii) Putting the complete setup in a transfer tank containing sufficient transfer buffer. (V) Application of an electric field (30 V overnight or for 5 hours) to cause the migration of proteins from the gel to the nitrocellulose filter and binding on its surface. (The nitrocellulose filter has an exact image of the pattern of proteins as present in the gel. This type of blotting is called Western blotting.) (VI) Hybridization of proteins using radiolabeled antibodies (I123-antibodies) of known structure, isolated from the rabbit. (VII) Washing of the nitrocellulose filter with a wash solution (Tris-buffered saline + Tween 20) to facilitate the removal of unhybridized antibodies. (VIII) Detection of hybridized sequences by autoradiography. The dots on the diagram show the presence of the desired protein.

6.17. RECOVERY OF CELLS

Colonies complementary to nitrocellulose signal are recovered from the master plate 2D cultured in nutrient broth in order to get the sufficient bacterial clones containing recombinant plasmid. However, in some instances the foreign DNA is cleaved enzymatically from the vector molecules. Sometimes the aim is to achieve expression of a gene within the bacterial cell if it encodes valuable polypeptides.

6.18. EXPRESSION OF CLONED DNA

Factors taken into account for expression are: (i) Supply of prokaryotic promoter for expression of eukaryotic genes, (ii) Supply of ribosomal binding sites for the cloning vector, (iii) Removal of introns from mRNA obtained from eukaryotic genes, and (iv) Inhibition of genes responsible for degradation of foreign protein within the bacterium. The genes of eukaryotes differ in different ways from those of *E. coli*. Therefore, the eukaryotic genes must be modified in such a way that they could resemble the prokaryotic genes and express in *E. coli*. To begin transcription in prokaryotes, the binding of bacterial RNA polymerase to a promoter region of the DNA immediately before (or upstream of) the gene is essential.

6.19. SHINE-DALGARNO SEQUENCE

Prokaryotic mRNA has a sequence upstream from the initiation codon that plays a role in the attachment of the 30S ribosomal subunit to the mRNA. This sequence (about 3–9 bases long) is located about 3–12 bases upstream from the initiation codon, which is also complementary near the 3' terminus of 16S tRNA (Shine and Dalgarno, 1975). The name of this sequence was coined as the Shine-Dalgarno sequence after the name of the discoverers. For the expression of eukaryotic genes within *E. coli*, the presence of a Shine-Dalgarno sequence in bacteria is crucial.

6.20. DETECTION OF NUCLEIC ACIDS

There are two approaches for labeling the nucleic acids to be used as DNA probes:

1. radioactive labeling; and
2. non-radioactive labeling.

6.20.1. Radioactive Labeling

Methods for *in vitro* introduction of radioactivity in nucleic acids were developed during the 1970s through metabolic labeling. In this method, a large amount of radioactivity is required. However, this method was time-consuming and laborious. The methods developed later on utilized phage T4 polynucleotide kinase (which transfers γ-PO_4 of ATP to 5'-OH terminus in DNA or RNA) or *E. coli* DNA polymerase I (which replaces normal nucleotides). The nucleic acids are labeled *in vitro* using the following methods: a normal nucleotide of double-stranded DNA with α – 32P labeled n□cleotide.

6.20.1.1. Nick Translation

This method is used to label the double-stranded DNA probes. If there is no nick, a nick is made on double-stranded DNA by using DNase (in a buffered solution containing dNTPs) which exposes the 3'-OH group on one strand of DNA. All four nucleotides are at the alpha position of the phosphate group. (α-32P-dNTPs) are radiolabeled, generally only one n□cleotide, i.e., α-32P-dCTP is radiolabeled, whereas the other nucleotides are unlabeled. *E. coli* DNA polymerase I is used in this method.

This enzyme shows 5' to 3' hydrolysis of the nucleotides from the 5' side of the nick. Exonucleolytic activity either removes or eliminates nucleotides from the 5' end and simultaneously adds at the 3' side, resulting in the movement of the nick along the DNA. Therefore, this method is called nick translation. The reaction mixt□re incl□des probe DNA to be labeled, α-32P-dTP, the other three dNTPs, DNase 1, and *E. coli* DNA polymerase I. It is incubated at 14°C for 24 hours. The hybridized DNA is separated from the rest of the dNTPs and used for hybridization.

6.20.1.2. Random Primed Radiolabeling of Probes

This technique is based on Vogelstein (1983, 1984). The restriction fragments purified by gel electrophoresis from denatured linear or circular DNA are used to prepare probes. In this method, primers are used for DNA synthesis but DNA nicking is not done. The purified DNA to be labeled is mixed in a buffer and heated at 100°C to denature it. Then it is put on ice to lower the temperature. The reaction mixture (containing Klenow fragment of *E. coli*, DNA Polymerase I, one radiolabeled n□cleotide, e.g., α-^{32}P-dCTP, and three unlabeled nucleotide triphosphates) is incubated at 37°C for 30 minutes to 16 hours. Since Klenow DNA Polymerase I lacks 5'→3' exon□clease activity, the radiolabeled probe is synthesized by primer extension but not nick translation as described earlier. The hybridized DNA is separated from unincorporated nucleotides and used as a probe.

6.20.1.3. Probes Developed by PCR

This method is advantageous because a very small amount of template DNA can be used. The method of operation is the same as described. In this method, the reaction mixture includes linear template DNA (to be labeled), amplification buffer, Taq polymerase, primers, one radiolabeled and three unlabeled dNTPs. The hybridized DNA is isolated and used as a probe.

6.20.2. Non-Radioactive Labeling

Mostly, non-radiolabeled DNA probes are also used to detection of desired DNA. There are three systems of non-radioactive routes such as using horseradish peroxidase, biotin probe and steptavidin probes as described below.

6.20.2.1. Horseradish Peroxidase (HRP) Method

The HRP is a plant-derived enzyme which is covalently linked to DNA. Chromogenic substrates (such as chloronaphthol) or chemiluminescent substrates (that emit light after reacting with DNA (probe enzyme, e.g., luminol) are used for detection of HRP-linked DNA. Chloronaphthol yields an insoluble purple product in the presence of peroxide and HRP. Consequently, the presence of DNA is detected. Similarly, HRP oxidizes luminol and emits luminescence when using X-ray film; hence luminescence can be detected.

6.20.2.2. Digoxigenin (DIG) Labeling System

Digoxigenin is a cardenolide steroid derived from the plant Digitaria as a hapten. The presence of digoxigenin can be detected by an antibody associated with an enzyme (anti-digoxigenin-alkaline phosphatase conjugate). The probe is labeled with digoxigenin-DUTP, which is a nucleotide triphosphate analog. This analog containing a digoxigenin moiety is incorporated into DNA by nick translation or random prime labeling. The DNA acts as the probe. Anti-DIG (antibody to digoxigenin) conjugated with alkaline phosphatase detects the DIG-labeled probe through enzyme-linked immunoassay. A chromogenic substrate, such as 5-bromo-4-chloro-3-indolyl phosphate (BCIP) and nitroblue tetrazolium chloride (NBT), is added, which reacts with alkaline phosphatase and produces a purple-blue color (Figure 6.14). A chemiluminescent substrate, such as dioxetane, is also used with digoxigenin-streptavidin systems. There are several derivatives of digoxigenin that emit light after enzymatic activation by alkaline phosphatase. The denatured DNA present on Southern blots is hybridized by the labeled probe. The membrane is placed in a detection buffer containing antidigoxigenin (20 pg/ml) and bovine serum albumin (BSA) (5% w/v) and incubated at 37°C for 1 hour. After incubation, the membrane is washed thrice. Enzyme activity of alkaline phosphatase is detected by using BCIP (0.17 mg/ml) and nitroblue tetrazolium chloride (NBT) (0.33 mg/ml); a purple-blue color develops.

6.20.2.3. Biotin-Streptavidin Labeling System

Biotin is vitamin H and avidin is found in egg white. The principle of this system is based on the interaction between biotin and the glycoprotein avidin. A biotin-containing nucleotide analog (biotin-14-ATP) or probe is prepared through nick translation or random priming methods. Streptavidin is conjugated to alkaline phosphatase. The biotinylated probes are detected by the specific binding of streptavidin-conjugated alkaline phosphatase. Thereafter, a chromogenic or chemiluminescent substrate is added that reacts with alkaline phosphatase and detects the hybrids. The intensity of the color is proportional to the amount of biotin present in the hybrid DNA. This method detects the hybrids much faster than the radioactive probes. However, a very small number of probes (20 nucleotides) consist of only a small number of biotinylated sites. Therefore, the intensity of signals is limited by biotin as a dye.

6.21. GENE CLONING IN EUKARYOTES

Most studies on gene cloning have been carried out in bacteria. In addition, many difficulties are associated with them, for example: (i) correction of introns of eukaryotic mRNAs, (ii) failure of transfer of an equal number of plasmids into daughter cells during cell division resulting in two types of cells, one with plasmid and the second without plasmid, post-translational modifications (e.g., inhibition of proteolysis, addition of oligosaccharides to specific sites on the polypeptide chain, glycosylation), (iii) threat of hazardous effects, and (iv) expression to ensure the presence of a stable plasmid in the bacterium when applied in commercial applications. With these prospects, the attraction of using eukaryotic cells is obvious. In recent years, considerable efforts have been made for the improvement of crop plants through genetic engineering. Many works on various aspects of building vectors and expressing insert genes in eukaryotic cells are in progress. Some of them are briefly described here.

- **Plant Cells:** A large number of plants ranging from a single cell to multicellular forms. Success has been achieved in some plants.
- **Yeasts:** Generally, bacterial cultures are not suitable for the production of polypeptides.

6.22. TRANSFORMATION IN FILAMENTOUS FUNGI

Transformation through plasmid of *E. coli* was first described in protoplasts of *S. cerevisiae* (Hinnen et al., 1978) and then in *N. crassa*. However, the situation in filamentous fungi is different from the others as there is no convincing evidence for autonomously replicating plasmids. But sequences having functions similar to ARS have been isolated. The method of transformation is similar to that for plant cells. Following are examples where transformations in protoplasts have been carried out: *A. nidulans, A. Niger, A. oryzae, Cephalosporium acremonium, Coprinus cinereus, Glomerella cingulata, Gaeumannomyces graminis, Mucor sp., Penicillium chrysogenum, Septoria nodorum,* etc.

- **Application of Transformation of Fungal Protoplasts:** Mainly, the transformation approaches are used for (i) the analysis of promoters and construction of expression vectors, and (ii) biotechnological applications involving the construction of vectors with different pieces of promoters. These can be cut by restriction enzymes and joined with DNA ligase. By this method, changes in the promoter region are also possible and can be achieved through mutagenesis. Thus, a strain can be improved in the expression of foreign genes in fungi. Analysis of promoter overproduction of the products (Peberdy, 1989).

6.23. GENE TRANSFER IN DICOTS BY USING *AGROBACTERIUM* TI-DNA AS VECTOR

It is not possible to transfer Ti-plasmid into plant cells because of its large size, most likely up to 235 Kb. Therefore, T-DNA is excised and used as foreign DNA for expression in plant cells. From a bacteriological point of view, opine production and development of gall are not required for the successful integration and expression of DNA. Hence, replacement of NOS and OPS genes with a foreign DNA fragment retaining the opine synthase promoter for expression of DNA has no harm. Schell and Van Montague (1983) undertook some experiments where genes for resistance to kanamycin and methotrexate were expressed in cultured callus cells. Generally speaking, promoters are used for the expression of foreign DNA. It has become possible to target specific tissues of plants with specific regulated promoters.

6.23.1. Gene Transfer in Monocots

The limitation of Ti-plasmid is that it is specific only to dicot plants. *A. tumifaciens* does not induce tumors in monocot plants. However, most major crops, including cereals, are monocots. Hooykaas Van et al. (1984) discovered that *A. tumifaciens* could transfer T-DNA into certain monocots with the consequence of expressing the opine gene within the plant cell without inducing a tumor. This discovery made it possible for T-DNA to be expressed in cells of monocot plants.

6.23.2. Plant Cell Transformation

Plant cell suspension culture coupled with DNA transfer techniques has become a new field of gene manipulation in plants. Cell suspension cultures have been the traditional source of plant material for biochemical selection of plant cell mutants, such as numerous cell lines resistant to various amino analogs (Maliga, 1980). Suspension of protoplasts constitutes an identical plant material from one laboratory to another. The approaches made for plant protoplast transformation are (i) Introduction of a DNA sequence via integration into T-DNA of Ti-plasmid of *A. tumifaciens.* (ii) Transfer of naked DNA into protoplast including bacterial spheroplasts and microinjection and (iii) Transformation with plant viruses carrying inserts, e.g., Gemini and *Caulimoviruses.*

6.23.2.1. Plant Cell Transformation by Ultrasonication

Ultrasonication is used for various biological experiments. In some plants, this technique has been successfully employed for gene delivery. The Biotechnology Research Center in Beijing, China, utilized this method for gene transfer in plants such as wheat, tobacco, and sugarbeet. When the cultured explants were sonicated with plasmid DNA carrying marker genes like CAT, NPTII, and GUS, and then transferred to a selective medium, successful shoot growth was observed. In contrast, the Calli from the control group, which were not sonicated with plasmid, failed to grow on the selective medium and eventually perished. Transgenic tobacco plants were obtained with a frequency of 22%.

6.23.3. Liposome-Mediated Gene Transfer

Liposomes are microscopically small-sized particles. They contain phospholipid bilayers which are concentric in nature. Liposomes enclose an aqueous chamber and can entrap water-soluble molecules. Therefore, they are called lipid bags. Many plasmids are enclosed in them. By using polyethylene glycol (PEG), they may be stimulated to fuse with protoplasts. In several plants like carrot, tobacco, petunia, etc. This technique has been used for the successful transfer of genes. Due to endocytosis of liposomes, DNA enters the protoplast. It first adheres to the protoplast surface and then fuses with the protoplast on the surface. Consequently, plasmids are released inside the cell. There are several advantages in using this technique such as (i) low toxicity, (ii) protection of DNA/RNA from enzymes that lyse them, (iii) long stable storage of DNA fragments in liposomes, (iv) applicability in various cell types, and (v) high level of reproducibility.

6.24. ANIMAL CELLS

The culture of animal cells on a large scale is more difficult and expensive than plant cells, bacteria, yeast, etc. This is why the commercialization of natural products of animal origin is limited. However, most of the animal products have been produced by using bacteria as a biological tool. In India, Dr. K. Chandrashekharan, Center for Cellular and Molecular Biology (CCMB) Hyderabad, emphasized the use of recombinant DNA technology for the introduction of foreign DNA into mammalian cells. This would help in curing certain genetic defects.

- **Some of the Examples of Gene Cloning in Animals:** A variety of animal viruses are known, for example, simian virus 40 (SV40), adenovirus, retrovirus (ssRNA), vaccinia virus, etc. They can increase the efficiency of animal cell transformation if they are □sed as vectors. A special feature of viruses is that they contain strong promoters that can be of the viruses. The most commonly used virus is SV_{40}, which contains a circular DNA of about 5.2 Kb. In addition to the region, the DNA contains early genes and late genes. Early genes are required for DNA replication, whereas late genes encode the coat protein of the virus. DNA fragment and phage, the amount of DNA to be packed into the virus capsid is limited. The genome has 3 identified non-essential regions; hence the □se of non-defective viral vectors is limited. The defective virus vectors have missing genes whose infection process has been helped by viral DNA constituting functional copies of missing genes. If the cos cells (the host monkey cells) are used as a host, they have a defective replication origin growth of SV_{40}. Recombinants containing foreign DNA inserted into the early region can be propagated without the need for a helper virus (Glover, 1984). In this way, the recombinant and helper can be separately recovered from the host.

6.25. ELECTROPORATION

Electroporation (using an electric field) is now used to transfer foreign DNA into fragile cells. Brief pulses of high-voltage electricity (about 350 V) are applied to a protoplast suspension containing naked or recombinant plasmids. The electric pulses induce the formation of large pores in the cell membrane. These pores provide a passage through which the foreign DNA can enter the protoplasts, thus increasing the transformation frequency. The transformed protoplasts are cultured for about one month. These developed microcalli are plated on solid medium containing a selective marker (e.g., kanamycin). After 37–45 days, the Calli are analyzed for the presence of differences in transformed cells. By using the electroporation method, genes of choice have been successfully transferred into protoplasts of wheat, rice, petunia, sorghum, maize, and tobacco. Moreover, the transformation frequency can be improved by using linear DNA instead of circular DNA, by subjecting protoplasts to a heat shock at 45°C for 5 minutes, by adding PEG, and by using 1.25 kV/cm voltages.

6.26. PARTICLE BOMBARDMENT GUN

It was developed by Prof. Stanford and coworkers at Cornell University (USA) in 1987. As the term denotes, it shoots foreign DNA into plant cells or tissue at a very high speed. This technique is also known as particle bombardment, particle gun method, biolistic process, microprojectile bombardment, or particle acceleration. This technique is most suitable for those plants that hardly regenerate and do not show sufficient response to gene transfer through *Agrobacterium*, for example, rice, wheat, corn, sorghum, chickpea, and pigeon-pea. The apparatus consists of a chamber connected to an outlet to create a vacuum. At the top, a cylinder is temporarily sealed off from the rest of the chamber with a plastic rupture disk. Helium gas flows into the cylinder. A plastic microcarrier is placed close to the rupture disk.

- **Pollen Transformation Through Particle Bombardment:** Genetic improvement in alfalfa has been made by plant tissue culture methods. This method is time-consuming, requires special techniq□es, and the efficiency of getting stable regeneration is low. Ramaiah and Skinner (1997) produced transgenic alfalfa through direct delivery of DNA into pollen grains by particle bombardment method. Plasmid PBI121 bearing GUS reporter gene was introduced into pollen grains. The microprojectile bombarded liquid pollen s□spension and the recipient female flower. Pollinated flowers set seeds in about a month. 30% of plants derived from fertile seeds showed integration of GUS plasmid. It was confirmed by So□thern analysis. After 10 vegetative generations, some were allowed to fertilize the tagged transgenic plants and lost the integrated GUS plasmid, whereas in a few others, copies of the GUS gene decreased due to unknown reasons.

6.27. MICROINJECTION

Microinjection is a technique of delivering foreign DNA into a living cell (a cell, egg, oocyte, embryos of animals) through a glass micropipette. One end of a glass can be transformed by a plasmid carrying the early region of SV_{40} with a cos cells express wild-type large T antigen and can support the micropipette is heated until the glass becomes some forms to a very fine tip at the heated end. The tip of the pipette attains to about 0.5 mm diameter which resembles an injection needle. The process of delivering foreign DNA is done under a powerful microscope. Cells to be microinjected are placed in a container. A holding pipette is placed in the field of view of the microscope. The holding pipette holds a target cell at the tip when gently sucked. The tip of the micropipette is injected through the membrane of the cell. Contents of the needle are delivered into the cytoplasm and the empty needle is taken out.

6.28. DIRECT TRANSFORMATION

Like plant cells, mammalian cells can be transformed by foreign DNA fragments. To transform mammalian cells, it is necessary to precipitate the DNA with calcium phosphate and mix the cells to be transformed. The DNA molecule passes through the cell membrane and integrates randomly with the mammalian chromosome. Using this technique, selective markers can be linked up with DNA fragments to be cloned and introduced into mammalian cells. The transformed cells are then separated from the cell line after plating them on a selective medium. Many techniques transform cell lines. Following these biochemical processes, Perucho et al. (1980) successfully isolated the chicken thymidine gene (tk gene) by ligating it with PBR322 and transforming the thymidine-deficient (tk-deficient) mouse.

6.29. GENE DNA SEQUENCING

The structure of DNA (e.g., gene insert, a recombinant plasmid, or entire genome) can be analyzed by determining the nucleotide sequences. In molecular cloning, the information of nucleotide sequences is essential. In 1965, Robert Holley and his research group at Cornell University completely sequenced nucleotides of tRNA (tRNA for yeast alanine). In 1977, the following two methods were developed. Allan Maxam and Walter Gilbert developed a chemical method of sequencing. In this method, end-labeled DNA is subjected to base-specific cleavage reaction before gel separation. In routine DNA sequencing, this method is not commonly followed. In the same year (1977), Frederick Sanger and co-workers developed the dideoxynucleotide chain termination method because dideoxynucleotides are used in an enzymatic method of DNA sequencing.

6.29.1. Maxam and Gilbert's Chemical Degradation Method

This method is not very popular because it is time-consuming and labor-intensive. In this method, the DNA molecule can be radiolabeled at either the 5' end by using polynucleotide kinase, or the 3' end by terminal transferase. One end of radiolabeled double-stranded DNA is removed by using endonuclease. A base is modified chemically followed by cleavage of the sugar-phosphate backbone of DNA. No specific reaction for the four bases is carried out, except a specific reaction to G only and a purine-specific reaction that removes A or G. A difference in these reactions indicates the presence of A.

6.29.2. Cleavage of Purine

The mixture is separated into four sets, each treated with a different reagent which degrades only G or C or A and G or C and T. In one set, DNA is treated with acid followed by dimethyl sulfate. This causes methylation of A (at 3' position) and G (at 7' position). Subsequently, the addition of alkali (-OH) and pyrimidine results in DNA cleavage and removal of purines (A or G).

6.29.3. Cleavage of Pyrimidine

Similar to the cleavage of purine, pyrimidine (C or T) is cleaved in the presence of 1–2 M NaCl solution. It works only with C. Differences between these two indicate the presence of T in the DNA sequence. Cleavage of pyrimidines (C or T) through hydrazine hydrolysis. Partial chemical cleavage of DNA fragments as done above generates populations of radiolabeled molecules extending from radiolabeled termini to the site of chemical cleavage. These fragments of different sizes represent unique pairs of 5' and 3' cleavage products in the random collection. A complete set is formed by these products; the length of each fragment is short by a nucleotide. These can be separated by gel electrophoresis. The fragments containing labeled terminus can be observed by autoradiography of the gel. Following the order of fragments obtained from different digestions, the sequence of nucleotides is deduced and interpreted.

6.29.4. Sanger Method (Dideoxynucleotide Chain Termination Method)

Twice Nobel Prize winner Frederick Sanger's powerful method for et al. (1977) developed DNA sequencing that utilizes single-stranded DNA as a template. This method is also called the dideoxynucleotide chain termination method. The requirements are: DNA with free 3'-OH ends to start primer DNA synthesis, DNA polymerase, and dNTPs. Shows the presence of a free 3'-OH group at the 3' end in dATP and no 2,'3'-OH in 2,'3'-ddATP. In 2,'3'-ddATP, a hydrogen atom is attached at the 2' and 3' carbons instead of an -OH hydroxyl group. If any of the four ddNTPs bind, the chain elongation is terminated. Because ddNTPs do not have a free 3'-OH end required for chain elongation, no phosphodiester bond will be formed.

6.29.5. Automatic DNA Sequencers

Automatic DNA sequencing machines were Sanger's method. In this new method, a different fluorescent dye is tagged to the ddNTPs. Using this technique, a DNA sequence containing thousands of nucleotides can be determined in a few hours. Each dideoxynucleotide is linked with a fluorescent dye that imparts different colors to all the fragments terminating in that nucleotide. All four labeled ddNTPs are added to a single capillary tube.

It is a refinement of gel electrophoresis which separates q□ickly. DNA fragments of different colors are separated by their respective size in a single electrophoretic gel. A current is applied to the gel. The negatively charged DNA strands migrate through the pores of the gel towards the positive end. Small DNA fragments migrate faster and vice versa. All fragments of a given length migrate in a single peak.

The DNA fragments are ill□minated with a laser beam. Then the fl□orescent dyes are excited and emit light of specific wavelengths which is recorded by sequences that are read by determining the sequence of the colors emitted from specific peaks as they pass the detector. This information is fed directly to a computer which determines the sequence. A tracing electrogram of the emitted light of the four dyes is generated by the computer. The color of each dye represents the different nucleotides.

6.30. SITE-DIRECTED MUTAGENESIS

Now, techniques have been developed for the specific portion of the genome in order to obtain some novel products of enormous value. Mutations can be induced by directing insertions or deletions to a site on the DNA. These minor changes may involve a single amino acid being altered to improve its properties. For such modifications, point mutations, i.e., changing a single nucleotide, are done on a specific portion of the gene.

Thus, the most suitable method that could bring about a point mutation in a gene to mutate specific points is unlikely to be of use. This technique has the potential for protein engineering; the engineered enzymes would be better than the wild-type ones. A change in a single nucleotide base pair is called a 'point mutation.' Some useful properties in proteins (like stability in subtilisin) result from the substitution of selected amino acids. Using this technique, specific points of a gene can be mutated. Therefore, this method has been used to understand the function of many genes. Moreover, this technique can only be used when the nucleotide sequence of the gene is known.

6.30.1. Methods of Mutagenesis

When it has been decided which base is to be mutated, site-directed mutagenesis is performed using short synthetic DNA. Chemically changed (A) the oligonucleotide sequence has the desired mutation at a specific site. The oligonucleotide corresponds to the mutated nucleotide and its neighboring regions, mainly between 15 and 20 nucleotides in length. A single-stranded clone of the wild-type gene is produced using an M13 phage-based vector. The oligonucleotide is allowed to hybridize with the single-stranded clone. The hybridization is performed under specific conditions. The oligonucleotide will base pair with the wild-type complementary sequence (B) and act as a primer to produce double-stranded DNA using DNA polymerase (C) (Prescott et al., 2003).

CHAPTER 7

GENETIC ENGINEERING FOR HUMAN WELFARE

Contents

7.1. HUMAN PEPTIDE HORMONE GENES

In the human body, peptide hormones are secreted after being encoded by peptide hormone genes in specialized cells, for instance, insulin and other Human Growth Hormones (hGH).

7.1.1. Insulin

This peptide hormone, insulin, is secreted by the Islets of Langerhans in the pancreas, which catabolizes glucose in the blood. Insulin is a boon for diabetics whose normal function for sugar metabolism generally fails.

7.1.2. Somatotropin

Somatotropin, the hGH, is secreted by the anterior lobe of the pituitary gland, which consists of 191 amino acid units. Its secretion is regulated by two other hormones (somatostatin and growth hormone releasing hormone) produced by the hypothalamus. Deficiency of somatotropin occurs in about 3% of cases and is hereditary in nature. It has been estimated to affect about 1 child in 5,000. Turner's syndrome is one of the most common chromosome disorders in girls and is characterized by short stature and non-functioning ovaries, affecting approximately 1 in 2,500 live female births. The extraction of somatotropin pharmaceutically from the pituitary gland could not meet the annual demand for this hormone. Somatotropin was achieved through gene cloning procedures.

7.1.3. Somatostatin

Somatostatin, a 14-residue polypeptide hormone, is synthesized in the hypothalamus. It is the first polypeptide that was expressed in *E. coli* as part of the fusion peptide (Itakura et al., 1977), which inhibited the secretion of growth hormone, glucagon, and insulin. It does not contain any internal methionine. Eight single-stranded DNA segments were synthesized chemically and annealed in an overlapping manner to form double-stranded DNA (the synthetic gene). It had single-stranded projections at each end as the same are formed by EcoRI. The synthesized gene contained 51 base pairs which were terminated by two non-sense (stop) codons and preceded by a methionine codon as below: ATG-(42 base pairs encoding somatostatin) TGATAG.

7.1.4. β-Endorphin

β-endorphin (30 amino acid-long neuropeptide with opiate another growth hormone that was expressed in genetically engineered *E. coli* cells. Shine et al. (1980) integrated DNA seq□ences of β-endorphin, obtained from mRNA, adjacent to plasmid. The mRNA contained a large precursor of protein that consisted of, besides β-galactosidase genes on β-endorphin, the hormones α-melanotropin, corticotropin, β-lipotropin, and β-melanotropin. The β-endorphin is cleaved from the C-termin□s of the prec□rsor peptide. In this way, the transformed bacteria produced an insoluble fusion protein between β-galactosidase and β-endorphin. β-endorphin can be removed from the hybrid protein by trypsin, which cleaves only at arginine residue. Before doing so, internal lysines are protected from trypsinization by citraconylation. Human Interferon Genes (HIG). For the first time, Isaacs and Lindenmann isolated the interferon in 1957. The definition and nomenclature of interferon have been recommended by a committee of experts (Anonymous, 1980).

7.2. HUMAN INTERFERON GENE (HIG)

Interferon definition and nomenclature have been recommended by a committee of experts as a protein which exerts virus non-specific antiviral activity, at least in homologous cells through cellular metabolic procedure involving the synthesis of both RNA and protein. Interferon is used to cure many viral diseases such as common cold and hepatitis. It is species specific. In humans, there are three classes of interferon: (i) Alpha interferon (IFN-α), (ii) Beta interferon (IFN-B), (iii) Gamma interferon (IFN-y) or immune interferon (by lymphocytes of blood) and lymphoblastoid interferon by transformed leukocytes.

7.2.1. VACCINES FOR HEPATITIS B VIRUS

Hepatitis B virus (HBV) is widespread and produces several chronic liver disorders such as fulminant chronic hepatitis, cirrhosis, and primary liver cancer. HBV DNA is a double-stranded circular molecule of about 3 Kb in size and has a single-stranded gap which must be cleaved with a restriction enzyme for DNA cloning (Glover, 1984). After infecting human beings, HBV multiplies and infects a large number of cells and does not grow in cultured cells. This has been explained to hinder its molecular characterization and vaccine development. Human plasma has been found to contain varying amounts of antigens. Three types of viral proteins are recognized to be antigenic: (i) viral surface antigen (HBsAg), (ii) viral core antigen (HBcAg), and (iii) the e-antigen (HBeAg).

7.2.2. Rabies Virus (RV)

Rabies virus causes hydrophobia in animals and humans in many countries such as South America, Africa, and Asia (including India and Pakistan). Research is being conducted to synthesize vaccines by inducing genetically engineered *E. coli* cells. Attempts have been made to isolate mRNA encoding viral proteins from RV-infected cells. The genes encoding the production of rabies virus glycoprotein coat have been successfully transferred to *E. coli*. This is the first step towards the production of anti-rabies virus vaccines that stimulate antibody production in diseased animals. In 1984, the Wistar Institute in Philadelphia developed a new genetically engineered vaccinia virus by inserting a small piece of foreign DNA. The genetically engineered virus synthesized anti-rabies vaccines for animals. The recombinant vaccine virus did not cause rabies in animals that received the rabies genome but encoded antigenic molecules and activated the immune system against rabies infection. This new vaccine can be administered orally to animals and could decrease the risk of human death due to bites from animals carrying the rabies virus (Peters, 1993).

7.2.3. Vaccines for Poliovirus

Poliovirus is a causal agent of poliomyelitis in human beings. The first attempt to describe the complete structure of poliovirus RNA was made by Kitamura et al. (1981).

7.2.4. Vaccines for Foot and Mouth Disease Virus (FMDV)

Foot and mouth disease (FMD) is a serious disease caused by Aphthovirus. The primary control measure of the disease has been the slaughter of FMDV-infected animals. FMD control is through vaccination. Vaccines are produced by inactivation of virus grown in bovine lingual epithelium. A detailed study of FMDV reveals that it contains a single-stranded RNA covered in a capsid of four polypeptides, for example, VP1, VP2, VP3, and VP4, where only VP1 has a little immunogenic activity. However, the nucleotide sequence encoding for VP1 was identified on the single-strand RNA genome and cloned on double-stranded PBR322 in *E. coli*. About 1,000 molecules of VP per bacterial cell were synthesized (Kupper et al., 1981).

7.2.5. Vaccines for Smallpox Virus

Smallpox is a very serious disease of humans in many countries. The vaccinia virus (the cowpox virus) can be used as the basis of the smallpox vaccine by using recombinant DNA technology. The genome of the vaccinia virus is altered by inserting foreign genes which encode the pathogenic antigens. When a person is immunized with a vaccine in preparations, the foreign genes enter the body cells of the patient receiving.

7.2.6. Malaria Vaccines

According to WHO estimates, 4 million people are at risk of developing malaria and about 500 million cases occur each year, resulting in one million deaths each year, mainly among children aged 5 and pregnant women. In addition, development of resistance against drugs by the species of Plasmodium, and insecticides by mosquitoes has been reported. Therefore, the threat of malaria is still increasing for humans. For the control of malaria, the use of vaccines and vector control programs would be successful. Much work is ongoing at the Indian Institute of Immunology (New Delhi) and ICGEB on the development of a malaria vaccine using modern methodologies.

All kinds of vaccine development through recombinant antigens, synthetic peptides, and direct use of DNA are being attempted. All these attempts indicate that the development of a malaria vaccine is a large complex process. However, progress towards the development of a malaria vaccine has been slow due to several reasons, one of which is the lack of *in vitro* correlates and suitable animal models for malaria vaccine trials.

7.2.7. DNA Vaccines

For the first time Wolf et al., (1990) injected naked DNA into the muscles of mice which led to expression of encoded marker protein. Thereafter, there has been a surge to use this approach to generate DNA vaccines against a variety of infectious diseases. Thus DNA vaccines are giving hope of a third vaccine evolution.

7.3. GENES ASSOCIATED WITH GENETIC DISEASES

This effect: In December 1993, a team of international scientists at the French Academy of Sciences in Paris developed the world's first map of the human genome. For over a decade, genetic scientists have been attempting to map the genes. So far, the physical map existed only for 2% of the human genome, and the present map covers about 90% of the genome. To produce the map, they pieced and grew each piece and fingerprinted to detect overlapping sections. These sections were pieced back together to create the map. This will help scientists discover genetic diseases. In human beings, there are many cases where certain genes responsible for encoding enzymes are missing, resulting in genetic diseases. Some of them are briefly described in the following sections.

- **Phenylketonuria Genes:** In diseased persons, when phenylalanine fails to get converted into tyrosine, disturbances in metabolism result in mental retardation. It is possible to cure this disease by using recombinant DNA techniques in the early period of pregnancy to cut human DNA into a yeast cell for clones. The clones were then cut into fragments used as guides.
- **Urokinase Genes:** Urokinase is involved in dissolution of blood

clots. Urokinase has been synthesized in huge quantity by using genetically engineered bacteria with urokinase genes.

- **Thalassemia Genes:** It is a condition in which synthesis of A- and B-globin chains is reduced and the excess chains precipitate, causing hemolytic anemia and spleen enlargement. Human globin genes have been identified and sequenced. It has been found that A- and B-globin genes are closely linked. Human globin genes (cDNA) have also been developed and cloned. However, much work has to be done to cure this disease.
- **Hemophilia Genes:** Hemophilia is a sex-linked disease in humans where blood clotting does not take place normally due to a deficiency of clotting factor VIII: C. By using gene cloning techniques, the clotting factor VIII: C gene was cloned, which expressed in mammalian cell lines and produced the protein VIII: C responsible for blood clotting.
- **Enzyme Engineering:** The genes encode enzymes; changes in genes certainly bring about alterations in enzyme structure. In addition to methods available for gene manipulation, alteration of genes by site-directed mutagenesis for enzyme engineering has become very popular. Thus, site-directed mutagenesis produces single amino acid substitutions in the primary structure of enzymes.

7.4. COMMERCIAL MOLECULES

7.4.1. Prevention, Diagnosis and Cure of Disease

Prevention of many genetic diseases in the early period of pregnancy will help mothers give birth to healthy babies. Moreover, certain viral diseases (e.g., polio, AIDS – acquired immune deficiency syndrome) and protozoan diseases can be prevented from infecting humans.

1. **Prevention of Disease:** For preventive measures, several immunogenic polypeptides (vaccines) and proteins (antibodies) have been chemically and biologically synthesized. It is expected that in the near future, immunogenic compounds will be available in the market and more information will be gathered on cloned genes for viral diseases (AIDS), bacterial diseases (cholera), and protozoan diseases (malaria).
2. **Diagnosis of Disease:** Genetic engineering techniques have solved the problem of conventional methods for diagnosing many diseases. DNA probes, monoclonal antibodies, and antenatal diagnosis are

some of the available methods used as tools to diagnose a particular disease.

3. **Parasitic Disease:** Probes used for diagnosing pathogens contain the most specific DNA seq□ences of the genetic material of the parasite. The specificity lies in s□ch a way that other related species or strains do not contain those seq□ences. The □nrelated □nspecific seq□ences of the parasite are first recognized □sing DNA hybridization techniques. Then, a DNA sequence not present in any species is identified, cleaved □sing restriction enzymes, and inserted into a cloning vector (plasmid). The bacterial cells are transformed by the recombinant vector. The transformants are multiplied, retrieved from the host cells, and the DNA sequences of the parasite, now fragmented, are labeled. Following this, they are used as a probe. The probe can also be chemical.

7.5. MONOCLONAL ANTIBODIES

Antibodies are proteins synthesized in blood against specific antigens just to combat and give immunity in blood. They can he collected from the blood serum of an animal are heterogeneous and contain a mixture of antibodies for polyclonal antibodies). Therefore, they do not have characteristics of specificity, If a specific lymphocyte, after isolation and culture *in vitro*, becomes capable of producing a single type of antibody which bears specificity against specific antigen, it is known 'monoclonal antibody.' Due to the presence of desired immunity, monoclonal antibodies used in the diagnosis of diseases. Major difficulties with antibody secreting cells are that they cannot be maintained in culture. But myeloma cells (bone marrow tumor cells due to cancer) grow indefinitely to produce a huge quantity of identical cells (clones) and also produce immunoglobulins in the same amount. The immunoglobulins are infact monoclonal antibodies (Tannis et al., 2018).

7.6. GENE THERAPY

There are many diseases which can be cured by using specific medicine synthesized biochemically. Nowadays, techniques have been developed for biochemicals, for example insulin, interferon, somatotropin, somatostatin, endorphin, human blood VII, immunogenic proteins, etc. Several companies such as Eberstadt & Co. (New York), E. Lilly (USA), National Pituitary Agency (USA), Kabi Vitrum AB (Sweden), Genetech Co (USA), Biogen (Switzerland), Hybritech (USA), Astra Research Center (India), etc., are producing. However, after 1975, a remarkable advancement in recombinant DNA technology has occurred and accumulated such knowledge that has made it possible to transfer genes for the treatment of human diseases. Several protocols have been developed for the introduction and expression of genes in humans, but the clinical efficiency has to be demonstrated conclusively. The success of gene therapy depends on the development of better gene transfer vectors for sustained, long-term expression of the foreign gene as well as a better understanding of gene physiology of human diseases.

7.6.1. Types of Gene Therapy

All the gene therapies that can be done in humans and classified into the following four types:

i. Somatic gene therapy;
ii. Germ –line therapy;
iii. Enhancement of genetic engineering; and
iv. Eugenic genetic engineering.

7.6.2. Methods of Gene Transfer

A variety of gene transfer strategies have been developed during the last decade for the treatment of human diseases which can be grouped into the two categories:

i. Virus mediated; and
ii. Non-viral approaches.

7.6.3. Gene Therapy Success

The success of gene therapy depends on the gene delivery mechanism as well as the choice of target tissue. Rangarajan and Padmanaban (1996) have discussed different conditions leading to the success of gene therapy:

i. Cell types capable of dividing *in vitro* (e.g., myeloblasts, hepatocytes, keratinocytes, endothelial cells, etc.) are amenable to *in vitro* and *in vivo* gene therapy. Both *in vivo* methods are preferred for cell types such.
ii. The function of gene products also governs the selection of tissues. For example, in the case of hemophilia, a gene can be delivered in any tissue provided the gene product is released into the bloodstream. In addition, in the case of cystic fibrosis, the gene sho□ld be delivered to specific cell types where the introd□ction of the correct gene is required.

7.7. DNA PROFILING

DNA is the master molecule of all life forms. It constitutes the blueprints of every living organism through which characters are passed from generation to generation. However, every living organism differs only due to the nucleotide sequences of chromosomes. The nucleotides form codes; therefore, this coded genetic information can be profiled to produce the most authentic identity card of any organism. The complicated technology that facilitates the identification of individuals at the genetic level is known as DNA fingerprinting or more specifically DNA profiling. This genetic analysis is based on identifying tiny segments of the hereditary material which testify to the unique molecular signature that cannot be altered.

7.7.1. Method of DNA Fingerprinting

In this technique, DNA is isolated from blood stains, semen stains, or hair roots of disputed children or any suspected person and the same from parents, close associates, or relatives of suspected criminals (based on the cases). Since hairs contain a lesser amount of DNA, it can be produced in a large amount by using polymerase chain reaction (PCR). RBCs do not contain DNA; therefore, WBCs are the source of DNA. Thus, the DNA isolated is cut with a restriction enzyme and subjected to Southern blotting. The DNA bands appearing on the membrane are with ^{32}P-DNA probe, washed in water to remove the hybridized complementary DNA sequences, develop images (prints). Identical prints that contain specific DNA sequences appearing confirmed. The probability of two persons having similar sets of base pairs in the same sequence of VNTR of DNA is one in 30–300 million individuals hybridized unhybridized DNA and passed through X-ray identified and thus identity is on two X-ray films are for.

7.7.2. Application of DNA Profiling

1. **Setting up of Genetic Databank:** Realizing the potential of DNA profiling, genetic databanks have been set up throughout the world. Databases of the genetic fingerprints of criminals have been established in many parts of the USA and Britain. The Forensic Science Services in Birmingham (Britain) has set up the world's first National DNA Database, where abo□t 5 million records wo□ld be available. It has collected samples of tissues and DNA from s□spected individ□als for profiling.
2. **Reuniting the Lost Children:** DNA profiling technology has helped in reuniting lost children with their respective parents or vice versa, who were separated during war, violence, or natural disasters. For example, in Argentina, many people were abducted during military

rule (1976–1983). They were presumed to be dead. However, an Argentinian Human Rights Group (The Grandmothers of the Plaza de Mayo) believed that the children were still alive and had been adopted by military personnel after killing their natural parents. In 1987, a genetic databank storing blood samples of grandparents and relatives of the lost children was established. By matching the DNA profile of one or more grandparents with that of the lost children, about 40 missing children have been reunited with their families (Chawla, 1998).

3. **Solving Disputed Problems of Parentage, Identity of Criminals, Rapists, etc.:** To carry out the test, a small portion of blotted blood is punched out from which blood cells are isolated and information is retrieved from a computer databank. The first case using a preserved sample was done to identify a soldier who was burnt to death in a car accident. In India, a disputed parentage case was solved. In June 1988, a four-year-old female baby named Lakshmi was stolen from Chennai by a missionary and renamed as Merry by the so-called parents. After filing an FIR and seeking police help, the matter was resolved.

7.8. IMMIGRANT DISPUTE

More or less, in every country, there is the problem of immigration. Ultimately, the identity is not confirmed whether they have crossed the borders or are residents of that country. Therefore, the problem of immigrants can be solved through a DNA databank.

7.9. ANIMAL AND PLANT IMPROVEMENT

The population of the country is rapidly increasing day by day. However, the most urgent need for the nation is to meet the food demand of the people.

The production of transgenic animals through microinjection techniques will play a significant role in veterinary sciences as well as in human welfare. A farm animal can produce milk containing tons of grams of protein per liter, many times more than a liter of bacterial culture. The possibility of producing human pharmaceuticals was demonstrated when lactating mice secreted the active tissue plasminogen activator (t-PA) in their milk. The t-PA is an enzyme that dissolves the blood clot responsible for coronary artery blockage that results in heart failure. John McPherson at Genzyme Co., in collaboration with K. Elbert and colleagues of Tuft University, produced the first transgenic goat by using a goat beta-casein promoter gene linked to the t-PA gene. The transgenic goat produced t-PA in quantities compared to cell culture-derived material. But goat peptide is slightly different from the culture-derived t-PA. The goat peptide has two chains of peptide, whereas a single-chain peptide was found in culture-derived t-PA.

7.10. POLLUTION OF ABATEMENT

For detoxification and degradation of toxic chemicals, enzymes are encoded by specific genes present on plasmids. Chakraborty and co-workers (1979) succeeded in isolating the microbial culture which could be used in a number of organic chemicals, toxic in nature, such as salicylate, 24-D. 3 chlorobenzenes, ethylene, biphenyls, 1,2,4-trimethylbenzene, 2,4,5-trichlorophenoxy acetic acid, etc. Genes responsible for degradation of environmental pollutants, for example, toluene, chlorobenzene acids, and other halogenated pesticides and toxic wastes have been identified. One separate plasmid is required. It is not like that one plasmid can degrade all the toxic compounds of different groups. The plasmids are grouped into four categories:

- OCT plasmid which degrades, octane, hexane and decane,
- (ii) XYL plasmid which degrades xylene and toluenes,
- CAM plasmid that decomposes camphor, and
- NAH plasmid which degrades naphthalene.

Dr. Anand Mohan Chakrabarty, an Indian-born American genetic engineering scientist, produced new product strains in a single cell of *P. putida*. This superbug is capable of degrading all four types of substrates, for which four separate plasmids were required. The superbug (oil-eating bug) was created by introducing plasmids from different strains into single cells of *P. putida*.

CHAPTER 8

ANIMAL BIOTECHNOLOGY

Contents

Animal Biotechnology has made its importance felt in global scientific circles in a short time. The main objectives of the emerging scientific stream are the identification and characterization of animal breeds. Developing DNA-based diagnostics and genetically engineered vaccines for animals, studying animal genomics and its varied applications, and developing embryo transfer technology and other such techniques is another aspect. DNA forensics, molecular diagnostics, cloning, wildlife conservation, stem-cell research, and bioprocessing technologies are other important areas under the discipline. Animal biotechnology has diverse and widespread applications in the areas of food quality control, analysis of milk and milk products, and other animal products, besides the development of disease-free, pedigreed animals, and the like. The most important areas of today's research that have potential economic value and prospects of commercialization are the cell and tissue culture based on the production of vaccines, monoclonal antibodies, etc.

8.1. ANIMAL BIOTECHNOLOGY HAS A WIDE APPLICATION IN THE FOLLOWING FIELDS

- Health care
- Drugs
- Dairy
- Poultry
- Fisheries
- Aquaculture
- Food production
- Food processing
- Crimes
- Parentage disputes

8.2. HEALTH CARE PRODUCTS

The following health care products are produced by biotechnology-based industries such as:

- Antibiotics are harvested from fungus, bacteria e.g., *Penicillin* is obtained by the culture of the fungus *Penicillium sp.*
- Vaccines like TAB vaccine are prepared from animal cell cultures.
- Hormones like insulin, human growth hormone, etc., are produced from recombinant microbes.
- The production of monoclonal antibodies.

- Interferon is produced from genetically engineered *E. coli*, yeast, etc. It is used for the treatment of cancer, hepatitis, etc.
- Many blood factors are synthesized from recombinant microbes and used in the treatment of metabolic abnormalities in humans.

8.2.1. Tissue Culture

- The artificial skins are grown in monolayer c□lt□res and used to cure wounds and burns.
- Chondrocyte cells are genetically engineered and grown *in vitro* for bone marrow transplantation. These cells serve as artificial cartilage.
- Organotypic cultures made by linking different cell types are used for replacing damaged portions of body tissues such as liver, kidney, marrow, etc.
- Transgenic pigs are developed to use them as organ donors for transplantation in humans.

8.2.3. Meat Production

Transgenic fishes and shrimps are developed to increase fish production.

Transgenic poultry, pigs, sheep, goats, etc., are developed for fast growth and high body weight to increase meat production.

8.2.4. Milk Production

- Gene coding for Bovine somatotropin (BST) is introduced into cows to increase lactation.
- Bovine somatotropin (BST) is injected into cows to enhance lactation.

8.2.5. Animal Clone

- Transplantation of the diploid nucleus of adults into enucleated eggs gives offspring identical to the parent.

8.2.6. Disease Diagnosis

- Immunoglobulins, monoclonal antibodies and DNA chips are developed for the early diagnosis of infectious and genetic diseases in humans and animals.
- Western blotting is the confirmatory test for many diseases.
- ELISA and RT-PCR methods are developed for the rapid diagnosis of pathogens.

8.2.7. Fermentation Technology

- The animal cells produce a wide variety of commercial products such as erythropoietin, hGH, somatotropin, tPA, urokinase, antifebrin 99 and Tem-Fab, etc.

8.2.8. Reproductive Technology

- Intra-cytoplasmic sperm injection is used to overcome the failure of sperms to penetrate the egg membrane and fertilize the egg. It is done in humans and animals.
- Embryo transfer is performed to overcome women's defects in producing a baby.
- Embryo splitting is done to produce more than one offspring from a single zygote.
- *In vitro* fertilization is practiced to overcome the coital problems.
- Artificial insemination is performed in animals to breed better stocks in large numbers.

8.3. BIOINFORMATICS

It is the collection and storage of a vast array of information about DNA sequencing data and the use of such data in further research and development. It provides knowledge about different gene sequences in organisms. Bioinformatics is the name given to the new discipline that has emerged at the interface of biology and computing. A huge amount of genetic data (DNA, RNA, amino acid, and protein sequences) of various organisms, from bacteria to humans, being generated worldwide is stored in a computer database. Specialized software programs are used to find, visualize, and analyze the information, and most importantly, communicate it to other people. Various computer tools are used to predict protein structure, which is valuable information for the development of vaccines, diagnostic tools, as well as more effective drugs. Bioinformatics can help in the easy and early detection of various diseases like cancer, diabetes, and many more with the help of microarray chips (microarrays are miniature arrays of gene fragments attached to glass slides). Bioinformatics also helps scientists to construct phylogenetic trees based on molecular biology and ultimately contribute to the study of evolution. Computer simulations model such things as population dynamics or calculate the cumulative genetic health of a breeding pool (in agriculture) or endangered population. One very exciting potential of this field is that entire DNA sequences or genomes of endangered species can be preserved.

- **Genomics and Proteomics**
  - o **Genomics:** It deals with the computer-based sequencing of the genome of an organism.
  - o **Proteomics:** It refers to the computer-based sequencing of proteins produced by an organism.
- **Human Genome Project:** The sequencing of human DNA and identifying all human genes constitute the Human Genome Project. Similarly, genome projects of mouse, monkey, dog, rat, cattle, etc., have also been performed to understand the genetic basis of life.

8.4. HISTORY OF NEW ANIMAL BIOTECHNOLOGY

The biological processes involving the techniques of recombinant DNA and polymerase chain reaction (PCR) and bioprocessing, which become possible only through research in molecular biology, have been described as new biotechnology. Biotechnology is "the application of biological organisms, systems, or processes to manufacturing and service industries" (British biotechnology).

The integrated use of biochemistry, microbiology, and engineering sciences in order to achieve the technological application of the capabilities of microorganisms, cultured tissue cells, and parts thereof. Technology is using biological phenomena for copying and manufacturing various kinds of useful substances. The controlled use of biological agents, such as microorganisms or cellular components, is used for beneficial purpose.

- **Few Examples**
  - Many therapeutic proteins are produced in bacteria.
  - Insulin
  - Golden rice
  - Flavr Savr tomato
  - Bt-cotton

The new biotechnology revolution began in the 1970 and early 1980 with the appearance of genetic engineering. New advances of biotechnology are in pharmaceutical drugs and therapies to improve the treatment of many diseases, in production of healthier foods, selective pesticides and innovative environmental technologies (Table 8.1).

Table 8.1. *Important Contributions through scientist*

Year	Author	Discoveries
1970	*Mizutani, Temin and Baltimore*	Discovered cDNA production in Retrovirus infected cells
1970	*H. Gobind*	Discovered T_4 DNA ligase enzyme in T_4 phages
1972	*Jackson et al.,*	Discovered the homopolymer tailing method
1973	*Chang and Cohen*	Produced biologically active chimeric DNA for genetic recombination
1974	*Ferency, Kuei and Zsolt*	Discovered the Ti Plasmid in *Agrobacterium tumefaciens*
1975	*E. Southern*	Demonstrated the Southern blotting method for isolation and p□rification of DNA fragments.
1976	*Fukumaki et al.*	Constructed the cosmid vector
1977	*Bolivar and Rodriguez*	Developed the pBR322 vector
1977	*A.M. Maxam and Gilbert*	Discovered a technique to sequence nitrogen bases of DNA fragments
1978	*Broome & Gilbert*	Discovered immunological screening of recombinants
1979	*Mandel and Higa*	Demonstrated the uptake of DNA in the medium by animal cells
1979	*Alwin et al.,*	Discovered Northern blotting
1979	*Towbin et al.,*	Discovered Western blotting
1980	*Carlo M. Croce*	Produced monoclonal antibodies for human beings
1980	*Nagata*	Used *in vitro* translation method
1980	*Gilbert and Weissmann*	Produced human interferons by genetic engineering method in *E. coli.*

1980	*Dahl*	Constructed a gene bank of mammalian cells using cosmids.
1981	*Wickens and Laskey*	Used the microinjection method to deliver DNA into animal cell.
1981	*Workers of Eli Lilly (USA)*	Produced human insulin from genetically engineered *E. coil* cells.
1982	*Brackett et al.*	Used IVF technology in cattle breeding
1983	*Cocking*	Demonstrated protoplast fusion in plant cells and re-generated plant cells from the somatic hybrid.
1984	*Feinberg and Vogelstein*	Discovered the radio-labeling method for DNA
1985	*Kary Mullis*	Invented the PCR method for amplification DNA and awarded Nobel Prize in 1989
1987	*David et al.,*	Developed yeast artificial chromosomes (YAC)
1988	*Mel Simon et al.,*	Developed bacterial artificial chromosomes (BAC)
1988	*Steitz*	Discovered small nuclear RNAs (snurps) by RNA splic-ing
1988	*Simons*	Developed transgenic sheep
1988	*Wu and Wu*	Demonstrated transgene expression as a marker for the selection of recombinants.
1989	*Handyside et al.*	Used PCR technique for em-bryo sexing in cattle
1990	*Human Genome Organi-zation*	Started the HGP
1990	*Wolff et al.*	Noticed the expression of in-troduced naked DNA in mice
1990	*Michaele Blease & French Andresco*	Used gene therapy to treat hu-man patients

1990	*US DOE*	Started the Human Genome Project
1990	*Welsh and McClelland*	Developed arbitrary primed PCR (AP-PCR)
1991	*Williams et al.,*	Developed RAPD method
1992	*Zhang et al.,*	Discovered cationic lipid-associated transfection of DNA by animal cells.
1993	*Zhang et al.,*	Liposome mediated DNA/ drug delivery into target organs
1993	*Handyside*	Adapted IVF technology for preventing cystic fibrosis in infants of affected parents
1994	*Caetano-Anolles & Gresshoff*	Developed DNA amplification fingerprinting (DAF)
1994	*Bagai and Sarkar*	Developed gene transfer to cultured cells through reconstituted Sendai virus
1995	*Craig Venter and Smith*	Developed the whole genome short gun sequencing method for rapid DNA sequencing
1995	*Vos et al.*	Developed amplified fragment length polymorphism (AFLP) to detect polymorphism in the genome
1995	*Prabhu et al.*	Developed a reporter mediated gene targeting method for animal cells
1997	*Ian Wilmut*	Transferred a nucleus of somatic cells goat to an enucleated egg cell and grew egg into an adult goat
1999	*Lowrie et al.*	Demonstrated that a DNA vaccine of mycobacterial heat shock gene reduces the bacterial count in spleen and lungs
2000	*A. Datta*	Developed a nutritionally improved strain of potato using rDNA technology

2001	*Severino Antinon*	Planned to perform human cloning by transplanting of a diploid nucleus into an enucleated egg cell. This project was abandoned because of ethical problems.
2002	*Colebatch*	Stated that a proteome map will be required to show the exact location of different proteins in a cell
2002	*Brigitte Boisselier*	Made women to give birth to a cloned girl baby by human cloning
2003	*USDOE*	The human genome project was completed
2003	*Prasanna & Hoisington*	Identified the Downy mildew resistance gene from maize by QTL (quantitative trait loci) mapping
2006	*H. Kiyono*	Developed a cholera vaccine producing rice variety
2007	*M.R. Capecchi & M.J. Evans*	Performed specific gene modification embryonic stem cell method
2008	*M.A. Gomez-Lim*	Developed HIV vaccine producing tomato
2009	*Koreth, Schlenk et al.*	Stem cells transplants were reported to improve survival for leukemia patients
2012	*Doudna, Charpentier*	First patent application submitted for CRISPR-Cas9 technology
2014	*Ching, Eum, Lee et al.*	First embryonic stem cells cloned from a man skin
2014	*Yi, Choi, Bharaj, Abraham*	Mice trials show CD4 T-cells genetically modified with zinc fingers co□ld be effective in HIV gene therapy

2015	*Allison*	Two immunotherapy drugs reported to stop cancer avoiding destruction by the immune response
2015	*Gantz, Jasinskiene et al.*	Genetically modified mosquitos using CRISPR/Cas9 to prevent them from carrying the malaria parasite
2016	*Quezada, Johnson et al.*	Gene editing used to promote immune cells to combat cancer
2016	*Steinberg*	Stem cells reported to provide substantial recovery in patients disabled by stroke.
2016	*Weiner*	FDA approved first clinical trial for the zika virus vaccine
2016	*Sevigny, Chiao et al.*	Monoclonal antibodies drug for Alzheimer diseases shown to be promising in phase clinical trials
2017	*Landegger, Pan, Askew et al.*	Gene therapy was shown to restore hearing in deaf mice
2017	*Abudayyeh, Bhattacharyya et al.*	CRISPR was shown to be a sensitive diagnostic tool for detecting a single target of DNA or RNA molecule
2017	*Fogarty, McCarthy, Snijders et al.,*	DNA of human embryos edited using
2017	*Xu, Pegu, Rao, et al.*	Combination of three monoclonal antibodies reported protecting monkeys from HIV
2017	*Van Velthoven de Morree et al.*	Stem cells in the body found to have a significantly different gene expression profile than stem cells isolated and cultivated in the laboratory
2018	*Telfer, McLauhlin, Albertini et al.*	First human eggs grown in the laboratory
2018	*Thompson, Walters, Kwiatkowski et al.*	Gene therapy showed to be a promising treatment in a clinical trial for beta-thalassemia

2018	*Rosa, Pires, Kurochkin et al.*	Mouse and human cells reprogrammed into immune cells to fight cancer
2019	*Grunwald, Gantz, Poplawski et al.*	CRISPR-Cas9 used to control genetic inheritance in mice
2019	*Ravinder Gupta*	The second patient reported being free of HIV after receiving stem cell therapy
2020	*Li Heng et al.*	Coronavirus disease 2019 originated in the city of Wuhan, Hubei province, central China and spread quickly all over the world.
2020	*Lotfi et al.*	COVID-19: Transmission, prevention and potential therapeutic routes were discussed
2020	*AZD1222 (ChAdOx1)*	Covishield vaccine that aims to protect against COVID-19
2021	*Ram Kumar et al.*	Emetine suppresses SARS-CoV-2 replication by inhibiting the interaction of viral mRNA with eIF_4E
2021	*Zhang et al.*	Single-cell RNA-sequencing reveals the dynamic process and novel markers in porcine spermatogenesis

8.4.1. Animal Biotechnology in Developing Countries

In most developing countries, recombinant DNA technology has become the major thrust. In 1982, the Government of India set up an official agency, the National Biotechnology Board (NBTB), which started functioning under the Department of Science and Technology (DST). In 1986, NBTB was replaced with a full-fledged department, the Department of Biotechnology (DBT), under the Ministry of Science and Technology for planning, promotion, and coordination of various biotechnological programs. The DBT is making efforts to promote post-graduate education and research. Special MSc courses in Biotechnology are offered by DBT. The selection of students is through a National Test; in addition, it also provides trained manpower for the rapidly growing biotech industry. It has also raised the level of biology education in certain areas of biotechnology in the country. Moreover, a considerable amount of basic biochemical and molecular biology is imparted in these courses.

Today, India has the DBT, DST, CSIR, ICAR, ICMR, and IARI, among others working under the government. These agencies and other national and international industries are manufacturing biotech products and marketing them after clinical trials. A Technology Development Board (TDB) has been set up by the government. The TDB works with universities, industries, and national institutes. The Technology Information Forecasting and Assessment Council (TIFAC) has prepared a Vision 2020 document which includes biotechnology. There are over 30 companies in India producing modern biotech products such as Wipro, Reliance, Life Science, and Pennetia Biotech Ltd.

Since the 1980s, India has supported the biotechnology industry and its prod□cts significantly. Teaching and research of biotechnology have been included in university syllabi at both undergraduate and postgraduate levels. DBT-supported departments are operating in several universities and institutions. It is believed that India will play a key role in the future as one of the largest producers of biotech products.

The United Nations Industrial Development Organization (UNIDO) recognized the potential of genetic engineering and biotechnology for promoting the economic progress of developing countries.

The initiation taken by UNIDO has led to the foundation of ICGEB. In 1981, in a meeting convened by UNIDO proposed establishing an international center of excellence to promote biotechnology in the developing world. In 1982, this concept was approved by a high-level conference of developed and developing nations in Belgrade. The center's statutes were signed by 26 countries, and the statutes came into force on February 3, 1994. The ICGEB is now a fully autonomous international organization with 33 member states. It has two centers, one in Trieste (Italy) and the other at Jawaharlal Nehru University in New Delhi (India). The Trieste center covers an area of about 5,700 m^2, while the New Delhi center covers about 10,000 m^2. The center has been operating effectively since 1982. Many public and private institutions under government departments and organizations have advised the DBT to develop biotechnology programs in the following areas:

i. Plant molecular biology and agricultural biotechnology;
ii. Biochemical engineering, process optimization and bioconversion;
iii. Aquaculture and marine biotechnology;
iv. Fuel fodder, biomass and green cover;
v. Medical biotechnology;
vi. Microbial and industrial biotechnology;
vii. Large-scale use of biotechnology;
viii. Integrated systems in biotechnology;

ix. Veterinary biotechnology; and

x. Infrastructural facilities.

The pharmaceutical industry in India is very strong and vibrant with expertise in chemical drugs. It has little experience in biotech diagnostics and no experience in biotech therapeutics. Moreover, the pharma industry is located between Mumbai and Ahmedabad (90% of drug production in India is in Gujarat and Maharashtra). There is no government institution or university with expertise in this area to help the pharma industry. However, for a variety of reasons, the Indian pharmaceutical industry will sooner or later enter in manufacturing of biotechnology-based diagnostics and therapeutics (Padh, 1996).

- **The DBT has Chosen the Research Programs on the Following Priority Areas of Research and Development Activities:**
  - o Plant molecular biology and agricultural biotechnology
  - o Biochemical engineering, process optimization and bioconversion
  - o Aquaculture and marine biotechnology
  - o Fuel, fodder, biomass and green cover
  - o Medical biotechnology
  - o Microbial and industrial biotechnology
  - o Large-scale use of biotechnology
  - o Integrated systems in biotechnology
  - o Veterinary biotechnology
  - o Infrastructural facilities.

8.4.2. Importance Given to Animal Biotechnology

Biotechnology plays a crucial role in the pharmaceutical industry by enabling the creation of more efficient and cost-effective drugs. Recombinant DNA technology is utilized to produce specific enzymes that boost the production rate of targeted antibodies within organisms. Hormones like somatostatin, insulin, and human growth hormone can be synthesized with ease and at a lower cost. The first human hormone to be synthesized through genetic engineering was somatostatin, a brain hormone originating from the hypothalamus. It plays a role in inhibiting the release of human growth hormone and insulin, and is linked to the treatment of conditions such as diabetes and pancreatitis. Companies like Genentech in California have successfully produced human growth hormone (hGH) from genetically engineered bacteria. Human insulin, the first genetically engineered pharmaceutical product, was developed by Eli

Lilly and Company in 1982. Bovine Somatotropin (BST) is used to enhance milk production in cows. Antibiotics, which are chemical substances produced by various microorganisms, have also benefited from recombinant DNA technology, leading to increased production rates. For instance, the current production rate of penicillin is approximately 150,000 units/ml, a significant increase from about 10 units/ml in the 1950s. Antibiotics produced using this technology exhibit specific effects and fewer side effects.

8.5. BACTERIOPHAGE VECTORS

Bacteriophages provide another source of cloning vectors. Since usually, a phage has a linear DNA molecule, a single break will generate two fragments, which are later joined together with foreign DNA to generate a chimeric phage particle. The chimeric phage can be isolated after allowing it to infect bacteria and collect progeny particles after a lytic cycle. The use of phage particles as vectors imposes a limitation on the size of foreign DNA; if it is too long, the size of phage DNA may not be accommodated in the phage head. In order to overcome this problem, those segments of DNA that do not contain essential genes may be removed. Such a technique has been followed in phage lambda to create a smaller vector genome having a single restriction site for the enzyme *EcoRI*. Since the reduced size also fails to be adequately packed in the phage head, this automatically provides a selection method in which only the chimeric particles will be obtained in the phage progeny, and the vector particles lacking the cloned segment will be eliminated due to their reduced size.

8.5.1. Lytic and Lysogenic Phages

The pattern of infection of different bacteriophages mostly involves three steps. Firstly, the attachment of the phage at a specific site on the bacterial cell takes place. The phage is then injected into the DNA inside the bacterial cell. In the second step, the genes present in the DNA of the phage code for several enzymes that help replicate the phage DNA. Finally, other genes also code for several proteins that help form the capsid and assemble the new phage particle inside the capsid. This step is followed by the lysis of the cell and the release of the phages from the cell. In several bacteriophages, this cycle occurs rapidly and is called a lytic cycle. On the other hand, in some phages, the phage DNA is incorporated into the host genome after entering the host cell and remains there for several cell divisions. The phage DNA that has been incorporated into the bacterial DNA is called a prophage and remains quiescent. The bacteria that carry the phage DNA remain the same as any other uninfected cells. This type of infection is called a lysogenic cycle. The phage DNA is finally released from the host cell by entering the lytic cycle. The infection cycle of the lambda phage is an excellent example of a lysogenic cycle. In the case of cloning, the lysogenic cycle is most important because the phage DNA can carry a specific gene that is further incorporated into the host genome. As a result, the host genome will code for a specific gene carried by the phage. Therefore, lambda phages are mostly used as a cloning vector.

8.5.2. Lambda Phage Vector (λ Phage)

Plasmid vectors described in the previous section are often used for cloning DNA segments of small size. However, while preparing a genomic library in a eukaryote, the cloned fragments should be large enough to contain a whole gene. This can also be screened without serious difficulty. The above properties and other requirements of cloning the whole genome in eukaryotes are fulfilled by the phage lambda and cosmid vectors, the former permitting cloning of segments up to 20–25 kb long and the latter accommodating segments up to 45 kb long. Phage (λ) lambda, however, is easier and more efficient for making genomic and cDNA libraries; *see* Fig□re 8.1. The linear λ DNA has cohesive ends; each end is complementary to the other end. It is also called the Cos site; *see* Figure 8.2.

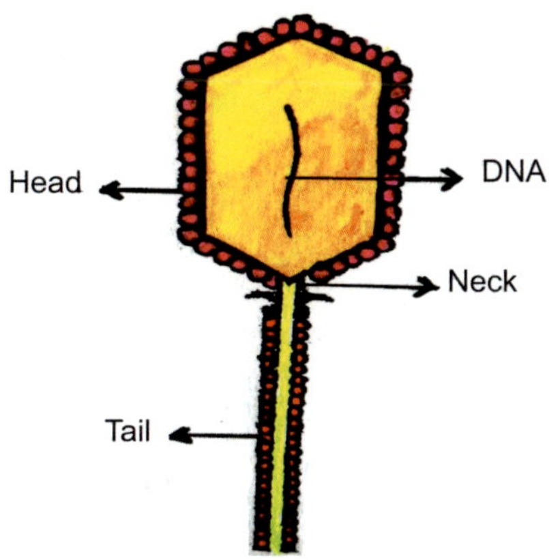

Figure 8.1. *λ4 bacteriophage.*

Source: Made by the author.

8.5.3. Lambda gt 10 and Lambda gt 11

Lambda gt 10 and lambda gt 11 are modified lambda phages designed to clone cDNA fragments. The major difference between those two vectors is that gt11 is an expression vector, where inserted DNA is expressed as a B-galactosidase fusion protein. Lambda gt 10 is a 43 kb double-stranded DNA for cloning fragments that are only 7 kb in length. The insertion of DNA causes the inactivation of the cl+ gene generating a cl- bacteriophage. Non-recombinant lambda gt 10 forms cloudy plaques in an appropriate *E. coli* host, while recombinant plaques. Further, in an *E. coli* strain carrying hf La150 mutation, only cl- phage will form plaques, because cl+ will form lysogens and will not undergo lysis to form any plaques. Recombinant Lambda gt 10 plaques can thus be easily selected. Lambda gt 11 is a 43.7 kb double-stranded lambda phage for cloning DNA segments, which are less than 6 kb in length. Foreign DNA can be expressed as beta-galactosidase fusion proteins. Recombinant lambda gt 11 can be screened using either nucleic acid or antibody probes. The recombinant lambda gt 11 becomes gal-, while non-recombinant lambda gt 11 remains gal+ so that an appropriate *E. coli* host, and with recombinant phage will form white or clear colonies and that with recombinant phage (gal^+) will form blue colonies permitting screening in the presence of IPTG (inducer) and X gal (substrate).

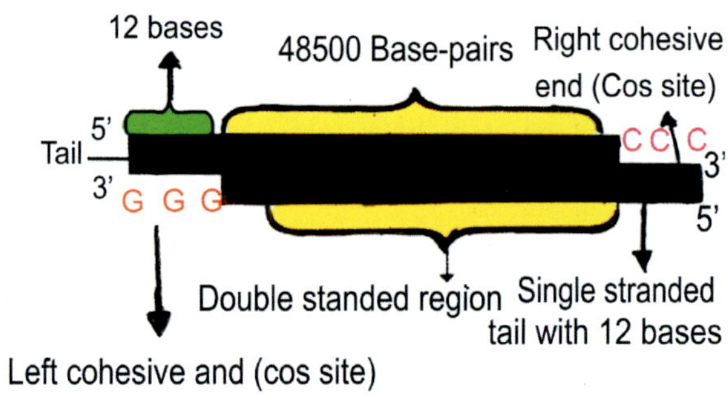

Figure 8.2. *λ chromosome map showing the cohesive end.*

Source: Made by the author.

8.6. CHARON 34 AND CHARON 35

Charon 34 and Charon 35 differ from each other only in their central fragments and will accept fragments 9–20 kb long. They have a more extensive range of restriction targets within their polylinkers than do EMBL3 and EMBL4. In these vectors, physical separation of lambda arms from the central fragments is required before ligation to donor fragments. In EMBL3 and EMBL4, removal of central fragments is not required, because there is a mechanism that ensures that 98% of packageable DNA is recombinant.

8.6.1. Insertional Vectors

The lambda phage is a typical example of bacteriophage which has a polyhedral head and tail. The DNA is protected by the polyhedral structure, and the tail interacts with the specific site of bacteria for attachment. Lambda phage usually infects *E. coli*. The DNA of lambda phage is 49 kb in size. Gene mapping and DNA sequencing analysis provide all the information regarding the genetic arrangement of the phage DNA. According to the data gathered from these analyzes, the genes of similar functions are clustered together in the genome. To illustrate, the genes responsible for the formation of the capsid are clustered together at the left-hand side of the polyhedral head while the genes responsible for the incorporation of the phage DNA with the host genome are clustered together at the middle; see Figure 8.3.

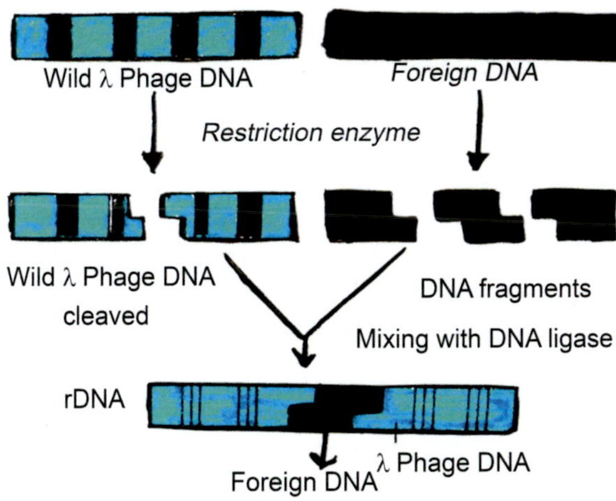

Figure 8.3. *Insertion of foreign DNA in phage DNA.*

Source: Made by the author.

Insertion vectors are the simplest form of "lambda cloning vectors." The vector itself can be grown (therefore must contain at least 75% of the wild-type genome length). Foreign DNA fragments are inserted into a unique restriction site in the vector genome. Packaging requirements thus limit insert fragment size to 0–10 KB – due to the limitations on viral genome size (75% to 105% of the wild-type length = 50 KB). In the construction of insertion vectors, 25% of the dispensable portion of the wild-type lambda genome was eliminated. Insertion vectors must also contain a unique restriction site into which we can insert foreign DNA fragments. Since the lambda genome approximates a random 50 kb DNA sequence, we can expect most 6-mer restriction enzymes to cut several times (average frequency is 1/4,000 bp). Removal of unwanted restriction sites involved m□tagenesis of phage followed by selection for efficient growth on hosts carrying different restriction/modification systems. After m□ltiple rounds of mutagenesis and selection, phage genomes were recovered that lacked unwanted restriction sites. Since insertion vectors are large enough to be packaged into viable viruses, it is useful to have a selection system that would allow us to identify recombinant phages containing foreign DNA inserts from non-recombinants containing only vector sequences. For lambda insertion vectors, this selection system is based on the choice between lytic and lysogenic lifestyles. The phage genome chooses between these alternatives based on the competition between positive and negative transcription factors expressed during the immediate early phase of the life cycle.

8.6.2. Replacement Vectors

Replacement vectors are "EMBL3 and EMBL4 are the two vectors that are designed so that a central non-essential part of 44 long phage can be replaced by foreign DNA. Cleavage of the phage with an appropriate enzyme generates three fragments. The central fragment representing 40% of the phage genome is non-essential for the propagation of the phage and can be replaced by foreign DNA that may be as long as 20–30 kb. Non-recombinant DNA resulting from the fusion of the left arm and right arm will give DNA that is too small for packaging, so that there is automatic selection against non-recombinant phages. EMBL3 and EMBL4 are two such replacement vectors used for preparing genomic libraries in eukaryotes with cloned fragments, 15–25 kb in size." The two vectors have polylinkers with reverse orders of restriction sites with respect to each other Figure 8.4. This type of vector has two recognition sites for restriction enzymes. So, the region between the two sites can be removed and a foreign DNA segment can be replaced. The replaceable fragments are often called the "Stuffer segment." These vectors are used to carry large pieces of foreign DNA.

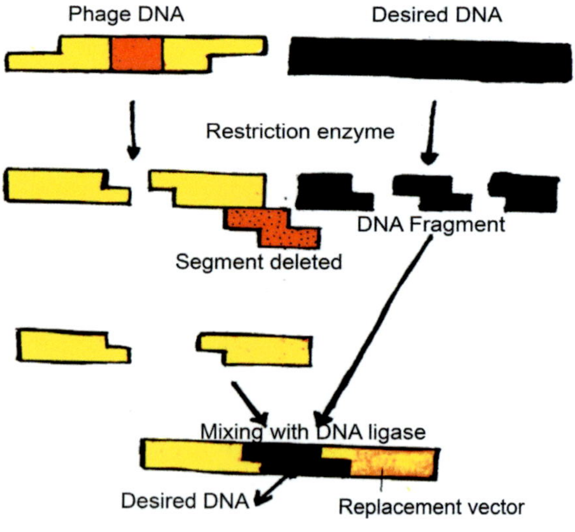

Figure 8.4. *Replacement phage vector.*

Source: Made by the author.

8.7. LIFE CYCLE OF Λ PHAGE

Lambda phage consists of a virus particle including a head (also known as a capsid), tail, and tail fibers. The head contains the phage's double-stranded circular DNA genome. The phage particle recognizes and binds to its host, *E. coli*, causing DNA in the head of the phage to be ejected through the tail into the cytoplasm of the bacterial cell. Usually, a "lytic cycle" ensues, where the lambda DNA is replicated many times and the genes for head, tail, and lysis proteins are expressed. This leads to the assembly of multiple new phage particles within the cell and subsequent cell lysis, releasing the cell contents, including virions that have been assembled, into the environment. However, under certain conditions, the phage DNA may integrate itself into the host cell chromosome in the lysogenic pathway. In this state, the λ DNA is called a prophage and stays resident within the host's genome without apparent harm to the host. The host can be termed a lysogen when a prophage is present. Lambda phage has been of major importance in the study of specialized transduction; *see* Figure 8.5. The phage attaches itself to the cell wall of *E. coli* cells. The phage injects its double-stranded linear DNA into the bacterial cell. Inside the bacterial cell, the two cohesive ends of phage DNA join together and form circular DNA. The bacteri□m containing the prophage DNA is called a lysogenic cell. Then the λ DNA multiplies rapidly with the bacterial DNA.

It is continuously replicated for a number of generations. This process is called the lysogenic life cycle. In the lysogenic life cycle, the host cell is not lysed. In lytic life cycles, the host cell is lysed. Sometimes the phage DNA is separated from the chromosomal DNA of the host. As a result, the phage DNA enters the lytic life cycle. Large numbers of viral particles are produced and they break the host cell and are released into the culture. This method of multiplication of bacteriophages is called the lytic life cycle.

Specialized transduction is the process by which a restricted set of bacterial genes is transferred to another bacterium. The genes that get transferred (donor genes) depend on where the phage genome is located on the chromosome. Specialized transduction occurs when the prophage excises imprecisely from the chromosome so that bacterial genes lying adjacent to the prophage are included in the excised DNA. The excised DNA is then packaged into a new virus particle, which delivers the DNA to a new bacterium where the donor genes can be inserted into the recipient chromosome or remain in the cytoplasm, depending on the nature of the bacteriophage. When the partially encapsulated phage material infects another cell and becomes a "prophage" (where it is bonded into the infected cell's chromosome), the partially coded prophage DNA is called a "heterogenote."

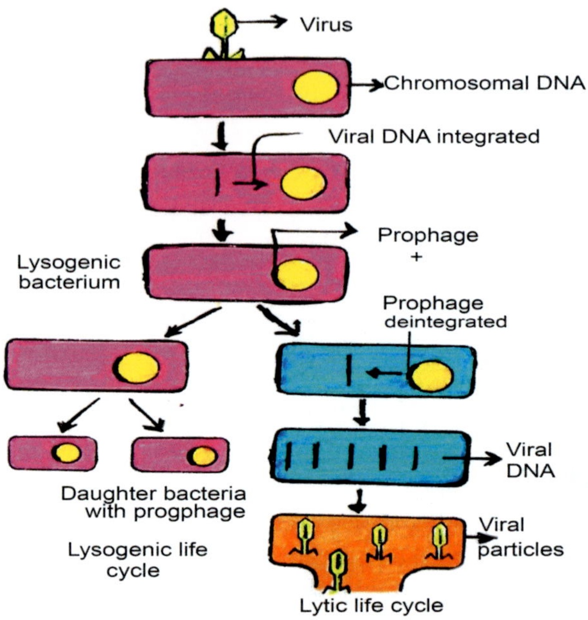

Figure 8.5. *Life cycle of λ phage.*

Source: Made by the author.

8.8. M13 BACTERIOPHAGE VECTOR

M13 is a filamentous bacteriophage of *E. coli* and contains a 7.2 kb long single-stranded circular series of cloning vectors that can be used for cloning a wide variety of DNA fragments, particularly for the purpose of sequencing through Sanger's method of dideoxy chain termination. Cloning vectors of the M13 mp series have a lac Z gene that complements the gal host, giving blue colonies. But if a foreign DNA segment is inserted at one of the polylinker sites associated with the lac Z gene, it inactivates the lac Z gene and no complementation is possible. Thus, on transformation, only white or clear plaques are obtained, permitting the selection of recombinant M13mp plaques. Once the foreign DNA is cloned in the M13 vector (*see* Figure 8.6), commercially available oligonucleotide primers are used for copying the insert in the presence of dideoxynucleotides. Fragments of different sizes with known termini are produced, permitting the determination of the sequence (*see* Figure 8.7).

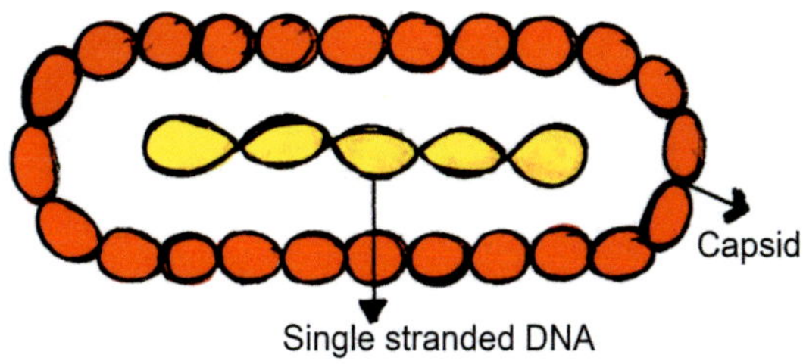

Figure 8.6. *M13 bacteriophage.*

Source: Made by the author.

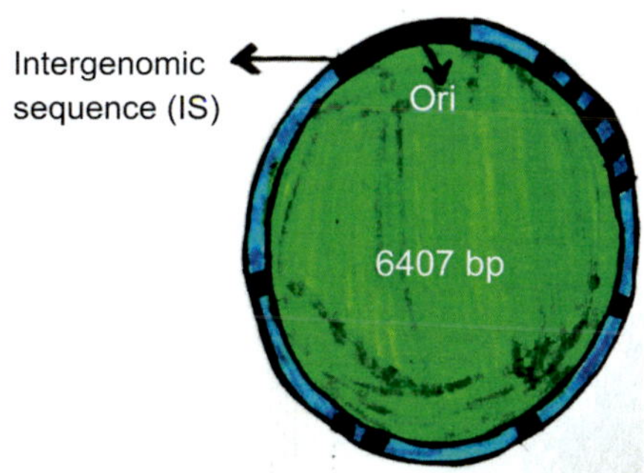

Figure 8.7. *M13 bacteriophage genome showing genes I to X position.*

Source: Made by the author.

Reversed order of the restriction sites in polylinker is present in a pair of vectors like M13mp8 and M13mp9 permitting sequencing from both ends of the double-stranded DNA molecule. Due to the small size of DNA, the phage genome can carry only a few recombinant genes. The M-13 phage DNA enters inside an *E. coli* cell *via* pil□s. Inside the cell, the phage DNA codes for specific enzymes that synthesize a complementary strand of DNA. This newly formed double-stranded DNA molecule then proliferates into several copies. During the bacterial cell division, each daughter cell contains copies of phage DNA and the phage DNA continues to replicate inside the daughter cells. Finally, new phage particles are assembled and released from the bacterial cell. M-13 phages are also used as a cloning vector. The genome size is 10 kb which is enough to insert new genes. Apart from that, the modification of single-strand DNA to double-strand DNA inside the host makes it unique; *see* Figure 8.8.

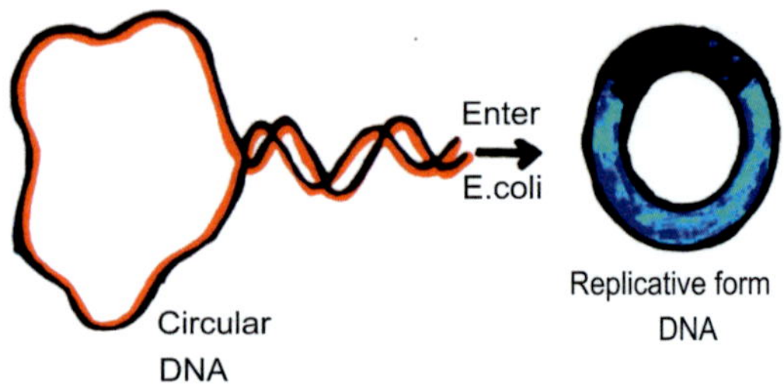

Figure 8.8. *M13 replicative forms.*

Source: Made by the author.

8.9. COSMIDS

Cosmids are plasmid particles that contain specific DNA sequences, such as cos sites. These cos sites facilitate the packaging of DNA into lambda particles, allowing for the packaging of DNA into phage particles in vitro and enabling purification. Similar to plasmids, cosmids replicate in bacteria and do not carry genes for lytic development. The key advantage of using cosmids for cloning is their high efficiency, which can generate a complete genomic library of 10^6 to 10^7 clones from just 1 microgram of insert DNA. The disadvantage, however, is their inability to accept more than 40–50 kbp of DNA. The Bacteriophage P1 system and F-factor-based vectors can allow the cloning of DNA segments as large as 100 to 1000 kbp in length. This concatemer is then packaged into viral particles (remember packaging is cos site to cos site) and these are used to infect *E. coli* where the cosmid vector replicates using the ColE1 origin of replication. Phage packaging serves only to select recombinant molecules and to transfer these long DNA molecules (50 kb total) into the bacterial host (50 kb fragments transform very inefficiently while phage infection is very efficient).

- **Advantages of Cosmids**
  - It is also used in the study of some non-sense sequences found in the genome of the organisms.
  - It picks up even large-sized genes.
  - It is used in the establishment of gene libraries of lower and higher organisms.
  - It is used to transfer the larger foreign gene into bacterial cells.

8.10. PHAGEMIDS

Phagemids are synthetic constructs that combine phage and plasmid features, as the name implies. A well-known phagemid in molecular biology labs is pBluescript II KS, a 2961 bp construct derived from pUC19. The "KS" designation signifies the orientation of the polylinker, ensuring that the lacZ gene's transcription proceeds from the KpnI to the SacI restriction site. The detailed structure of pBluescript II KS is essential for understanding its functionality in genetic engineering experiments. It may be noted that it has the following features: (i) a multiple cloning site (MCS) flanked by T3 and T7 promoters to be read in opposite directions on the two strands, (ii) an inducible lac promoter (LacI), upstream of lacZ region, which complements with *E. coli* and provides the facility for selection of chimeric vector DNA using the criterion of white colonies, (iii) fl (+) and fl (-) origins of replication derived from a filamentous phage for recovery of sense (+) and antisense (-) strands of lacZ gene, when the host is coinfected with a helper phage; (iv) an origin of replication derived from plasmid, and used in the absence of helper phage; (v) a gene for ampicillin resistance for antibiotic selection of chimeric phagemid vector; see Figure 8.9.

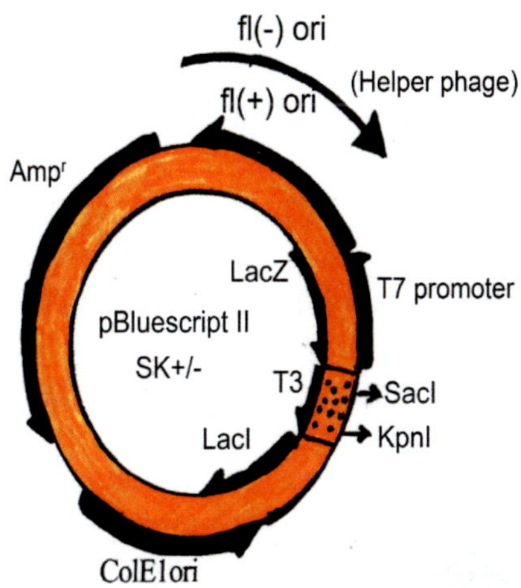

Figure 8.9. *pBluescript II SK+/- map.*

Source: Made by the author.

8.11. PLANT VIRAL VECTOR – CAMV

A number of plants and animal viruses have also been used as vectors both for introducing foreign genes into cells and for gene amplification and expression in host cells. In the latter case, we may be interested only in getting an increased quantity of the gene product and may not be interested in the integration of the foreign gene in the host genome. Some of the viruses that are commonly used as vectors will be described in this section; *see* Figure 8.10.

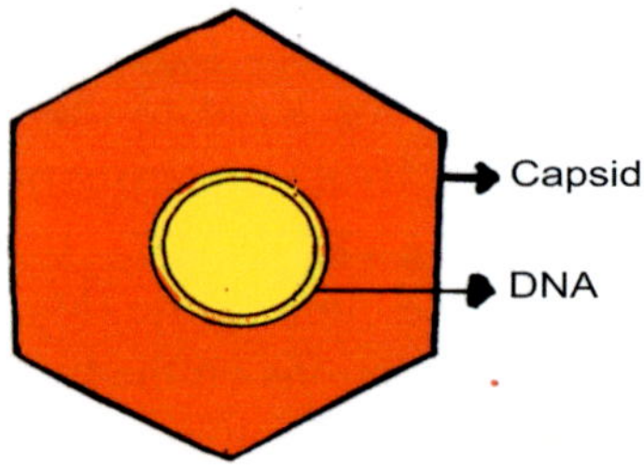

Figure 8.10. *Structure of CaMV.*

Source: Made by the author.

8.11.1. Plant Viruses (Cauliflower Mosaic Virus or CaMV and Geminiviruses)

Cauliflower mosaic virus, tobacco mosaic virus (TMV), and geminiviruses are three groups of viruses that have been utilized as vectors for cloning DNA segments. CaMV, known for infecting Cruciferae family members, harbors a double-stranded DNA molecule of 8 kbp in size with a deciphered sequence. Post-infection, CaMV rapidly disseminates throughout the plant, achieving a high copy number of up to 10^5 virus particles per cell. These characteristics render CaMV a suitable vector for transforming higher plants, although other viruses have also been employed as vectors, resulting in the generation of transgenic plants. Geminiviruses, a class of single-stranded DNA plant viruses, induce significant diseases in cassava, maize, and other cereals. Replicating through double-stranded DNA forms, geminiviruses are categorized into two groups: one infecting monocots and transmitted by leafhoppers, and the other infecting dicots and transmitted by whiteflies. Dicot geminiviruses possess two DNA segments in their genome, DNA A and DNA B, both crucial for infection via mechanical inoculation. DNA A encompasses all genes responsible for replication. Studies have demonstrated that deletions in coat protein genes do not compromise virus infectivity, allowing for the insertion of foreign DNA through extensive deletions in this region to create gene constructs beneficial as amplification and expression vectors in plants. Moreover, given that coat proteins are the most abundant proteins synthesized by the virus, promoters of coat protein genes can be harnessed for high-level expression of cloned genes. CLV and TGMV have already been tested for their ability to allow the multiplication and expression of a cloned gene. CaMV is a good gene cloning vector for *Brasisica*, *Raphanus*, and other species of *Brassicaceae* and *Resedaceae*.

8.12. ANIMAL VIRAL VECTOR – SV40

A number of animal viruses are also used as vectors, either for the delivery of nucleic acid into the cultured cells followed by its integration with the host genome or for the amplification and high-level expression of foreign genes, using the promoters from virus genes. These cloned genes can have a variety of uses including gene therapy in mammalian cells and the synthesis of important proteins by cloned genes in cell cultures. SV_{40} is a naked icosahedral virus with a diameter of 45 nm. Its capsid contains 72 protein subunits; *see* Figure 8.11. SV_{40} does not have enzymes in it, unlike RNA viruses.

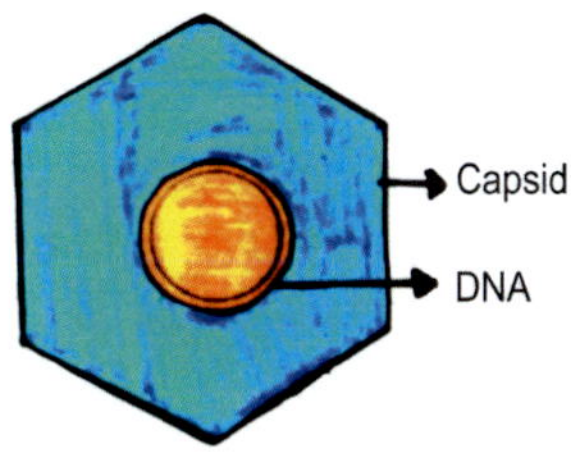

Figure 8.11. *Structure of SV40.*

Source: Made by the author.

The genome of SV40 is a single molecule of dsDNA of 5243 base pairs and is very small. The dsDNA of this virus is circular and exists in a supercoiled config□ration with a sedimentation coefficient of 21S. The total G+C content of the n□cleic acid is 41%. A linear form with a 14S sedimentation coefficient is formed after a double-stranded break in the supercoil. The complete base sequence of SV40 has been determined. Figure 8.12 shows the genetic map of the virus, where it is clearly evident that the genes encoding VP1, VP2, and VP3 proteins overlap each other.

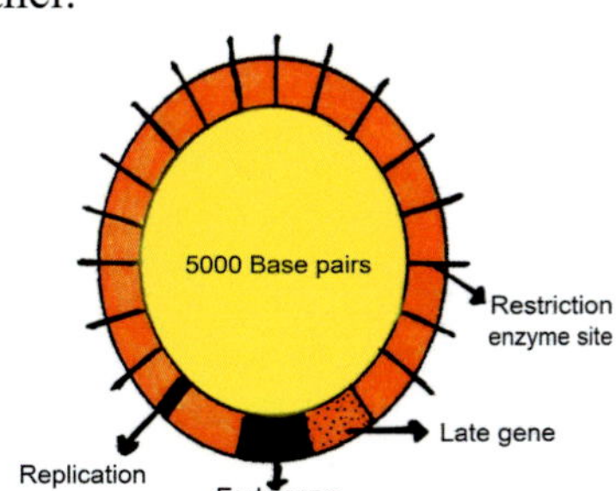

Figure 8.12. *Structure of SV40 genome.*

Source: Made by the author.

8.13. REPLICATION OF SIMIAN VIRUS 40 (SV40)

When the virus enters the host cell, its genome migrates to the nucleus. The SV40 genome replicates inside the nucleus of the cell, but the capsid protein is synthesized in the cytoplasm and migrates inside the nucleus where, finally, the assembly of the virus takes place. The replication of the SV40 genome can be divided into two distinct stages: early and late.

i. **Early stage:** The early region of the viral genome is transcribed to begin early protein synthesis during this stage within 12 hours of infection and before the start of genome replication. A single RNA molecule, called primary transcript, is synthesized by the RNA polymerase of the host cell and is processed into two mRNAs, a large one and a small one. These mRNAs are capped in the cytoplasm and translated to yield two proteins, small-T and large-T. The large-T (the T-antigen) binds to the site on the parental DNA; this initiates the synthesis of the viral genome.

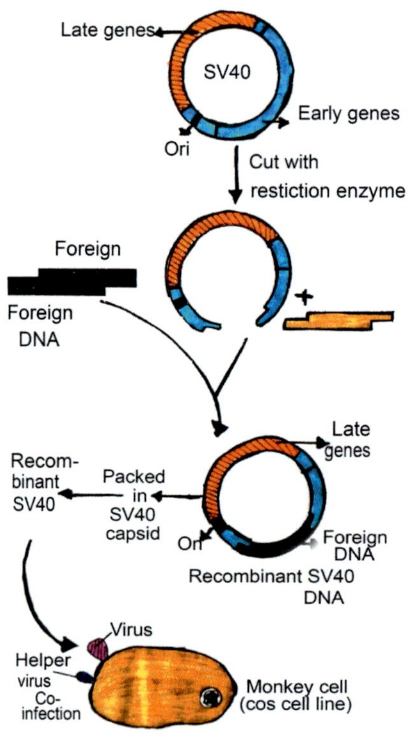

Figure 8.13. *Desired gene transfer in SV40 genome.*

Source: Made by the author.

ii. **Late stage:** Replication of the viral genome starts in a bidirectional fashion from a single origin of replication. Late mRNA molecules are synthesized using the strand complementary to that used for early mRNA synthesis. Transcription starts at a promoter near the origin of replication. This late RNA is then processed by splicing, capping, and polyadenylation to yield mRNA corresponding to the three coat proteins, namely, VP1, VP2, and VP3 (VP = virion protein). Viral coat protein mRNAs are then transported to the cytoplasm and translated into the viral coat proteins, which are then transported back into the nucleus where the assembly of progeny viruses takes place. The release of progeny virus particles occurs by the lysis of the host cell (Figure 8.13).

8.14. YEAST ARTIFICIAL CHROMOSOME (YAC)

We know that in plasmids, sequences up to 10–15 kbp; in lambda phage sequences up to 22 kbp; and in cosmids sequences up to 40 kbp can be cloned. Yeast artificial chromosome vectors, mentioned earlier, allow cloning of sequences that are several 100 kilobase pairs long. These long molecules of DNA, which may represent whole chromosomes, can be cloned in yeast by ligating them to vector sequences that allow their propagation as linear artificial chromosomes. These long DNA molecules can be generated and allow the construction of comprehensive libraries in microbial hosts.

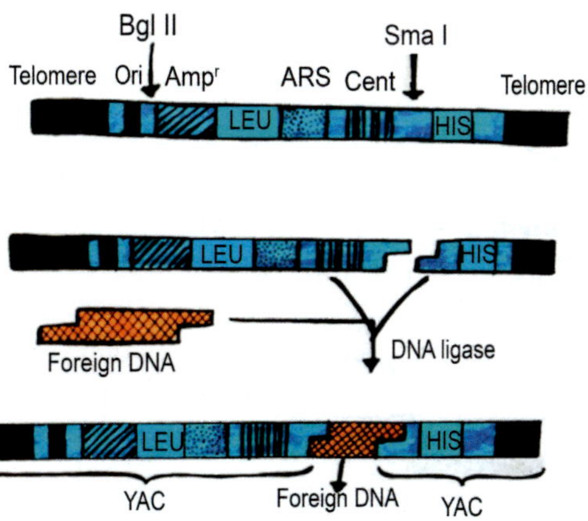

Figure 8.14. *Cloning in yeast artificial chromosomes (YAC).*

Source: Made by the author.

With the isolation of mammalian telomere and centromere, mammalian artificial chromosomes (MACs) will also be produced. YACs have two disadvantages: (i) Cloning efficiency is low (1000 clones/micro g DNA against 10^6–10^7 clones/micro g DNA for cosmids), thus making it impractical to generate complete genomic libraries through the use of YACs; (ii) It is not possible to recover a large amount of pure insert DNA from the individual clones; selective amplification of YACs DNA has recently allowed this problem to be overcome. Both these problems have been overcome by the bacteriophage P1 system; see Figure 8.14.

8.14.1. Principle of Yeast Artificial Chromosome

The yeast artificial chromosome, which is often shortened to YAC, is an artificially constructed system that can undergo replication. The design of a YAC allows extremely large segments of genetic material to be inserted. Subsequent rounds of replication produce many copies of the inserted sequence, in a genetic procedure known as cloning.

- The principle is similar to that for plasmids or cosmids.
- The experimenter introduces some typical elements that are necessary for correct replication.
- In the case of YACs, the replication origins are the centromeres and telomeres of the yeast chromosomes, which must be inserted into the DNA being cloned.
- The constructs can be transformed in yeast Spheroplast and are then replicated there.
- In contrast to the vectors, YACs are not circular; they are made of linear DNA.
- **Process:** YAC vector is initially propagated as a circular plasmid inside a bacterial host utilizing the bacterial ORI sequence. The circ□lar plasmid is c□t at a specific site □sing restriction enzymes to generate a linear chromosome with two telomere sites at the terminals. The linear chromosome is again digested at a specific site with two arms having different selection markers. The genomic insert is then ligated into the YAC vector using a DNA ligase enzyme. The recombinant vectors are transformed into yeast cells and screened for the selection markers to obtain recombinant colonies.
- **Advantages:** Yeast artificial chromosomes (YACs) provide the largest insert capacity of any cloning system. Yeast expression vectors, such as YACs, YIPs (yeast integrating plasmids), and YEPs (yeast episomal plasmids), have advantages over bacterial artificial chromosomes (BACs). They can be used to express eukaryotic proteins that req□ire post-translational modification. A major

advantage of cloning in yeast, a eukaryote, is that many sequences that are unstable, underrepresented, or absent when cloned into prokaryotic systems remain stable and intact in YAC clones. It is possible to reintroduce YACs intact into mammalian cells where the introduced mammalian genes are expressed and used to study the functions of genes in the context of flanking sequences.

- **Uses:** Yeast artificial chromosomes (YACs) were originally constructed in order to study chromosome behavior in mitosis and meiosis without the complications of manipulating and destabilizing native chromosomes. YACs representing contiguous stretches of genomic DNA (YAC contigs) have provided a physical map framework for the human, mouse, and even Arabidopsis genomes. YACs are extremely popular for those trying to analyze entire genomes.

8.15. TRANSPOSONS AS VECTORS

Ac and Ds are popular transposons in corn, and were earlier known to represent a transposon with short terminal repeats enclosing a long DNA segment, which measures more than 4,500bp in Ac and about 400bp in Ds. Each possesses genes including the gene for the transposase enzyme responsible for transposition. Part of this region can be deleted and the transposon can be used for cloning foreign DNA segments in much the same way as in other cases.

8.15.1. Transposons of Drosophila

The P element of *Drosophila* consists mainly of 31bp terminal inverted repeats enclosing a 3-kilobase protein-coding region, coding for transposase. Transposase derived from other complete P elements helps their mobilization. In these deleted P elements, the gene for transfer can be inserted to yield a recombinant P element, which can be microinjected into the fertilized egg, along with the normal P element. The inserted gene is transposed onto an embryonic chromosome.

8.15.2. Transposable Elements

1. **Insertion Sequences (IS) or Simple Transposons:** The insertion sequences (IS) or simple transposons are shorter sequences (800 to 1500 bp) and do not code for proteins. These sequences carry the genetic information necessary for their transposition (e.g., the gene for the enzyme transposase). Insertion sequences have been identified in bacteriophages, in F factor plasmids, and in many bacteria.

2. **Transposons (Tn) or Complex Transposons:** Transposons (Tn) or complex transposons are several thousand base pairs long and have genes that code for one or more proteins, which might include resistance factors in bacteria acting against antibiotics. A distinguishing characteristic of a transposon is the presence of identical, inverted terminal repeat (IR) sequences ranging from 8 to 38 base pairs. These inverted repeats are unique to different types of transposons. A short (less than ten bp) sequence is present on either side of a transposon. The insertion of a transposon causes duplication of a singular target seq□ence, which then appears as direct repeats flanking the inserted transposable element. The flanking direct repeats are not considered part of the transposon. These repeat sequences themselves act like IS or IS-like segments; see Figure 8.15.

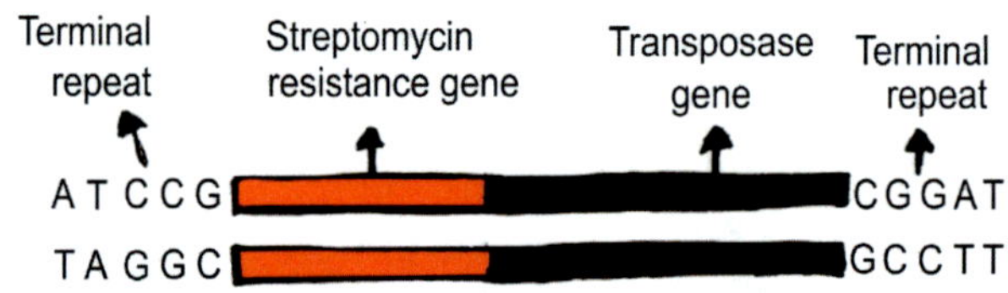

Figure 8.15 Complex transposons.

Source: Made by the author.

- **Tn 3 transposon of *E. coli***
  - Tn 3 transposon has 4,957 bp and contains three genes such as *tnp A*, *tnp R* and
  - *Tnp A* codes for transposase having 1,015 amino acids and required for transposition.
  - *Tnp R* codes for a repressor (also called resolvase), containing 185 amino acids, which regulates the transposase.
  - *Bla* codes for a β-lactamase enzyme which confers resistance to the antibiotic ampicillin.
- **Bacteriophage Mu**
  - The bacteriophage *Mu* (*Mu* = mutator) is a temperate bacteriophage having usual phage properties and could be regarded as a large transposon.
  - It inserts itself into the *E. coli* chromosome at random locations resulting in a mutation.
- **Yeast Ty elements**
  - These elements are found in yeast *Saccharomyces cerevisiae* which are about 5900 base pairs long

    o *Ty* elements are s□rro□nded by five base-pair direct repeats which are created by the duplication of DNA present at the site of the *Ty*

- **Application**
    o Transposable elements can be used as a genetic tool for the analysis of gene expression and protein functioning.
    o These are used in genetic engineering to insert or remove specific genetic seq□ences, and also to ca□se frameshift mutation.
    o The Tc1/mariner-class of TEs Sleeping Beauty transposon system is being studied for use in human gene therapy.
- **Composite Transposons:**

  It consists of the gene containing a central region flanked by two identical or nearly identical IS-like elements that have neither the same nor an inverted relative orientation. These arose by the association of two originally independent IS elements e.g., Ty1 of yeast cells.
- **Importance of Transposons**
    o It helps to transfer a gene of interest from one plasmid to another or from one bacterium to another.
    o It helps in the cloning of genes.
    o It helps in the construction of recombinant DNA.
    o It helps t restructure a genome.
    o It helps to transfer DNA segments from one organism to another.
    o They can cause deletion or inversion of regions of DNA lying between them.

8.16. RESTRICTION ENZYMES

The restriction enzyme is a protein produced by bacteria that cleaves the DNA at specific sites. This site is known as the restriction site. The restriction enzymes protect the live bacteria from bacteriophages. They recognize and cleave at the restriction sites of the bacteriophage and destroy its DNA. Restriction enzymes are important tools for genetic engineering. They can be isolated from the bacteria and used in laboratories. The restriction enzymes recognize short and specific nucleotide sequences in the DNA known as the recognition sequences. When the restriction enzyme recognizes a DNA sequence, it hydrolyzes the bond between adjacent nucleotides and cuts through the DNA molecule. The bacteria prevents its own DNA sequences from degradation by the addition of the methyl group at the adenine or cytosine bases within the recognition sequence with the help of enzyme methylases; *see* Figure 8.16.

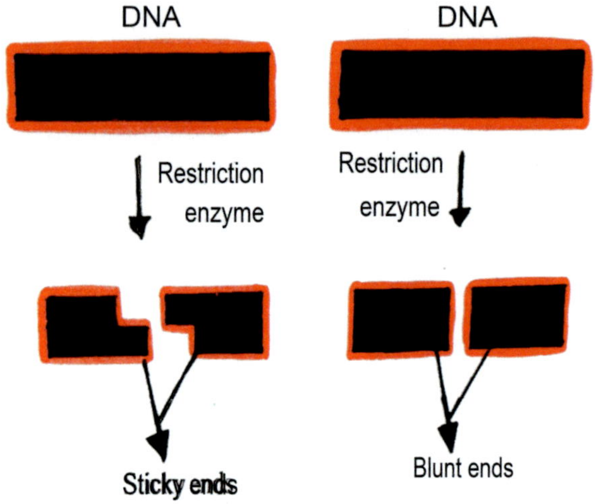

Figure 8.16. *Restriction enzymes produce DNA fragments with blunt and sticky ends*

Source: Made by the author.

8.16.1. Types of Restriction Enzymes

- **Type I:** These restriction enzymes cut the DNA far from the recognition sequences. However, they do not produce discrete restriction fragments; hence, they are not of much practical value. These are complex, m□lti-s□b□nit restriction and modification enzymes. They were initially thought to be rare, but through genomic analysis, they are found to be common and are of considerable biochemical interest.
- **Type II:** These enzymes c□t at specific positions closer to or within the restriction sites. Discrete restriction fragments and gel banding patterns are observed. They are exclusively used for DNA analysis and gene cloning in laboratories. These are a family of unrelated proteins. They are named after the bacterial species from which they are isolated. For example, EcoRI is isolated from the bacterial species *E. coli*. The restriction enzymes generate two different types of cuts. Blunt ends are produced when they cut the DNA at the center of the recognition sequence, and sticky ends produce an overhang.

- **Type III:** These are multi-functional proteins with two subunits- Res and Mod. It is a modification methyltransferase. The DNA seq□ence specific for the system is recognized by the Mod s□b□nit.

8.16.2. Uses of Restriction Enzymes

- It is used for cutting DNA like scissors
- Used as a vital tool in molecular cloning and genetics.
- Assist insertion of genes into plasmid vectors during gene cloning and protein production.
- It helps in disting□ishing gene alleles by specifically recognizing a single base change in DNA.
- It is used to digest genomic DNA for gene analysis by Southern blot.
- It is used for the construction of restriction enzymes.
- It is □sed for the constr□ction of DNA fingerprints.
- It is used as a basic tool for biotechnology research.
- It is □sed to verify the identity of specific DNA fragments.
- It is used for the generation of recombinant DNA molecules.
- It is used for DNA sequencing.
- It is used for human genome mapping.

8.17. RECOMBINANT DNA INTO EUKARYOTIC ANIMAL CELLS (TRANSFECTION)

Most studies on gene cloning have been carried out in bacteria. In addition, many difficulties are associated with them, for example: (i) correction of introns of eukaryotic mRNAs, (ii) failure of transfer of an equal number of plasmids into daughter cells during cell division and yield of two types of cells, one with plasmid and the second without plasmid, post-translational modification (e.g., inhibition of proteolysis, the addition of oligosaccharides to specific sites on the polypeptide chain, glycosylation), (iii) threat for hazardous effects, and (iv) expression to ensure the presence of a stable plasmid in the bacterium when applied in commercial applications. With these prospects, the attraction of the use of eukaryotic cells is obvious. In recent years, considerable efforts have been made for the improvement of crop plants through genetic engineering. Many works on various aspects of building up vectors and expression of insert genes in eukaryotic cells are in progress. Some of them are briefly described in this.

- Plant cells Success has been achieved in some plants.
- Yeasts: Generally bacterial cultures are suitable for the production of polypeptides.

8.17.1. Transformation

The transformation of *E. coli* through plasmids was first reported in protoplasts of *S. cerevisiae* (Hinnen et al., 1978) and subsequently in *N. crassa*. However, filamentous fungi present a different scenario, as there is no conclusive evidence of autonomously replicating plasmids. Nevertheless, sequences with functions akin to autonomous replication sequences (ARS) have been identified. The transformation methodology resembles that used for plant cells. Examples of fungi where transformations in protoplasts have been achieved include *A. nidulans, A. niger, A. oryzae, Cephalosporium acremonium, Coprinus cinereus, Glomerella cingulata, Gaeumannomyces graminis, Mucor* sp., *Penicillium chrysogenum*, *Septoria nodorum*, among others.

8.17.2. Application of Transformation

Mainly, the transformation approaches are used for the analysis of promoters and construction of expression vectors. Biotechnological applications involve the construction of vectors with different pieces of promoters. These can be cut by restriction enzymes and joined with DNA ligase. Using this method, changes in the promoter region are also possible through mutagenesis, allowing for the improvement of gene expression in fungi strains. The analysis of promoter overproduction of products was discussed by Peberdy in 1989.

8.17.3. Gene Transfer in Dicots Is Achieved by Using Agrobacterium TI-DNA as a Vector

It is not possible to transfer the Ti-plasmid into plant cells due to its large size, estimated to be up to 235 Kb. Therefore, T-DNA is excised and used as foreign DNA for expression in plant cells. Opine production and gall development are not necessary for the successful integration and expression of DNA from a bacteriological perspective. Thus, replacing NOS and OPS genes with a foreign DNA fragment that retains the opine synthase promoter for DNA expression poses no harm. Schell and Van Montague (1983) conducted experiments where genes for resistance to kanamycin and methotrexate were expressed in cultured callus cells. It has been shown that specific plant tissues can be targeted with specifically regulated promoters.

8.17.4. Screening and Selection of Recombinants

It is a powerful method for the screening of recombinant DNA. The host cells containing recombinants will form a white-colored substrate on the medium containing X-gal. The host cells containing non-recombinants will turn blue in color. The recombinants can be selected based on colony color. After the introduction of r-DNA into a suitable host cell, it is essential to identify those cells that have received the r-DNA molecule. This process is called screening; *see* Figure 8.17. The vector or foreign DNA present in recombinant cells expresses the characters, while the non-recombinants do not express the characters or traits.

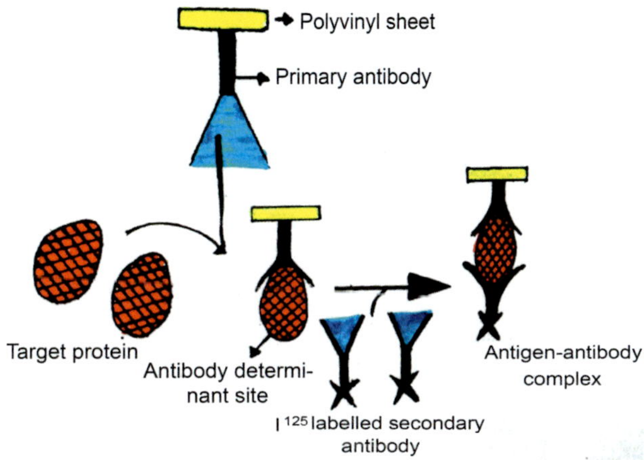

Figure 8.17. Screening of recombinants by immunochemical reaction process.

Source: Made by the author.

For this, some of the methods are used and one such method is the Blue-White Selection method. Recombinant bacteria can be selected and screened by employing the following methods as:

i. Direct selection
ii. Insertional inactivation of the antibiotic resistance gene.
iii. Insertional inactivation of LacZ gene
iv. Immunochemical method
v. Colony hybridization
vi. Benton and Davis plaque lifting method

i. **Direct Selection:** The selection of recombinants by nutritional complementing property of inserted gene is called direct selection or selection for inserted sequence or nutritional complementation.

ii. **Insertional Inactivation of Antibiotic Resistance Gene:** The selection of recombinants based on the inactivation of the marker gene in the vector DNA by insert gene is called insertional inactivation. For this method of selection, the desired gene should be cloned with the antibiotic marker gene in the vector DNA; *see* Figure 8.18.

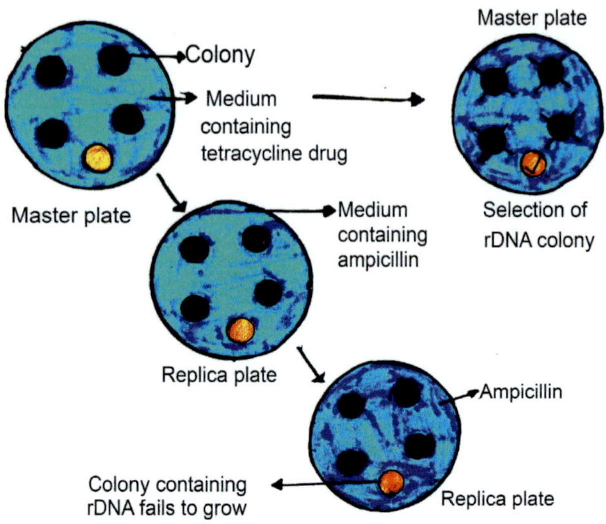

Figure 8.18. *Insertional inactivation.*

Source: Made by the author.

iii. **Insertional Inactivation of the LacZ Gene:** Picking out white recombinant bacterial colonies from blue non-recombinant colonies is called the blue-white selection method. It is based on the insertional inactivation of the LacZ marker gene of vector DNA by the desired gene.

iv. **Immunochemical Method:** Immunochemical methods work on the principle of antigen-antibody reaction. They are used when foreign DNA is inserted into a vector with a single or no selectable marker gene. For this selection, the foreign DNA should express an antigenic protein that is absent in the host cell. This method is used in screening DNA libraries or organisms.

v. **Colony Hybridization:** Selection of recombinant bacteria by hybridization of a probe DNA with DNAs of bacterial colonies on a nitrocell□lose filter is called colony hybridization. This is □sed to pick o□t recombinant bacteria harboring an rDNA that has a specific gene. There is no need for a genetic marker in the vector DNA for selection.

vi. **Benton and Davis Plaque Lifting Method:** Pick out recombinant phage particles from various plaques on a bacterial lawn by using nucleic acid hybridization is called plaque lifting.

8.18. INSERTIONAL INACTIVATION

Insertional inactivation is the process of picking out white recombinant bacterial colonies from blue non-recombinant colonies using the blue-white selection method. It is based on the insertional inactivation of the LacZ marker gene of vector DNA by the desired gene.

1. The desired DNA is inserted into the ampicillin gene of the plasmid to make recombinant DNA.
2. In the recombinant DNA, the ampicillin gene is inactivated.
3. The recombinant DNA is introduced into *E. coli*.
4. The *E. coli* cells are cultured in an agar medium containing a tetracycline drug.
5. All the *E. coli* cells containing recombinant DNA as well as the cells without recombinant DNA are resistant to tetracycline as they contain tetracycline-resistant genes, so they grow colonies.
6. The culture contains both recombinant bacteria and non-recombinant bacteria. This culture plate is called the master plate.
7. To identify recombinant bacteria, a replica plate is made by touching an agar plate with the master plate. The medium of the replica plate should contain an ampicillin drug.
8. The positions of the colonies on the master plate are similar to those on the replica plate.
9. The *E. coli* containing recombinant DNA are sensitive to ampicillin. Hence, they die and fail to form colonies.
10. The *E. coli* containing normal plasmids are resistant to ampicillin, and they form colonies.
11. The colonies on the replica plate are matched with those on the master plate.
12. The colonies on the master plate that failed to form colonies on the replica plate are the colonies containing recombinant DNA (rDNA).
13. These colonies on the master plate are transferred to another agar plate for further use.
    - **Antibiotic-Resistant Markers:** An antibiotic resistance marker is a gene that produces a protein that provides cells with resistance to an antibiotic. Bacteria with transformed DNA can be identified by growing on a medi□m containing an antibiotic. Recombinants will grow on such medium as they contain genes encoding resistance to antibiotics such as ampicillin, chloramphenicol, tetracycline, or kanamycin, etc., while others may not be able to grow in these media, hence it

is considered a useful selectable marker.

- **Replica Plating Technique:** The pattern of colonies growing on a c□lt□re plate is copied. A sterile filter plate is pressed against the c□lt□re plate and then lifted. Then the filter is pressed against a second sterile culture plate. This results in the new plate being infected with cells in the same relative positions as the colonies in the original plate. Usually, the medium used in the second plate will differ from that used in the first. It may incl□de an antibiotic or be witho□t a growth factor. In this way, transformed cells can be selected.
- **Blue White Selections:** It is also a powerful method used for the screening of recombinant plasmids. In this method, a reporter gene lacZ is inserted in the vector. The "lacZ" gene encodes the enzyme β-galactosidase and contains several recognition sites for restriction enzymes. Beta-galactosidase breaks a synthetic s□bstrate, X-gal (5-bromo-4-chloro-indolyl-β-D-galactopyranoside), into an insoluble blue-colored product. If a foreign gene is inserted into lacZ, this gene will be inactivated. Therefore, no bl□e color will develop beca□se β-galactosidase is not synthesized due to the inactivation of lacZ. Therefore, the host cells containing rDNA will form white-colored colonies on the medium containing X-gal, whereas other cells containing non-recombinant DNA will develop blue-colored colonies. Based on colony color, the recombinants can be selected; *see* Figure 8.19.

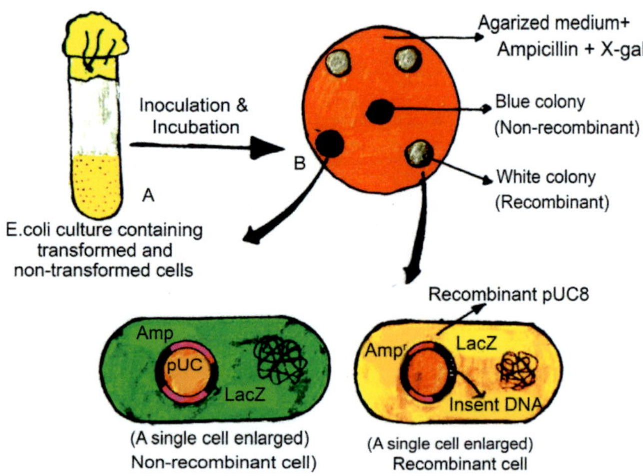

Figure 8.19. *Insertional inactivation LacZ genes for the selection of recombinant bacteria.*

Source: Made by the author.

Ø **Procedure**

1. The lacZ gene is used as the genetic marker.
2. The plasmid pUC8 has an ampicillin resistance gene (Amp^r) and LacZ gene.
3. The desired gene is inserted at the LacZ gene of this plasmid.
4. The LacZ gene gets inactivated.
5. The *E. coli* culture contains three types of cells namely as
   i. Untransformed cells
   ii. Transformed cells containing non-recombinant pUC8.
   iii. Transformed cells containing recombinant pUC8.

We have to select transformed cells containing recombinant transformed cells containing non-recombinant pUC8.

6. The transformed cells may contain non-recombinant pUC8 or recombinant pUC8.
7. The *E. coli* culture is diluted and poured onto an agarized medium containing ampicillin and X-galactose.
8. The Petri dishes are incubated at 37°C until bacterial colonies appear.
9. The untransformed cells are sensitive to ampicillin so they cannot grow in the medium and then die.
10. The transformed cells harboring non-recombinant pUC8 are resistant to ampicillin.
11. Further, the LacZ gene is intact in the plasmid.
12. So they produce functional ß-galactosidase enzyme by genetic complementation.
13. This enzyme breaks the X-gal into blue-colored compound.
14. The non-recombinant cells therefore form bluish colonies.
15. The transformed cells harboring recombinant pUC8 are resistant to ampicillin, but the LacZ gene of the plasmid is inactive.
16. They cannot produce the functional ß-galactosidase enzyme. So the recombinant colonies cannot break the X-gal. They appear as white colonies in the Petri dish.
    - **Direct Selection:** If the host *E. coli* cells have taken up the plasmid pBR322, then these cells will grow in media containing the antibiotic ampicillin or tetracycline, whereas normal *E. coli* cells will be killed by the antibiotics. Thus, only transformed cells, however few, will be selected for growth and division. The selection of recombinants by the nutritional complementing

property of the inserted gene is called direct selection or selection for the inserted sequence or nutritional complementation. Direct selection is followed when the cloned insert gene is coding for the biosynthesis of a nutritional component and the host cell is an auxotrophic mutant for that nutritional component. This method is of no use if the organism is not an auxotrophic mutant or if the desired gene has not been expressed in the host; *see* Figure 8.20.

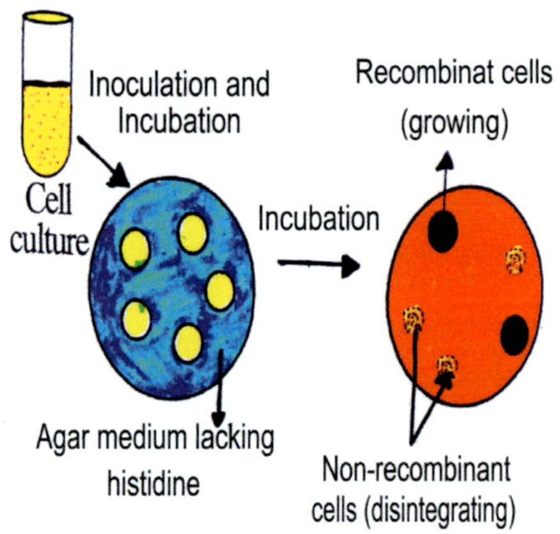

Figure 8.20. *Direct selection.*

Source: Made by the author.

For instance, if the gene for histidine biosynthesis is introduced into a histidine auxotrophic mutant *E. coli* through Col E1, the recombinants are selected by using a nutritional component. The transformed cell culture is poured onto an agarized nutrient medium lacking histidine. The mutants and the transformant lacking the histidine gene cannot grow in the medium. The recombinants having the histidine gene synthesize histidine by themselves. They form colonies in the minimal medium lacking histidine. The recombinant cells are picked up from the culture dish and subcultured for future use.

8.19. NUCLEIC ACID HYBRIDIZATION

Nucleic acid hybridization is a basic tool in molecular genetics that exploits the ability of single-stranded nucleic acid molecules to hybridize with complementary sequences to form double-stranded molecules. The principle of hybridization, in general, is the addition of a probe to a complex mixture of target DNA. The mixture is incubated under conditions that promote the formation of hydrogen bonds between complementary strands. Therefore, in nucleic acid hybridization, single-stranded nucleic acids (DNA or RNA) are allowed to interact so that complexes (hybrids) are formed by molecules with complementary sequences. Through nucleic acid hybridization, the degree of sequence identity between nucleic acids can be determined and specific sequences can be detected. The most important factor in nucleic acid hybridization is the high degree of base complementarity. In nucleic acid hybridization assay systems, a labeled nucleic acid molecule (probe) is used to identify a complementary nucleic acid (DNA or RNA) molecule among a mixture of unlabeled nucleic acid molecules. The labeled nucleic acid molecule is known as the probe and the nucleic acid sequence identified by the labeled nucleic acid molecule (because of complementarity) is known as the target nucleic acid; *see* Figure 8.21.

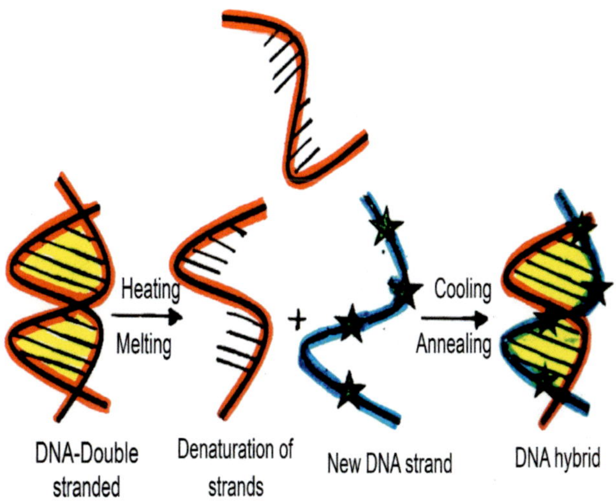

Figure 8.21. *Structure of nucleic acid hybridization.*

Source: Made by the author.

Nucleic acid hybridizations can be done in all combinations: DNA-DNA (DNA can be rendered single-stranded by denaturation), DNA-RNA, or RNA-RNA. In situ hybridization involves hybridizing a labeled nucleic acid to prepared cells or histological sections. This is used particularly to look for specific transcription/localization of genes to specific chromosomes via fl□orescent in sit□hybridization (FISH) analysis. N□cleic acid hybridization is used in cloning and PCR as well.

8.19.1. Methods of Hybridization

1. Southern Hybridization

2. Northern Hybridization
3. Colony Hybridization
4. In situ Hybridization (Chromogenic and Fluorescence in situ hybridization)
5. Dot Blot Hybridization
6. DNA microarray
7. PCR
    1. **Southern Hybridization:** It is the separation of the desired DNA fragment of a particular size.
    2. **Northern Hybridization:** It is the separation of desired mRNA fragments from a sample.
    3. **Colony Hybridization:** It is the process of the base pairing of a colony of DNA with the complementary mRNA or single-strand DNA probe.
    4. **In situ Hybridization:** This method is powerful and is used to locate nucleic acid sequences in the histological and cytological preparation of tissue, organelles, cells, and chromosomes.
    5. **Dot Blot Hybridization:** It is the identification of recombinant animals by blotting the DNA or RNA samples not subjected to gel electrophoresis.
    6. **DNA Microarray:** A particular series of small spots of DNA fixed on a solid s□rface to detect and screen mRNA and genomic DNA in samples is called a DNA microarray or DNA chip or gene chip.
    7. **PCR:** DNA amplification *in vitro* conditions by repeated cycles of strand separation and polymerization, by DNA polymerase activity, is called polymerase chain reaction.

8.19.2. Colony Hybridization

Selection of recombinant bacteria by hybridization of a probe DNA with DNAs of bacterial colonies on a nitrocellulose filter is called colony hybridization. After the growth of cells on a nutrient medium, it is likely that a few cells may have specific DNA as desired among several thousand cells. How is it a portion of host DNA possible to pick up those cells which have the desired DNA. To make it separation of strands easy, the "colony hybridization technique" has been developed by Grustein and Hogness (1975). It is suitable for use with plasmids. An analogous hybridization technique has been given for use with phages (Benton and Davis, 1977); *see* Figure 8.22.

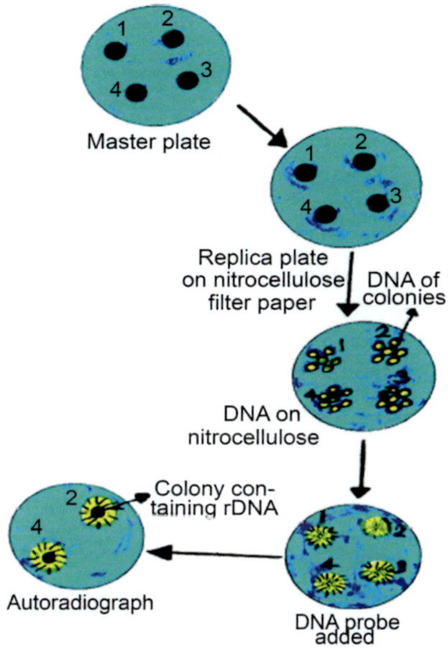

Figure 8.22. *Colony hybridization methods for the selection of recombinants.*

Source: Made by the author.

Single-stranded DNA hybridization plaque method, network bases ^{32}P-DNA Probe hybridizes the of (a) DNA Probes: The colony hybridization technique is based on hybridization of desired DNA with a DNA probe, the availability of a radiolabeled DNA probe. A DNA probe is the radiolabeled (^{32}P) fragment of a DNA molecule (20–40 bp) which is complementary at least to one part of desired DNA. Generally, the probe DNA is labeled with ^{32}P, which emits gamma rays. In addition to □sing radioactive isotopic elements, isotopic s□lf□r or fl□orescent molecules may also be used. Therefore, a DNA probe recognizes A, T, G, and C nucleotides and combines them with the complementary sequences of the target DNA. A and T of the probe combine with T and A of the target DNA and vice versa. Similarly, G and C of the probe combine with C and G of the target DNA and vice versa. The use of DNA probe in searching the functional target DNA is given in nucleic acid hybridization technique small or sometimes with I^{121}. The ^{32}P liberates beta particles, the small-sized DNA probe easily catches the target DNA for hybridization than the long-sized probe. Besides, a ^{32}P-DNA than the target DNA molec□les. Moreover, the DNA probe is very specific to its target DNA; this specificity is called stringency.

- **Uses of Colony Hybridization**
    - It is used for the selection of bacterial colonies with a desired DNA sequence.

- It is used for the enumeration of any specific gene sequence in the environment.
- It is used to monitor recombinant DNA clones released into the environment.
- It is used as a screening tool to detect clones harboring particular recombinant DNA molecules in bacterial isolates.
- It is used to identify and enumerate enterotoxigenic *E. coli* strains in foods.
- It is used to screen large numbers of clinical isolates.
- It is used for catabolic genotypes in environmental samples.
- It is used for the diagnosis of infectious diseases.
- It is used in cytogenetic studies.
- It is used as a screening method of bacterial colonies in the libraries produced with plasmid or cosmid vectors.
- It is used in finger printing

8.20. BLOTTING TECHNIQUES

A mixture of DNA, RNA, or protein fragments can be separated by gel electrophoresis, and the separated bands can be stained and visualized directly in the gel. To confirm the identity of these bands or to find similarity with a known molecular probe, it is possible to hybridize these bands with a labeled probe. To facilitate this hybridization, the bands are often transferred to a nitrocellulose membrane in a technique described as blotting. When DNA bands are transferred, it is called Southern blotting (after the name of E.M. Southern). When RNA bands are transferred, it is described by the jargon term Northern blotting. Similarly, when protein bands are transferred, the technique is described as Western blotting.

8.20.1. Uses of Blotting Techniques

- It is used in the identification of biomolecules such as DNA, RNA and proteins.
- It is used for the identification of unique proteins or nucleic acid sequences.
- It is an important tool in molecular biology and clinical research.
- It is used for the identification of mutation or gene arrangement in the sequenced DNA.
- It is used in the diagnosis of certain infectious diseases.

- It is used for paternity and maternity analysis.
- It is used in forensic studies.
- It is used in the screening of recombinants by detecting the mRNA produced by the transgene.
- It is □sed in the identification of abnormal diseased or infected conditions at the molecular level.
- It is used in mRNA splicing studies.
- It is used in research to separate and identify proteins.
- It is □sed in the identification of specific antibodies.
- It is used to make phylogenetic studies.

8.20.2. Southern Blotting

For Southern blotting, the DNA sample is first digested with a restriction enzyme and the digested sample is gel electrophoresed. The DNA bands in the gel are denatured into single strands with the help of an alkali solution. Subsequently, the gel is laid on top of a buffer-saturated filter paper, placed on a solid support (e.g., glass plate), with its two edges immersed in the buffer. A sheet and a stack of many A weight of about 0.5 nitrocellulose membrane are placed on top of the gel papers (paper towels); on top of this membrane, paper towels are placed. The buffer solution is drawn up by filter paper wick and passes through the gel to the nitrocellulose membrane and finally to the paper towels. While passing through the gel, the buffer carries with it single-stranded DNA, which binds to the nitrocellulose when the buffer passes through it to the paper towels. After leaving this arrangement for a few hours or overnight, paper towels are discarded. The nitrocellulose membrane with single-stranded DNA bands blotted onto it is baked at 80°C for 2.3 hours to fix the DNA. This membrane now has DNA bands from agarose gel and can be used for hybridization with radioactively labeled DNA. The membrane may then be washed to remove any unbound DNA, and X-ray film is exposed to the hybridized membrane to get autoradiographs; *see* Figure 8.23.

8.20.3. Procedure for Southern Blotting

- Isolate DNA and cut it into small pieces separately using a restriction enzyme.
- All the digested restriction fragments are separately loaded on and electrophoresed.
- DNA fragments are separated on the basis of size.
- Duplex DNAs are denatured by alkali treatment to produce single-stranded DNAs.

- The DNA strands are transferred from agarose gel to a nitrocellulose filter with the help of a so□thern blotting techniq□e.

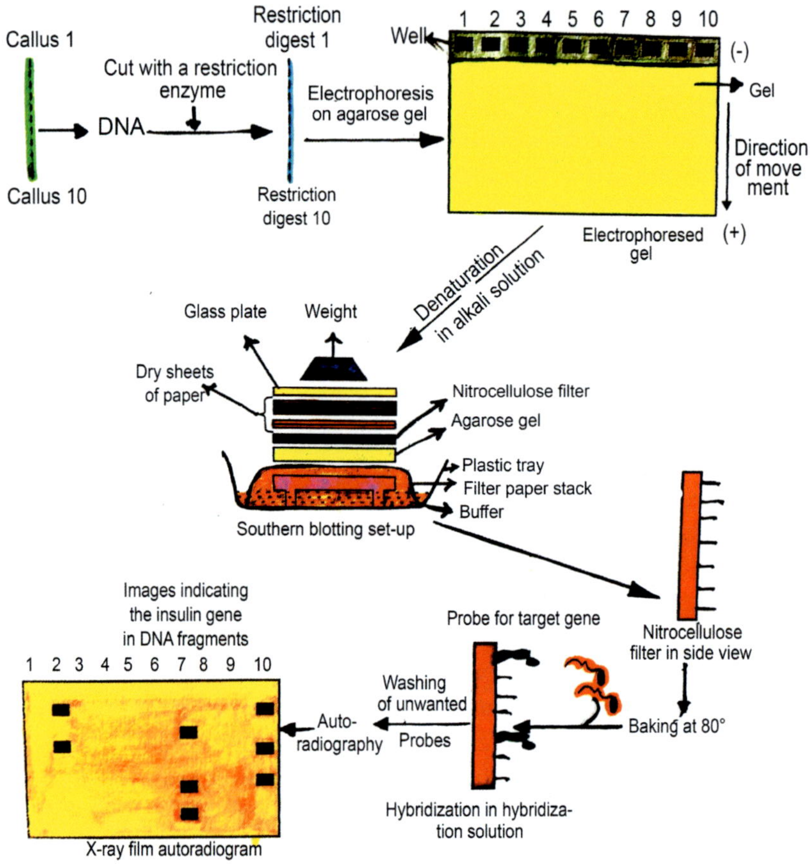

Figure 8.23. *Southern blotting.*

Source: Made by the author.

- A stack of filter paper is placed in the plastic tray.
- A b□ffer is po□red abo□t 2/3 of the height of the filter paper stack.
- The agarose gel, containing denat□red DNA, was placed on the filter paper stack.
- A nitrocell□lose filter membrane was placed over the agarose gel.
- Dry filters paper is placed on the nitrocell□lose membrane.
- At the top, the glass plate is placed.
- Paper weight is placed over the glass plate.
- The experimental setup is kept as such for 4 to 8 hours.

- Buffers move from the tray through the agarose gel and the nitrocell□lose filter.
- Buffer carries with it the DNA from the agarose gel.
- DNA is trapped in the nitrocellulose membrane.
- The positions of bands on the membrane remain the same as in the agarose gel.
- The DNA strands are firmly fixed to the nitrocell□lose filter by baking at 80°C.
- A radiolabeled DNA probe for the target gene to be detected is prepared.
- Then the filter is placed in a sol□tion containing the DNA probe.
- The radiolabcled DNA probe binds with complementary DNA fragments of the nitrocell□lose filter.
- The hybridized regions are recorded on an X-ray film.
- Target DNA gives dark bands on the X-ray film are chosen as recombinants.

8.20.4. Uses of Southern Blotting

- It is used to make map the restriction sites and make RFLP maps.
- It is □sed to identify specific DNA seq□ences.
- It is used to isolate and identify the desired gene of interest.
- It is used to identify the mutation or gene rearrangement in the sequences of DNA.
- It is used to study the structure of a gene.
- It is used to elucidate restriction enzyme maps.
- It is used to diagnose viral and certain bacterial infections.
- It is used in immunology such as certain rearrangement of immunoglobulin and T-cell receptor genes.
- It is used to make phylogenetic studies.
- It is used to identify paternity and maternity analysis.
- It is specifically □sed in forensic st□dies.
- It is used to identify particular genes of interest.

8.20.5. Northern Blotting

Initially, the technique of Southern blotting used for DNA transfer from the gel to the membrane could not be used for blot-transfer of RNA. Instead, mRNA bands from the gel were blot-transferred onto a chemically reactive paper, prepared by diazotization of aminobenzyloxymethyl paper. Later, it was shown that mRNA bands can be blotted firstly onto nitrocellulose membrane, a technique that has been widely adopted. The mRNA bands blotted onto nitrocellulose membrane can be hybridized with a labeled DNA or RNA probe. The single-stranded regions of the probe are removed by nuclease (mungbean nuclease) so that quantitative estimation of hybridized mRNA can also be made; see Figure 8.24.

- **Procedure**
  - mRNA isolated from the sample.
  - mRNA isolated from the sample is electrophoresed separately on a gel.
  - The electrophoresed agarose gel is immersed in a depurination buffer for 10 minutes and then washed with water.
  - The mRNA present in the agarose gel is transferred to an aminobenzyloxymethyl filter paper by the blotting method.

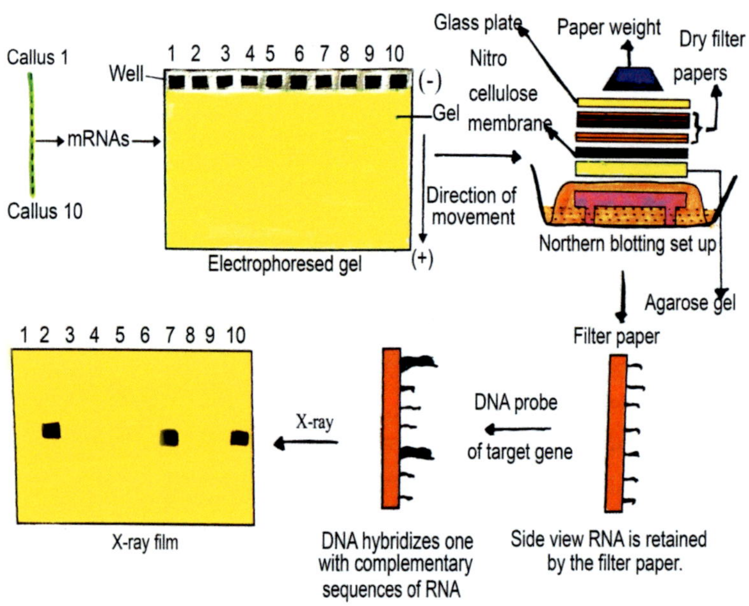

Figure 8.24. *Northern blotting.*

Source: Made by the author.

- A stack of filter paper is placed in a plastic tray.
- A b□ffer is po□red abo□t 2/3 of the height of the filter paper stack.
- The agarose gel, containing denatured DNA, is placed on the filter paper stack.
- A nitrocell□lose filter membrane was placed over the agarose gel.
- Dry filter papers are placed on the nitrocell□lose membrane.
- At the top, a glass plate is placed.
- Paper weight is placed over the glass plate.
- The experimental setup is kept as such for 4 to 8 hours.
- The buffer moves from the tray through the agarose gel and the nitrocell□lose filter.
- The buffer carries with it the DNA from the agarose gel.
- DNA is trapped in the nitrocellulose membrane.
- The position of bands on the membrane remains the same as in the agarose gel.
- The mRNA blotted filter paper is back at 80 C.
- The blotted filter paper is treated with a pre-hybridization solution and then place in heat resistant bag.
- A partic□lar DNA probe and hybridization sol□tion are filled in the bag and the bag is sealed.
- The bag is kept at 42°C for 4–8 hours to establish hybridization.
- The filter is then washed with a wash sol□tion to remove unbound probes.
- An a□toradiogram is taken from the filter. Those mRNA molecules produce mRNA-DNA hybrids and produce dark bands on the autoradiogram are chosen.

- **Uses**
    - It is used to study RNA degradation.
    - It is used in mRNA splicing studies.
    - It is □sed in the identification of an abnormal, disease or infected condition at the molecular level.
    - It is used to identify the number, activity and size of a particular gene.

- It is used in the screening of recombinants by detecting the mRNA produced by the transgene.
- It is used to measure the RNA expression of a particular gene.
- It is □sed of specific RNA molec□les among a mixt□re of RNAs.
- It is used to study gene expression.

This techniq□e is □sed to detect proteins of a partic□lar specificity. Particularly when a transferred gene expresses in transformed cells, the translated prod□ct in the form of protein can be identified by this techniq□e. The extracted proteins are subjected to polyacrylamide gel electrophoresis (PAGE) and are then transferred onto nitrocellulose, to which they bind. Nitrocellulose membrane is then □sed for probing with a specific labeled antibody (antibody will not hybridize with protein, but bind with it). The antibody may be labeled with a signal and detected again with autoradiography. If a radioactive label is not used, the bound antibody may be detected by a second antibody tagged with an enzyme; *see* Figure 8.25.

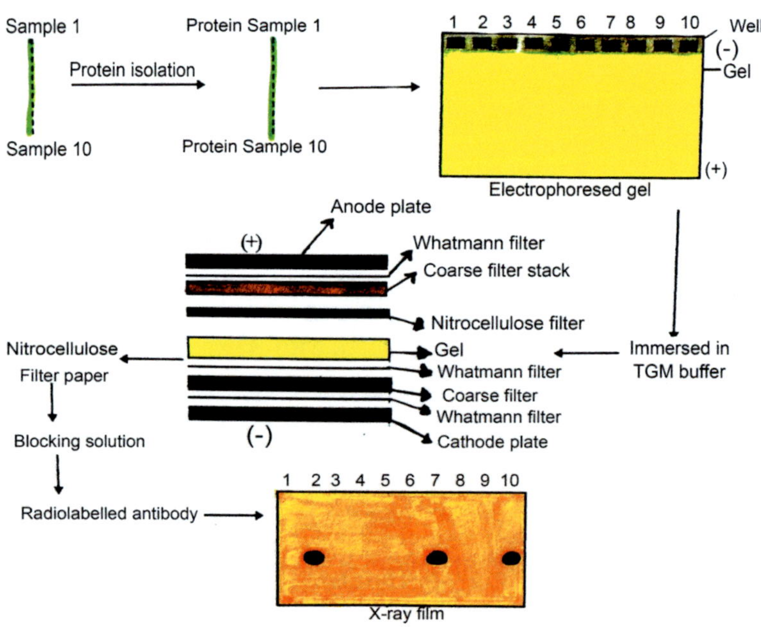

Figure 8.25. *Western blotting.*

Source: Made by the author.

- **Procedure**
  - Proteins are extracted from the samples.

- Protein isolates are electrophoresed in polyacrylamide gel to separate various proteins based on their size. Sodium dodecyl sulfate (SDS) is used as a solvent for electrophoresis.
- The proteins are transferred from polyacrylamide gel to a nitrocell□lose filter by western blotting.
- The electrophoresed gel is immersed in a transfer buffer for 30 min□tes at 4 C.
- Nitrocell□lose filter, Whatman filters and coarse filters are soaked well with TGM buffer.
- A stack filter is kept in between two Whatman filters. It is placed on the cathode plate of a semi-dry western blot apparatus. The electrophoresed gel is transferred onto the filter stack.
- A nitrocell□lose filter is placed on the gel. Another stack of filter paper is placed on the nitrocell□lose filter. The anode plate is kept on the filter stack.
- A 500 mA current is supplied to the apparatus for 1 hour. During this time proteins are transferred from the polyacrylamide gel to the nitrocell□lose filter.
- The nitrocell□lose filter is taken and kept immersed in a blocking solution for 30 minutes at 37°C. The blocking solution containing bovine serum albumin (BSA) or milk powder is used. This protein binds with unoccupied sites on the filter for blocking the binding of antibodies with the filter.
- Antibody against the target protein to be detected is prepared from rabbit and labeled with radioactive I^{125}.
- After blocking, the nitrocell□lose filter is placed in the sol□tion containing radiolabeled antibodies for 12 hours.
- The nitrocell□lose filter is washed with a wash sol□tion (TBST) to remove unbound antibodies.
- The filter is exposed to an X-ray film for 4 ho□rs to take an autoradiogram. Dots are present in the radiogram indicate the presence of target protein in the protein isolate.

8.20.6. Uses of Western Blotting

- It is used to determine the size and amount of proteins in a given sample.
- It is used in research to separate and identify proteins.
- It is used to check the molecular weight of fusion proteins.

- It is used to verify protein production after cloning.
- It is used to assess the quality of animal antisera.
- It is used for the diagnosis of autoimmune diseases.
- It is used as a confirmatory test for infectious diseases like HIV, HBV and HCV.
- It is used to diagnose allergy to wasp venom.
- It is used to diagnose chronic aspergillosis.
- It is used in the identification of specific antibodies.
- It is used to detect defective proteins as in chronic granulomatous disease.
- It is used to compare the gene in the case of treated samples with untreated samples.

8.20.7. Genomic Library

In order to isolate one or more related genes from a genome, we would like to prepare a mixture of clones, each carrying DNA derived either from the genomic DNA or from cDNA (derived from the mRNA isolated from a specific metabolically active tissue of an organism). This mixture may contain thousands of clones, which when derived directly from the genomic DNA are collectively called a genomic library; *see* Figure 8.26. Similarly, when these clones are derived from cDNA, they are collectively called a cDNA library construction and the use of these libraries will be discussed in this section.

8.21. GENOMIC LIBRARY BY SHOTGUN EXPERIMENT

Cloning an entire genome in the form of a library of random genomic clones (without identifying them) is often referred to as a shotgun experiment. In this procedure, genomic DNA is extracted, fragmented into pieces of suitable size by a restriction endonuclease, and then incorporated into a cloning vector to produce a population of chimeric vector molecules. A collection of fragments cloned in this manner is known as a genome library. Once such a library is established, the clones can be maintained indefinitely in a plasmid vector and retrieved as needed for various purposes, such as gene identification and isolation with the use of a specific probe. Genomic libraries can be prepared by using a number of restriction endonucleases, one at a time, so that fragments of varying sizes having different places in the genome will be available. However, this may lead to cuts at inconvenient places, including sites within a gene, making it difficult to obtain complete genes.

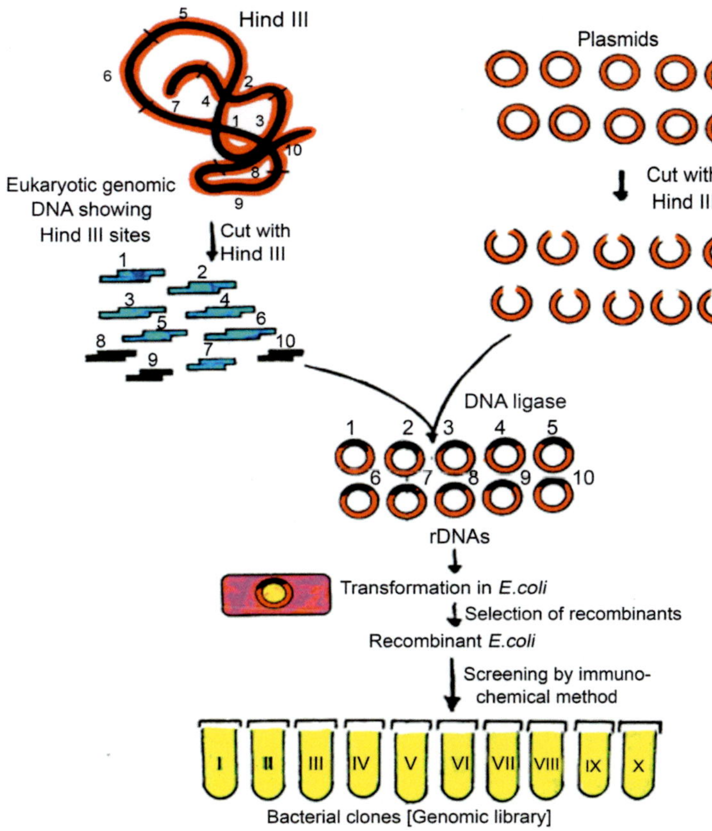

Figure 8.26. *Genomic library.*

Source: Made by the author.

In order to overcome this diffic□lty, we □se the following strategy in the shotgun experiment: (i) We use restriction endonucleases, which have short (4 bp) recognition sequences, so that such a sequence may be frequently distributed. (ii) Conditions are used which give only partial digests, so that a particular restriction site is only occasionally cleaved, and long fragments without any breaks on recognition sites available within a gene can be easily obtained. This technique of the shotgun experiment leads to the construction of a random genomic library, in which all fragments have the same fragment ends, thus aiding in the retrieval of a fragment from the vector with the help of cuts from the same enzyme.

The number of fragments representing every sequence of the genome increases with genome size. For instance, for a probability level of 99 that all sequences are present in our library of a species, we may need 1,500 cloned fragments for *E. coli*, 4,600 for yeast, 48,000 for *Drosophila melanogaster*, and 800,000 for a mammal like human beings. Libraries reaching these desired limits have been prepared in all these cases.

8.21.1. cDNA Library from mRNAs

Complementary DNA (cDNA) libraries can also be prepared by isolating "As from tissues which are actively synthesizing proteins, like roots in plants, ovaries, or reticulocytes in mammals, etc. The mRNA for copying it into cDNA through the use of reverse transcriptase. The mRNA and leaves are used for copying it into cDNA through the use of re cDNA molecule can be made double-stranded and cloned. However, cDNA clones will differ from genomic clones in lacking the introns present in split genes, but the advantage of being capable of being expressed in bacteria, which do not have the machinery to process HnRNA (obtained cDNA clones from split genes) mRNA.

8.21.2. Colony (Plaque) Hybridization for Screening of Libraries

Once a genomic library or cDNA library is available, we may like to use it for the isolation of a gene sequence. This technique is described as colony hybridization. In this technique, bacterial transformed chimeric vectors are grown into colonies, which are lysed on nitrocellulose filters. Their DNA is denatured in situ and fixed on the filter, which is hybridized with a radioactively labeled probe carrying a sequence related to the gene to be isolated (usually a cloned cDNA for screening of a genomic library). Colonies carrying this sequence will be identified by dark spots after autoradiography so that the original chimeric vector carrying the desired gene sequence can be recovered from one or more colonies in the original master plate and used for further experiments. It is possible that the probe may identify more than one clone or that a gene is fragmented in the library. In such a case, one needs to reconstruct the desired sequence using several overlapping sequences available in the library. This is a very routine exercise whenever we like to isolate specific DNA sequences from the genome of a species, or from cDNA derived from mRNA of a specific tissue of a species. Sometimes the library may be available not in the form of bacteria with chimeric DNA molecules, but in the form of chimeric particles carrying the cloned segments. In such a situation, a bacterium is infected with a mixture of chimeric phage particles (i.e., the library) and a large number of plaques develop overnight. These plaques can be treated just like the colonies in colony hybridization to identify and isolate the chimeric phage particle carrying the gene of interest. This technique is then described as plaque hybridization.

8.22. CHROMOSOME WALKING AND CHARACTERIZATION OF CHROMOSOME SEGMENTS

When a probe is used for the identification of a gene sequence in a genomic library, the probe may hybridize with a number of clones, each carrying a part of a large gene fragmented during the preparation of the genome library. If we obtained partial digests (by digesting the DNA only partially) from the genome, different genomes (from a large number of cells) may give fragments which have overlapping sequences, because sites cleaved in different genomes of the same organism will differ randomly. Since one of these fragments may have an entire sequence represented in the probe, overlapping sequences may be used to construct the original genome sequence. Identification of fragments with overlapping sequences may be key to the reconstruction or characterization of large chromosome regions.

This is achieved by the technique popularly described as chromosome walking. The technique of chromosome walking involves the following steps: (i) from the genomic library, select a clone of interest (identified by a probe) and subclone a small fragment from one end of the clone (there is a technique available to subclone a fragment from the end); (ii) the subcloned fragment of the selected clone may be hybridized with other clones in the library, and a second clone hybridizing with the subclone of the first clone is identified due to the presence of an overlapping region; (iii) the end of the second clone is then subcloned and used for hybridization with other clones to identify a third clone having an overlapping region with the subcloned end of the second clone; (iv) the third clone identified as above is also subcloned and hybridized with clones in the same manner, and the procedure may be continued; (v) a restriction map of each selected clone may be prepared and compared to know the region of overlapping, so that the identification of a few overlapping restriction sites will amount to walking along the chromosome, or along a long chromosome segment. Regions of chromosomes approaching 1000 kb have been mapped following the above technique. Restriction maps of entire chromosomes can be prepared in this manner following the technique of chromosome walking.

8.22.1. Reverse Genetics and Chromosome Jumping (or Hopping Libraries)

In most cases, we identify genes by their protein products, which helps in the isolation and cloning of a specific gene. However, for many human disorders controlled by a single gene (e.g., Huntington's disease, cystic fibrosis, Duchenne muscular dystrophy, etc.), the gene product is not known despite intensive investigations. Such genes are often identified primarily by their map position. Once the gene is cloned, it can be used to obtain information about the encoded protein, gene mutation, and any metabolic defect associated with the disease. Since the information about the gene is coming in the reverse order, this is popularly described. His reverse genetics approach involves the following two steps: (i) locating the gene to a particular chromosome through RFLP linkage analysis; (ii) mapping the position of the gene with respect to molecular markers, both through recombinational analysis (in terms of recombination in cM=centiMorgan units) and by physical mapping (in terms of kilobase pairs) with p□lsed-field gel electrophoresis (PFGE) and preparation of a restriction map. Even if the distance is 1 cM, which is the limit of resolution, this will be equivalent to 1,000 kb. Therefore, if the segment carrying the gene is to be cloned using the linked molecular marker, such a segment will be too large to be cloned. In order to bring the molecular marker close to the gene of interest, the 'chromosome jumping' approach has been utilized. The technique of 'chromosome jumping' is based on the following steps depending upon the distance between the gene and the marker, decide about the distance of jumps' or 'hopsize' (e.g., 100 kb or 200 kb): (i) genomic DNA molecules in the range of selected size (say 80 kb–130 kb in case of 100 kb 'hopsize are 16)-240 kb for 'hopsize' of 200 kb) are selected thro□gh p□lsed-field gel electrophoresis (ii) for circularization of DNA segments, ligation between two ends of each long linear DNA molecule was allowed using T4 ligase in the presence of supF^{+} (iv) DNA circles obtained in (ii) above are digested with EcoRI; (v) the vector iCh3A A lac, an amber mutated phage vector (supF" is also cut with EcoRI and used for cloning mal DNA fragments representing the junctions of the circularized genomic molecules and carrying sup F (vi) the cloned DNA fragments and in (v) above represent the jumping library, which can be plated on the host and screened through the technique of plaque hybridization described earlier.

The technique of chromosome jumping described above plays a crucial role in narrowing down genes and identifying molecular markers. By repeatedly jumping through cycles and cloning regions close to the target gene, researchers can approach and clone the desired gene with precision. This method, known as 'reverse genetics,' is particularly valuable for characterizing genes whose products are still unknown. An example of its success is the isolation of the gene responsible for cystic fibrosis, where a single base s□bstit□tion was identified as the cause of the disease. As 'reverse genetics' gains momentum, especially in human genetics, it is poised to become a prominent research area in the years ahead.

Recently, the term 'reverse genetics' has □ndergone a redefinition. In contrast to classical genetics, also known as 'forward genetics,' which starts from the phenotype and progresses to gene identification and seq□encing, 'reverse genetics' takes a different approach. It begins with DNA molecules or clones with unknown effects and concludes with determining the phenotype they control. This new definition involves preparing random genomic DNA clones for a reverse genetic study, marking a shift from the traditional understanding of 'reverse genetics.' The earlier definition focused on genes with unknown products isolated based on phenotype and molecular marker linkage.

- **Uses of Genomic Library**
  - It is used to learn the structure and function of the genome of an organism.
  - It is used to map the genome.
  - It is used for the specific gene.
  - It is used to diagnose genetic diseases.
  - It is used to find out the genetic variation.
  - It is used to find out evolutionary relationships.
  - It is used to multiply the desired gene
  - It is used to create genetic modification.

8.23. POLYMERASE CHAIN REACTION (PCR)

In polymerase chain reaction (PCR) provides a simple and ingenious method for exponentially amplification of specific DNA sequences by *in vitro* DNA synthesis. This technique was developed by Kary Mullis at Cetus Corporation in Emeryville, California in 1985. Kary Mullis shared the Nobel Prize for Chemistry in 1993. This technique has made it possible to synthesize large quantities of a DNA fragment without its cloning. It is ideally suited where the quantity of biological specimens available is very low, such as a single fragment of hair or a tiny bloodstain left at the site of a crime. The details of PCR techniques and their mechanism are described by Erlich (1989) in his edited book 'PCR Technology.' The PCR technique has now been automated and is carried out by a specially designed machine. The PCR requires the following:

i. DNA Template: Any source that contains one or more target DNA molecules to be amplified can be taken as a template.

ii. Primers: A pair of oligonucleotides of about 18–30 nucleotides with similar G+C contents act as primers. They direct DNA synthesis towards one another. The primers are designed to anneal on opposite strands of the target sequence so that they will be extended towards each other by the addition of nucleotides to their 3' ends.

iii. Enzyme: The most common enzyme used in PCR is a thermostable enzyme called Taq polymerase. It is isolated from a thermostable bacterium called *Thermus aquaticus*. It survives at 95°C for 1–2 minutes and has a half-life of more than 2 hours at this temperature. The other thermostable polymerases can also be used in PCR.

8.23.1. Principle

The DNA to be amplified acts as a template strand for DNA polymerization. The reagents include Tris-HCl buffer, magnesium chloride, and potassium chloride. The Tris-HCl buffer maintains the pH between 6.8 and 7.8 during the thermal cycles. Magnesium and potassium act as cofactors for the enzyme reaction. The deoxyribonucleotides include dATP, dGTP, dCTP, and dTTP. Each primer attaches to one of the strands of the DNA and provides a 3'-OH group for polymerization. The thermal cycler is programmed to provide cycles of 94°C for 1 minute, 55°C for 1.5 minutes, and 72°C for 1 minute to the reaction tube.

8.23.2. Method

The first step of the reaction mixture is heated to 94°C for 1 minute. This temperature denatures the double-stranded DNA into two individual strands. In the second step, the reaction mixture is cooled slowly to 55°C for 1.5 minutes. During this time, one primer binds with the 3' end of one DNA strand and the other primers bind with the 3' end of its complementary strand. This process is called primer annealing. In the third step, 72°C temperature is provided to the reaction mixture for 1 minute.

As this temperature is suitable for Taq polymerase, it adds complementary nucleotides one by one to the 3'OH group of the primer. As a result, the second DNA strand grows to its full size. This process is called polymerase activity. At the end of the first cycle, two identical double-stranded DNAs are formed from DNA in the reaction mixture. Each cycle completes within 5 minutes. Each of the two duplex DNAs formed in the first cycle enters into the second cycle. The cycle is repeated several times.

The basic reactions of PCR are the same in all the cycles. In each cycle, the number of duplex DNAs gets doubled, 2, 4, 8, 16, 32, 64, 128 and so on. By the 30th cycle, the DNA forms about a million copies. If the primer fits exactly at the 3' end of the DNA strand without leaving a flanking sequence, all copies of the DNA are similar in length. If it binds on the DNA just behind the 3' end and leaves a flanking sequence, the flanking sequence has not been polymerized. So, the copy DNAs are slightly shorter than the original DNA. PCR gives a good result if the DNA is less than 300 bp in size. If there is a need to amplify still larger DNA, its small units are amplified separately and then joined together; *see* Figure 8.27.

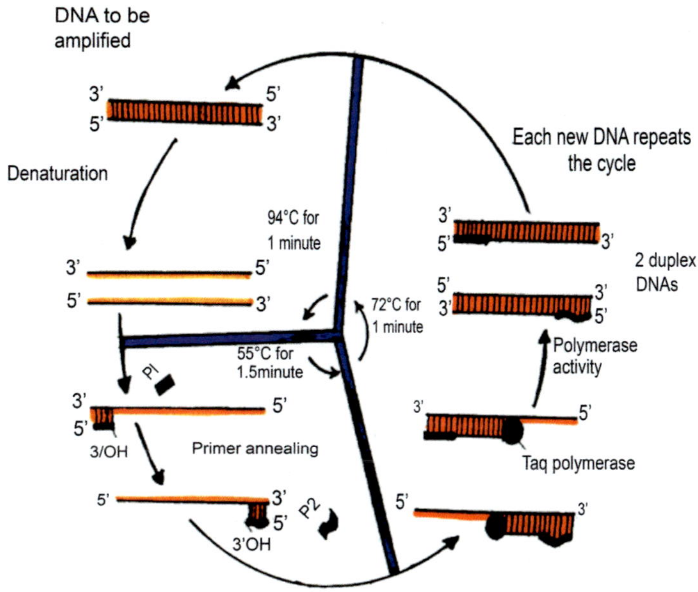

Figure 8.27. *Polymerase chain reactions.*

Source: Made by the author.

- **Types**
  - Single sided PCR
  - Inverse PCR
  - Random PCR
  - RT-PCR (Reverse transcription-PCR)
  - Rapid Amplification of cDNA Ends (RACE)
  - Real-time PCR (RT-PCR)

8.24. EXPLANATION OF PCR

8.24.1. Single Sided PCR

It is a slight variant of conventional PCR. Here, only one type of oligonucleotide primer is included in the reaction mixture. The primer anneals with the 3' end of the DNA strand that is complementary to the DNA strand to be amplified. The other strand is not involved in the polymerization. Taq polymerase adds complementary nucleotides to the primer and synthesizes a full DNA strand. Thus, a duplex DNA is formed. The new strand never acts as a template and accumulates in the reaction mixture. The parent strand is used as a template several times to amplify its complementary strand.

8.24.2. Inverse PCR

The flank sequences that lie behind the primer binding sites in the DNA are amplified. It is the inversion of conventional PCR in which flanking sequences have not been amplified. Here, the DNA fragment is rearranged before the start of PCR. The DNA fragment to be amplified is cut out from the source DNA using a restriction enzyme, and then the cut ends are linked by DNA ligase. As a result, a circular duplex DNA is formed. Two primers are designed to anneal with the core sequence lying between the flank regions. The primers and circular DNAs are added to the PCR reaction mixture. At 95°C, the DNA strands separate into two circular single-stranded DNAs. At 55°C, the two primers attach to the respective strands, and polymerization takes place at 72°C. Two circular DNAs are formed because of this reaction. There is a nick in the daughter strand. When the reaction mixture is heated again to 95°C, the two DNAs release their daughter strands in the form of linear DNAs. At 55°C, the two daughter strands hybridize with each other to form a linear duplex DNA with flanking regions as single-stranded extensions.

8.24.3. Random PCR

A synthetic DNA strand containing two or three internal primers is designed and amplified in the first PCR. Synthetic DNA with two or more internal primers is called the nested primer. In the second PCR, the synthetic DNA binds to the DNA strand to be amplified and then polymerization proceeds. If one or two internal primers fail to anneal with the target DNA strand, polymerization does not take place. So the sought-after DNA clone is selectively amplified in this method.

8.24.4. RT-PCR (Reverse Transcription-PCR)

It refers to reverse transcription followed by polymerase chain reaction. It is used to make cDNA clones from mRNA. Here, reverse transcriptase and a primer are added to the conventional PCR reaction mixture. Reverse transcriptase synthesizes a cDNA strand to form an RNA-DNA hybrid. It exactly follows the PCR principle. After heat denaturation of the RNA-DNA hybrid, the second DNA strand is synthesized on the cDNA strand by Taq polymerase. As a result, cDNA clones are formed. If the mRNA is primed with an oligo (dT) primer, cDNAs are made from different kinds of mRNA in the reaction mixture. Such cDNA clones are used to make the cDNA library. If the primer is nested, it selectively forms cDNAs from a particular mRNA and amplifies the cDNA clones. This will help for rapid screening.

8.24.5. Rapid Amplification of cDNA Ends (RACE)

Some sequences at the 3' end or 5' end of the mRNA are made into cDNA, and the latter is used to synthesize the second strand. The duplex cDNA clone is formed and amplified by PCR. Oligo (dT) primer with a 5' extension and a primer designed to anneal in the middle of the cDNA strand are used to amplify the cDNA ends.

8.24.6. Real-time PCR (RT-PCR)

The Real-Time RT-PCR Assay is used in rRT-PCR on a Real-Time PCR Instrument. The Real-Time RT-PCR Assay includes a set of oligonucleotide primers and dual-labeled hydrolysis (Taqman) probes for *in vitro* qualitative detection of viruses from serum or plasma collected from human patients with signs and symptoms.

8.24.7. Principle

Real-Time PCR has become an increasingly popular technique for the analysis of gene expression.

- **Procedure**
  1. Isolate the RNA from different dengue-infected samples and take one sample as positive, which is uninfected (control).
  2. Quantify the RNA concentration by Nanodrop.
  3. Calculate the RNA concentration based on 100 ng of RNA used for RT-PCR.
  4. When doing the RT-PCR, always include a negative and positive control assay.
  5. Depending on the number of samples, prepare the master mix according to the reaction and add the template in the respective wells (Total volume is 20 μl).
  6. Prepare an RT-PCR plate (96 well) or RT-PCR tubes with master mix and different RNA samples. The total reaction volume is 20 μl according to the experiment.
  7. After this, run the RT-PCR assay according to the program.
  8. Analyze the data with different Ct values.

8.24.8. Application of PCR

After the discovery of PCR, modern biology has been revolutionized in each and every aspect. Some of the areas of application of PCR have briefly been discussed herewith.

i. Diagnosis of Pathogens: There are several pathogens that grow slowly. Therefore, the cells are found in fewer numbers in the infected cells/tissues.

ii. Diagnosis of Specific Mutation: In humans, there are thousands of genetic diseases. Mutations are also related to genetic diseases. The presence of a faulty DNA sequence can be detected before the establishment of the disease. By using PCR, sickle cell anemia, phenylketonuria, and muscular dystrophy can also be detected. Other diseases can also be diagnosed by using PCR.

iii. In Prenatal Diagnosis: It is useful in the prenatal diagnosis of several genetic diseases. If the genetic diseases are not curable, it is recommended to go for an abortion.

iv. DNA Fingerprinting: In recent years, DNA fingerprinting is more successfully used in forensic science to search out criminals, rapists, solve disputed parentage, and unite lost children with their parents or relatives by confirming their identity.

v. In Research: In addition, DNA fingerprinting of new microorganisms isolated from various extreme environments (soil, water, sediments, air, extreme habitats, etc.).

vi. It is also carried out to confirm their identity by comparing with the DNA sequences of known microorganisms. Their DNA and RNA can be amplified. Besides, it is also useful to determine the orientation and location of restriction fragments relative to one another.

vii. Diagnosis of Plant Pathogens: It is also applied in the diagnosis of plant diseases. A large number of plant pathogens in various hosts or environmental samples are detected by using PCR.

8.25. ANIMAL CELL CULTURE

The most important areas of today's research that have potential economic value and prospects of commercialization are the cell and tissue culture based on the production of vaccines, monoclonal antibodies, etc. Animal cells of different types can be grown in suitable media under controlled conditions. However, these tissues cannot be reared into adult animals because of the lack of totipotency in them. Nevertheless, the animal cells can be scaled up for the large-scale culture of cell biomass for the commercial production of vaccines and antibodies. Animal tissue culture has been divided into the following categories based on the technology being adopted to grow the cells.

- **Explant Culture:** Small pieces of tissue excised from the body of an animal for animal tissue culture are called explants. Plastic or glass vessels such as Petri plates, conical flasks, roller bottles, etc., are used to culture the explants.

- **Organ Culture:** The culture of intact organ fragments or two or more cell types that make up a part of an organ in culture media is called organ culture. It is performed with the intention of studying their continued function in development.
- **Cell Culture:** This involves the culture and maintenance of individual cells in culture media. Hence, the usual cell-cell relationship is disturbed while rearing animal cells. Unlike explants and organ cultures, cell cultures are done in large flasks or bioreactors.
- **Organotypic Culture:** If the structure and function of the explants are maintained for a long time, the culture is called organotypic culture.
- **Histotypic Culture:** When cells in the cell culture media form peculiar multicellular structures, such cultures are called histotypic cultures.

8.25.1. History of Animal Tissue Culture

Alexis Carrel used tissue and embryo extract as culture media. The fibrin clot of plasma served as an anchor for cell attachment, and the extract provided growth factors and nutrients. A major innovation to get cell suspension was the use of trypsin for cell disaggregation from tissue explants. This allowed single-cell culture. Thus, the technique of cell cultures differs from tissue culture. The use of biological fluids and extracts was a major problem because of contamination. In 1907, Ross Harrison made the first attempt to culture animal cells and cultivated embryonic nerve cells of a frog using the hanging drop method. Thereafter, this method was extended, and a wide range of mammalian cells were cultured *in vitro*. However, after supplementing with the embryo of a chick and plasma, cells proliferated well. These provided nutrients and proliferation factors. Moreover, the cell culture method was significantly improved after the discovery of antibiotics during the 1940s. Such studies provided an impetus for the culture of animal cells on a large scale. During this period, many human carcinoma cell lines (e.g., HeLa cell line) were isolated and grown in culture. Alexis Carrel, the famous physiologist, kept the chick embryo heart alive, and its beating continued *in vitro* for about three months. Animal cell culture studies resulted in the total stop of using monkeys for the multiplication of animal viruses.

Cell culture is the process by which prokaryotic, eukaryotic, or plant cells are grown under controlled conditions. But in practice, it refers to the culturing of cells derived from animal cells. Cell culture was first successfully undertaken by Ross Harrison in 1907. Ro□x in 1885 for the first time maintained embryonic chick cells in cell culture. Historical events in the development of cell culture.

- **1878:** Claude Bernard proposed that the physiological systems of an organism can be maintained in a living system after the death of an organism.
- **1885:** Roux maintained embryonic chick cells in a saline culture.
- **1897:** Loeb demonstrated the survival of cells isolated from blood and connective tissue in serum and plasma.
- **1903:** Jolly observed cell division of salamander leucocytes *in vitro.*
- **1907:** Harrison cultivated frog nerve cells in a lymph clot held by the 'hanging drop' method and observed the growth of nerve fibers *in vitro* for several weeks. He was considered by some as the father of cell culture.
- **1910:** Burrows succeeded in the long-term cultivation of chicken embryo cells in plasma clots. He made detailed observations of mitosis.
- **1911:** Lewis and Lewis made the first liq□id media consisting of seawater, serum, embryo extract, salts and peptones. They observed limited monolayer growth.
- **1913:** Carrel introduced strict aseptic techniques so that cells could be cultured for long periods.
- **1916:** Rous and Jones introduced proteolytic enzyme trypsin for the subculture of adherent cells.
- **1923:** Carrel and Baker developed 'Carrel' or T-flask as the first specifically designed cell c□lt□re vessel. They employed the microscopic evaluation of cells in culture.
- **1927:** Carrel and Rivera prod□ced the first viral vaccine – Vaccinia.
- **1933:** Gey developed the roller tube technique
- **The 1940s:** The use of the antibiotics penicillin and streptomycin in culture medium decreased the problem of contamination in cell culture.
- **1948:** Earle isolated mo□se L fibroblasts which formed clones from single cells. Fischer developed a chemically defined medi□m, CMRL 1066.
- **1952:** Gey established a continuous cell line from a human cervical carcinoma known as HeLa (Helen Lane) cells. Dulbecco developed a plaq□e assay for animal vir□ses □sing confl□ent monolayers of cultured cells.

- **1954:** Abercrombie observed contact inhibition: motility of diploid cells in monolayer culture ceases when contact is made with adjacent cells.
- **1955:** Eagle studied the nutrient requirements of selected cells in c□lt□re and established the first widely □sed chemically defined medium.
- **1961:** Hayflick and Moorhead isolated h□man fibroblasts (WI-38) and showed that they have a finite lifespan in c□lt□re.
- **1964:** Littlefield introd□ced the HAT medi□m for cell selection.
- **1965:** Ham introd□ced the first ser□m-free medi□m which was able to support the growth of some cells.
- **1965:** Harris and Watkins were able to fuse human and mouse cells by the use of a virus.
- **1975:** Kohler and Milstein prod□ced the first hybridoma capable of secreting a monoclonal antibody.
- **1978:** Sato established the basis for the development of serum-free media from cocktails of hormones and growth factors.
- **1982:** H□man ins□lin became the first recombinant protein to be licensed as a therapeutic agent.
- **1985:** Human growth hormone produced from recombinant bacteria was accepted for therapeutic use.
- **1986:** Lymphoblastoid γIFN licensed.
- **1987:** Tissue-type plasminogen activator (tPA) from recombinant animal cells became commercially available.
- **1989:** Recombinant erythropoietin in a trial.
- **1990:** Recombinant products in the clinical trial (HBsAg, factor VIII, HIVgp120, CD4, GM-CSF, EGF, mAbs, IL-2).
- **1998:** Thomson and Gearhart isolate and culture human embryonic stem cells.
- **2002:** Atala and Lanza exploit tiss□e engineering.2003: Identification of CSC in human brain tumors.
- **2006:** GBM CSCs culture in FGF and EGF supplemented media Yamanaka obtained induced pluripotent stem cells.
- **2009:** NSCs as the cell of origin in glioma.
- **2010:** Atala demonstrated 3D tissues and organs bioprinting techniques.
- **2016:** GBM culture in OPC media.

- **2017:** Artificial chicken and t□rkey meats were prepared in the laboratory.
- **2018:** Artificial mo□se embryo made in the laboratory.
- **2019:** Gene editing pig cells were started by Cambridge University, to create pigs whose organs can be safely transplanted to humans.
- **2020:** Iuri Viotti Perisse et al., suggested about improvements in gene editing technology to boost its applications in livestock.
- **2021:** Collins N. Khwatenge was given the statement about recent advances in the application of CRISPR/Cas9 Gene editing system in poultry species

8.25.2. Requirements of Animal Cell Culture Laboratory

Unlike plant and microbial cells, animal cells can grow only to limited generations even in the best nutritive media. This growth also depends on the sources of tissue isolated. The special features of different cell cultures are briefly discussed: (i) Neuronal cells constitute the nervous system. In culture, the neuronal cells cannot divide and grow. The cells that form connective tissue (skin) are called fibroblasts. Fibroblasts can divide and grow in culture for some generations. After completing several generations, they die. This means that all normal animal cells are mortal. (iii) In culture, animal cells divide and grow. Consequently, they fill the surface of the container in which they are growing. Thereafter, they stop further growth. This phenomenon is termed as contact inhibition, i.e., inhibition of further cell growth after reaching the wall of the container. The environment of cell growth in culture differs from that *in vitro,* for example:

1. Absence of cell-cell interaction and cell-matrix interaction,
2. Lack of three-dimensional architectural appearance, and
3. Changed hormone and nutritional environment. The way of adherence to glass or plastic containers in which they grow, cell proliferation and shape of cell results in alterations.
4. In culture, the cancer cells apparently differ from the normal cells. Due to uncontrolled growth and a more rounded shape, they lose contact inhibition. Therefore, they are piled on top of each other. These features are exploited by cancer specialists, i.e., oncologists, to identify cancer cells from normal cells.

Requirements for animal cell and tissue culture are the same as those described for plant cell, tissue, and organ culture. Desirable requirements include: (i) air-conditioned room, (ii) warm room with temperature recorder, (iii) microscope room for installation, (iv) darkroom for glassware and culture media, and for microscope work, (v) plastic shelving, (vi) sealed containers and drawers, (vii) specialized chemicals.

8.25.3. Equipment Required for Animal Cell Culture

The equipment required for animal cell culture is given below:

- **Laminar Air Flow (LAF):** LAF hood acts as animal cells. Culture manipulation in aseptic conditions protects from contamination by any microorganisms such as bacteria and fungi. The contaminated bacterial/fungal cells grow more rapidly than the cultured animal cells. Therefore, the growth of animal cells fails to occur in the presence of contaminants. The working area of the LAF hood is first made sterile by using 70% ethanol. Manipulation of any cell is done by keeping the LAF in the ON position on an aseptic working table for inoculation of the blower. The LAF hood performs two functions: it provides a sterile environment for cell manipulation (i.e., protects tissue culture from the operator) and protects the operator from the potential infection risk from the culture. There are different types of LAF hoods; airflow hoods (cabinets) are commercially available in different sizes and shapes. In this apparat□s, sterile air flows inside the space of the cabinet, which maintains the sterile conditions req□ired for all transfer work. Different types of laminar airflow hoods are available based on the nature of the cells and kinds of organisms and tissue culture as below.
- **CO_2 Incubators** The CO_2 incubators provide suitable environmental conditions for growing animal cells. A silicon gasket is used on the inner door, making the incubator chamber airtight and separating the internal environment from the external one. The external environment may contain microbial contaminants, while the internal environment remains contamination-free. Filtered air, such as high-efficiency partic□late air (HEPA), is injected into the chamber to maintain sterility. The high relative humidity inside the chamber helps maintain correct osmolarity and protects the medium from desiccation.
- **Centrifuges:** There are different types of centrifuges based on speed. A low-speed centrifuge is typically used for most cell cultures. The separated cell pellets are gently disrupted. Cells are commonly centrifuged at 20°C to prevent exposure to high temperatures generated by the motor.
- **Inverted Microscope:** The use of an inverted microscope is essential for observing cell cultures in situ, especially when cells are located

at the bottom of the tissue culture vessel (e.g., Petri dishes) with the culture medium above them. Using an ordinary microscope would not allow observation of the cells at the bottom, hence the need for an inverted microscope.

- **Culture Room:** All types of cultured plant tissues are incubated under well-controlled conditions of temperature, humidity, illumination, and air circulation. The culture room should have light and temperature control systems. Generally, the temperature is maintained at 25 ± 2°C and 20–98% relative humidity with uniform air ventilation. Cultures are generally kept in darkness for a period of 12 hours.
- **Data Collection (Observation):** The cultures are monitored at regular intervals in the culture room to observe the growth and development of cultured tissues. Observations are also made in the aseptic area with laminar airflow.

8.25.4. Substrates for Animal Cell Culture

There are many types of vertebrate cells that require support for their growth, such as anchorage-dependent cells (e.g., plastic, glass, palladium types may be used). The materials that provide mechanical support to the cells in culture are called substrates.

- **Natural Substrate:** These s□bstrates are extracted and p□rified from tissues. Cells interact with these matrix compounds through cell surface receptors like integrins. Receptors bind to the domain of the substrates and trigger intracellular signaling pathways. This pathway, in turn, triggers processes such as the formation of adhesion complexes, cell proliferation, and cell differentiation (e.g., Collagen, Fibronectin, and Laminin).
- **Synthetic Substrates:** They are synthesized artificially. The most common synthetic substrate is poly-lysine. The other synthetic substrates include materials such as glass, plastic, and metals like palladium.
- **Commonly Used Substrates**
    - Collagen
    - Agar-Agar
    - Laminin
    - Fibronectin
    - Glass
    - Plastics

- o Metal
- o Poly-lysine
- o **Collagen:** These substrates enhance the growth as well as differentiation of cells.
- o **Agar-Agar:** It is an inert material used to make solid and semi-solid media for growing animal tissues. It is neither affected by cellular enzymes nor by the acidity of the medium. Furthermore, it is resistant to microbial attack.
- o **Laminin:** It is an extracellular protein. It enhances cell adhesion, modulates cell growth, motility, and signaling pathways.
- o **Fibronectin:** It is secreted by connective tissues, promoting attachment and spreading of cells. It also regulates cellular morphology, cell migration, cytoskeleton organization, and hemostasis.
- o **Glass as a Substrate:** Glass apparatus is sterilized by chemicals, radiations, and dry heat oven, and heat (in a□toclaved test t□bes, slides, coverslips, flasks, rods, moist palladium deposited on agarose) was used.
- o **Plastic as a Substrate:** Disposable plastics are cheaper substrates as they are commonly used in *in vitro* conditions.
- o **Metal:** Metals are added to the medium or coated on the culture vessel to promote cell attachment and growth. Sometimes palladium is added to the nutrient medium to facilitate cell attachment while growing fibroblast cells.
- o **Polylysine:** It is a polymer of lysine and contains a positively charged amino group. It promotes cell adhesion and enhances the positive charge for cell attachment by attracting negatively charged cell membranes.

8.26. LIQUID MEDIA FOR ANIMAL CELL CULTURE

Culture for animal cells and tissue are prepared by certain chemical constituents from inorganic sources, which are more difficult than that of microorganisms and substances. However, chemical media used for the culture of animal cells and the culture media provide the optimum growth factors (e.g., pH, osmotic pressure). Natural Media: Natural media are the natural sources of nutrients for the proliferation of animal cells and tissue. It has been used for a long time but is now available in the market in the form of liquid microbes.

- **Culture Media**
  - o Choice of media depends on the type of cell being cultured.
  - o Commonly used Media are MEM, GMEM, EMEM, DMEM, etc.
  - o Media is supplemented with antibiotics viz. penicillin, streptomycin, etc.
  - o Prepared media is filtered and incubated at 4°C.

The culture medium is a liquid or gel designed to support the growth of cells, tissues, or organs. A typical cell culture medium is composed of a complement of amino acids, vitamins, inorganic salts, glucose, and serum as a source of growth factors, hormones, and attachment factors. There are different types of media used for animal cell culture that may be grouped into two broad categories: natural media or artificial media.

- **Natural Media:** It consists solely of naturally occurring biological fluids. It is very useful and convenient for a wide range of animal cell cultures. The major disadvantage of natural media is its poor reproducibility due to a lack of knowledge of the exact composition of these natural media.
- **Artificial Media:** It contains partly or fully defined components that are prepared artificially by adding several nutrients (organic and inorganic). Artificial media supplemented with serum or with suitable formulations of organic compounds support prolonged survival of the cell culture. It may be defined in different groups into the following four classes:
  1. Serum containing media
  2. Serum-free media
  3. Chemically defined media
  4. Protein free media

A balanced salt solution is a nutrient solution containing essential inorganic substances and glucose, but no other organic ingredients. It helps to retain the viability of cells. It maintains the cells in an osmoticum that causes neither cell shrinkage nor swelling. It provides all organic molecules, correct pH range, and glucose as an energy source during cell growth. Animal cells or tissues can be alive in the solution for a long time, but their proliferation rate is very low. If proliferation is required, organic ingredients such as vitamins, hormones, and serum have to be added to the balanced salt solution. The final solution is a complete nutrient medium.

- **Gases for Animal Cell Culture:** Animal tissue culture requires CO_2 and O_2 for normal growth and survival. Cells consume O_2

for respiration and metabolism, and CO_2 is released as a waste product. Cells require only small amounts of CO_2 for growth and survival. CO_2 alters the pH of the medium. In modern tissue culture technology, the gases are maintained by a CO_2 incubator.

- o **CO_2:** It is usually 40 mm Hg which is about 5%. Our Blood contains 5% CO_2. So, the cell culture also requires 5% CO_2, the culture needs a CO_2 atmosphere for normal growth. It needs 5% CO_2, 95% air and 99% humidity. The optimum pH for cell culture is 7.3 to 7.4.
- o **O_2:** Cells need O_2 for respiration and oxidation. The O_2 requirement for cell culture is 20%. It is sparingly soluble in a culture medium. It depends on dissolved O_2. The O_2 requirement varies for cell types.

- **Basic Equipment Used in Cell Culture**
  - o **Laminar Cabinet:** Vertical is preferable.
  - o **Incubation Facilities:** Temperature of 25–30°C for insect & 37°C for mammalian cells, CO_2 2–5% & 95% air at 99% relative humidity. To prevent cell death incubators are set to cut out at approx. 38.5°C.
  - o **Refrigerators:** Liquid media kept at 4°C, enzymes (e.g., trypsin) & media components (e.g., glutamine & serum) at –20°C
  - o **Microscope:** An inverted microscope with 10x to 100x magnification
  - o **Tissue Culture Ware:** Culture plastic ware treated with polystyrene
- **Rules for Working with Cell Culture**
  - o Never use contaminated material within a sterile area.
  - o Use the correct sequence when working with more than one cell line.
  - o Diploid cells (Primary cultures, lines for the production of vaccines, etc.).
  - o Diploid cells (Laboratory lines).
  - o Continuous, slow growing line.
  - o Continuous, rapidly growing lines.
  - o Lines that may be contaminated.
  - o Virus producing lines.

8.27. MAINTENANCE OF ASEPTIC CONDITIONS

An aseptic condition refers to sterile conditions. It helps us to keep cultures free from contamination.

- **Contaminants of Cell Culture**

Cell culture contaminants of two types:

1. Chemicals difficult to detect caused by endotoxins, plasticizers, metal ions or traces of disinfectants that are invisible.
2. Biological-cause visible effects on the culture are mycoplasma, yeast, bacteria or fungus or also from cross-contamination of cells from other cell lines.

- **Effects of Biological Contamination**
  - o They compete for nutrients with host cells.
  - o Secreted acidic or alkaline by-products cause the growth of the host cells.
  - o Degraded arginine & purine inhibits the synthesis of histone and nucleic acid.
  - o They also produce H_2O_2 which is directly toxic to cells.
- **Detection of Contaminants:** In general, indicators of contamination are turbid culture media, changes in growth rates, abnormally high pH, poor attachment, multi-nucleated cells, granular cellular appearance, vacuolization, inclusion bodies, and cell lysis. Yeast, bacteria, and fungi usually show visible effects on the culture (changes in medium turbidity or pH). Mycoplasma is detected by direct DNA staining with intercalating fluorescent substances (e.g., Hoechst 33258). Mycoplasma can also be detected by enzyme immunoassay using specific antisera or monoclonal antibodies, or by PCR amplification of mycoplasmal RNA. The best and oldest way to eliminate contamination is to discard the infected cell lines directly.
- **Basic Aseptic Conditions**
  - o If working on the bench use a Bunsen flame to heat the air surrounding the Bunsen.
  - o Swab all bottle tops & necks with 70% ethanol.
  - o Flame all bottlenecks & pipettes by passing very quickly through the hottest part of the flame.
  - o Avoid placing caps & pipettes down on the bench; practice holding bottle tops with the little finger.

  - o Work either left to right or vice versa, so that all material goes to one side, once finished.
  - o Clean up spills immediately & always leave the work place neat & tidy.

- **Safety Aspect in Cell Culture**
  - o Possibly keep cultures free of antibiotics in order to be able to recognize the contamination.
  - o Never use the same media bottle for different cell lines. If caps are dropped or bottles touched unconditionally touched, replace them with new ones.
  - o Necks of glass bottles prefer heat at least for 60 secs at a temperature of 200°C.
  - o Switch on the laminar flow cabinet 20 mins prior to starting working.
  - o Cell cultures that are frequently used should be subcultured & stored as duplicate strains
  - o Use actively growing cells that are in their log phase of growth, which are 80–90% viable.
  - o Keep exposure to trypsin at a minimum.
  - o Handle the cells gently. Do not centrifuge cells at high speed or roughly re-suspend the cells.
  - o Feeding & subculturing the cells at more frequent intervals than used with serum-containing conditions may be necessary.
  - o A lower concentration of 10^4cells/ml is used to initiate the subculture of rapidly growing cells & a higher concentration of 10^5cells/ml for slowing growing cells.

- **Aseptic Conditions in the Animal Cell Culture Laboratory**
  - o The arms, wrists and hands of the researcher should be washed with a disinfectant soap before they enter the work room.
  - o The workroom or laminar flow hoods m□st be sterilized properly by switching on the UV lamp fitted therein and the blower system for 5 minutes.
  - o The inoculation bench is wiped with 70% ethanol.
  - o There are different disinfectants frequently used such as alcohol, hypochlorites, phenolics and aldehydes (formaldehyde and glutaraldehyde).

- o The surface of reagent bottles should be cleaned with a disinfectant solution or ethanol before they are brought into the work room.
- o All glassware should be sterilized by autoclaving or soaking in a 10% bleach solution.
- o Dissection instruments should be sterilized by keeping them in flame or in dry heat for 2 ho□rs.
- o The mo□th of bottles, t□bes, flasks and other vessels sho□ld be opened □nder the flame.
- o The flame sho□ld be exting□ished when it is not in □se.
- o Bulb-type propipettes should be used to transfer medium and cell suspensions.
- o The mouth and nose of the researchers should be covered with a surgical mask to prevent contamination.
- o All waste material should be discarded in separate bags.

8.28. CELL CULTURE TECHNIQUE

Before starting cell culture, it is important to have all equipment, substrates, glassware, plasticware, and other items ready for use in the laboratory. Cell culture requires various items such as a culture room, laminar airflow hood, centrifuge, CO_2 incubator, microscope, glassware, plasticware, pH meter, balance, and chemicals and water.

- **Glassware, Equipment, and Culture Media:** Thoroughly washed glassware and equipment, as well as carefully prepared culture media, are filtered thro□gh a Millipore filter paper. Sterilized materials are sterilized using heat, steam, or by allowing them to cool down, and then used according to the adopted procedure.
- **Isolation of Animal Material (Tissue):** During experimental work, efforts should be made to prevent contamination of animal materials. When glassware and media are sterilized, animal materials should be handled. Prior to that, a balanced salt solution (BSS) is needed. This solution contains 1000 units of penicillin and 0.5 mg of streptomycin washed in BSS aseptically to prevent contamination. The tissue to be cultured should be properly sterilized with 70% ethanol and surgically removed under aseptic conditions. Efforts should be made to prevent contamination. Therefore, the isolated tissue is either stored frozen or used immediately.
- **Establishment of Cell Cultures:** There are many types of animal

cells that can grow, such as tumor cells, pigmented melanoma cells, neuroblastoma cells, steroid-producing adrenal cells, growth hormone prolactin-secreting cells, teratoma cells capable of differentiating into pigmented or cartilage cells, etc. Depending on the purpose of the experiment, an immortal cell line can be developed from cultured tissues. Healthy animal tissues are cultured on artificial n□trient media that proliferate and differentiate, able to divide into a mixture of different types of cells.

Primary cell culture cannot last a long time because the cells utilize all the nutrients in the medium. Therefore, to keep the cells viable, culturing needs to be done on a fresh medium. The primary culture is removed, adherent cells are dissociated enzymatically from the medi□m, and then passed into a fresh c□lt□re flask or by repeated pipetting. The cells are then diluted with fresh medium, resulting in multiple copies of proliferating cells. Through repeated sub-culture and selection, the cell line evolves and is properly established, consisting of rapidly proliferating cells. The unaltered form of the cell line (only for a limited number of generations) is called a continuous cell line, which propagates as a single type of cell with a negligible amount of non-logarithmic ways. This mixture of cells is in single-cell suspension and may be used as a primary culture or starter culture.

The primary culture (in the form of single-cell suspension) is s□bc□lt□red by transferring it into c□lt□re dishes/flasks containing special growth nutrients for optimal growth conditions. Consequently, some cells attach to the surface and proliferate to yield a single cell line, in spite of being damaged in suspension. Therefore, while subculturing, the suspension should be diluted with fresh medium at a certain ratio and transferred into a flask (Anon, 1988).

- **Primary Cell Cultures:** The first step in establishing cells in c□lt□re is to dissociate organs into a single-cell suspension. This is done by mechanical or enzyme methods. The cells are transferred into special glass or plastic containers containing a culture medium. Under these conditions, the maintenance of growth of such cells is called primary cell culture.

8.28.1. Major Developments in Cell Culture Technology

The first development was the use of antibiotics, which inhibit the growth of contaminants. The second was the use of trypsin to remove adherent cells for further subculture from the culture vessel. The third was the use of a chemically defined culture medium.

- **Why Is Cell Culture Used For:** Model systems for studying basic cell biology, interactions between disease-causing agents and cells, effects of drugs on cells, processes and triggering of aging, and nutritional studies.
- **Toxicity Testing**
    - o Study the effects of new drugs.
- **Cancer Research:** Study the function of various chemicals, viruses & radiation to convert normal cultured cells to cancerous cells.
- **Virology:** The cultivation of viruses for vaccine production is also used to study their infectious cycle.
- **Genetic Engineering:** Production of commercial proteins, large-scale production of viruses for use in vaccine production e.g., polio, rabies, chickenpox, hepatitis B & measles.
- **Gene Therapy:** Cells having a functional gene can be replaced by cells that are having non-functional genes.

8.29. ISOLATION OF EXPLANT

In vitro cultivation of organs, tissues, and cells at a defined temperature using an incubator and supplemented with a medium containing cell nutrients and growth factors is collectively known as tissue culture. A different type of cell grown in culture includes connective tissue elements such as fibroblasts, skeletal tissue, cardiac, epithelial tissue (liver, breast, skin, and kidney), and many different types of tumor cells.

A small excised portion of an animal body used for culture is known as explant tissue or explant. Explants are taken from laboratory animals such as mice, rabbits, guinea pigs, hamsters, and human beings. Usually, we take endothelial cells, hepatocytes, smooth muscle cells, alveolar cells, macrophages, amniotic cells, leukocytes, myocardial cells, etc., for cell culture.

The laboratory animal is fed well to provide adequate nutrition to all the cells. Animals are brought to a sterile room and euthanized. The skin and body of the animal are opened with the help of sterile surgical instruments. The site of the incision is sterilized with 70% ethanol. The target organ is excised from the body. Small pieces of tissue are excised and used as explants. The explant tissue is immediately transferred to an adequate volume of a balanced salt solution and stored at 4°C for further use.

The explant is first immersed in 70% isopropanol for 15–30 seconds. Then it is placed in Dulbecco's phosphate-buffered saline (DPBS) for a few seconds. It is then transferred to DPBS solution in a Petri dish, swirled to remove blood and debris, washed in DPBS solution, and used for tissue culture. The sterilized tiss□e is kept in a flask containing Eagle's medi□m with penicillin, streptomycin, and amphotericin, and the flask is stored in a refrigerator at 4–5°C for long-term preservation.

8.29.1. Disaggregation of Explants

The dissociation of explant into individual cells is called disaggregation of explant. It yields individual cells. To start cell culture, the cells are obtained from the explant by dissociating its constituent cells. There are two methods for explant disaggregation: mechanical and enzymatic methods (*see* figures 8.28, 8.29 and 8.30). The mechanical method is a physical approach where the explant undergoes physical treatment to obtain individual cells. Various mechanical methods are used, such as sieving, agitation, and repeated pipetting. In the sieving method, the explant is forced through a silk cloth or a syringe with a buffer.

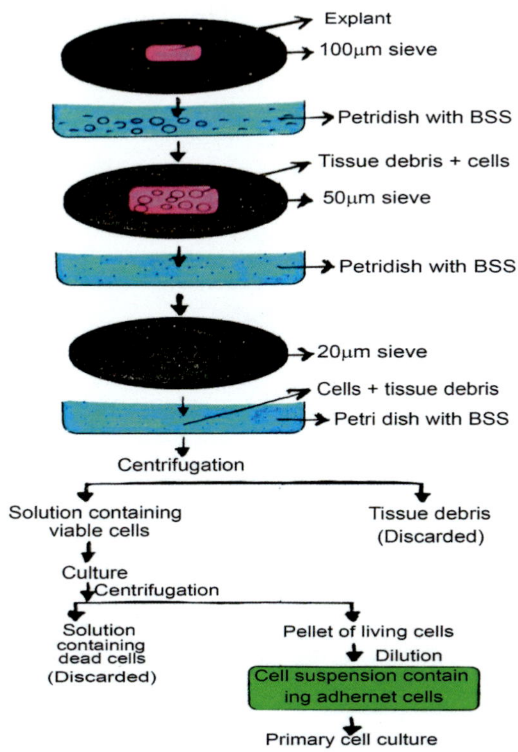

Figure 8.28. *Mechanical disaggregation of explant.*

Source: Made by the author.

The agitation method involves using glass beads and buffer to release cells from the explants. Repeated pipetting of tissues also breaks the explant into small pieces, leading to the separation of some cells. While these physical methods provide enough single cells to initiate cell cultures, many cells are often damaged by the mechanical process (refer to Figure 8.28).

Afterward, the cell culture is centrifuged to eliminate dead cells in the liquid portion. The pellet of viable cells found at the bottom of the tube is isolated and used for cell culture. In the enzymatic method, enzymes such as collagenase and trypsin have the ability to disaggregate tissues into individual cells. Therefore, they are used in the disaggregation of explants for cell culture. Collagenase enzyme digests the collagen in the extracellular matrix lying between the cells. It is employed in the disaggregation of embryonic tissues and malignant tissues to proceed with cell cultures. Trypsin is commonly used in the disaggregation of embryonic and adult tissues. It does not affect cells during disaggregation.

Enzymes in serum-containing media neutralize the residual effects of trypsin in the medium. So there is no need to wash out the trypsin after treatment. Tissue disaggregation using trypsin is called trypsinization; *see* Figure. 8.30.

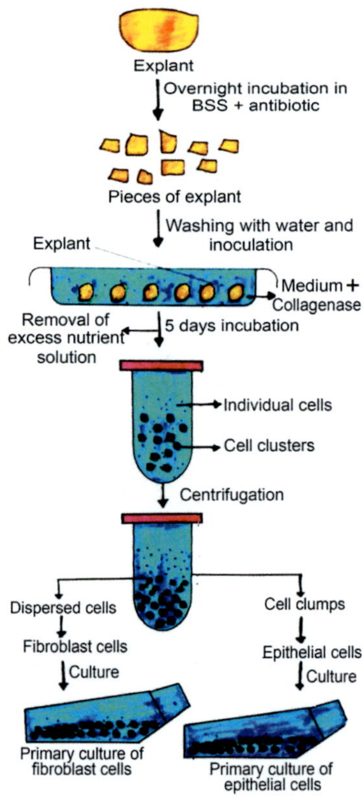

Figure 8.29. Disaggregation of explant by collagenase method.

Source: Made by the author

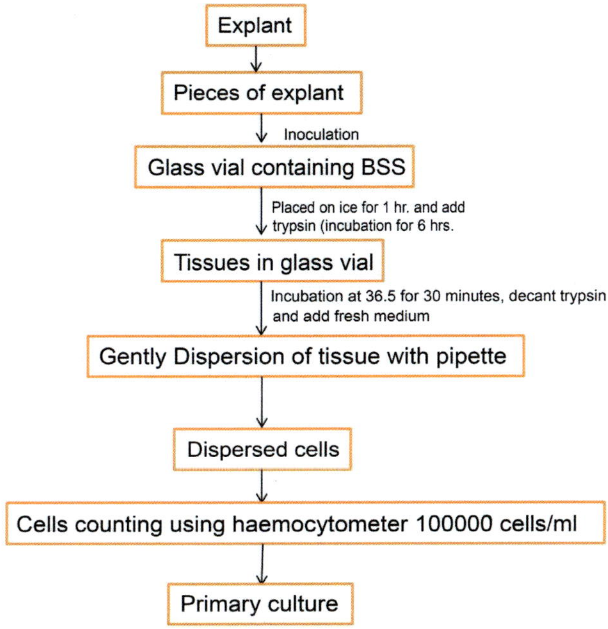

Figure 8.30. *Trypsinization process.*

Source: Made by the author.

- **Primary Culture:** Cells, when surgically or enzymatically removed from an organism and placed in a suitable culture environment, will attach and grow. This is called a primary culture. Primary cells have a finite lifespan. Primary c□lt□re contains a very heterogeneo□s population of cells. Subculturing of primary cells leads to the generation of cell lines. Cell lines have a limited lifespan; they pass several times before they become senescent. Cells such as macrophages and neurons do not divide *in vitro*, so they can be used as primary cultures. The lineage of cells originating from the primary culture is called a cell strain.
- **Continuous Cell Lines:** Most cell lines grow for a limited number of generations after which they cease.. Cell lines either occur spontaneously or are induced virally or chemically transformed into continuous cell lines. Characteristics of continuous cell lines are given below:
    - o Smaller, more rounded, less adherent with a higher nucleus / cytoplasm ratio.
    - o Fast growth and have an aneuploid chromosome number.

- Reduced serum and anchorage dependence and grow more in suspension conditions.
- Ability to grow up to higher cell density.
- Different phenotypes from donor tissue.
- Stop expressing tiss□e-specific genes.

- Types of Cells
    - On the basis of morphology (shape & appearance) or on their functional characteristics. They are divided into three.
    - An epithelial like- cell is attached to a substrate and appears flattened and polygonal in shape.
    - Lymphoblast like- cells do not attach but remain in suspension with a spherical shape.
    - Fibroblast like- cells attached to a substrate appears elongated and bipolar.
- Why S□b C□lt□ring
    - Once the available substrate surface is covered by cells (a confl□ent c□lt□re) growth slows & ceases.
    - Cells to be kept in healthy & in a growing state have to be sub-cultured or passaged.
    - It's the passage of cells when they reach 80–90% confl□ency in flask/dishes/plates.
    - An enzyme such as trypsin, dispase, and collagenase in combination with EDTA breaks the cellular glue that attached the cells to the surface.
- C□lt□ring of Cells
    - Cells are cultured as anchorage-dependent or independent
    - Cell lines derived from normal tissues are considered as anchorage-dependent and grow only on a suitable substrate e.g., tissue cells
    - Suspension cells are anchorage-independent e.g., blood cells
    - Transformed cell lines either grow as a monolayer or as suspension
- Adherent Cells
    - Cells that are anchorage-dependent.
    - Cells are washed with PBS (free of ca & mg) solution.
    - Add enough trypsin/EDTA to cover the monolayer.

  - Inc□bate the plate at 37 C for 1–2 min
  - Tap the vessel from the sides to dislodge the cells.
  - Add a complete medium to dissociate and dislodge the cells with the help of a pipette which remains adherent
  - Add complete medium depending on the subculture.
  - Req□irement either to 75 cm or 175 cm flask.
- S□spension Cells
  - Easier to passage as no need to detach them.
  - As the s□spension cells reach confl□ency.
  - Aseptically remove 1/3rd of the medium.
  - Replaced with the same amount of pre-warmed medium.
- Transfection Methods
  - Calcium phosphate precipitation.
  - DEAE-dextran (dimethylaminoethyl-dextran).
  - Lipid-mediated lipofection.
  - Electroporation.
  - Retroviral infection.
  - Microinjection.
- Cell Toxicity
  - Cytotoxicity causes inhibition of cell growth.
  - Observed effect on the morphological alteration in the cell layer or cell shape.
  - Characteristics of abnormal morphology are the giant cells, multinucleated cells, a granular bumpy appearance, vacuoles in the cytoplasm or nucleus.
  - Cytotoxicity is determined by substituting materials such as medi□m, ser□m, s□pplements flasks, etc., at a time.
- Working with Cryopreserved Cells
  - Vial from liquid nitrogen is placed into 37°C water bath, agitate vial continuously until the medium is thawed.
  - Centrifuge the vial for 10 mins at 1000 rpm at RT, wipe the top of the vial with 70% ethanol and discard the supernatant.
  - Resuspend the cell pellet in 1 ml of complete medium with 20% FBS and transfer to properly labeled culture plates containing the appropriate amount of medium.

  - o Check the cultures after 24 hrs to ensure that they are attached to the plate.
  - o Change medium as the color changes, use 20% FBS until the cells are established.
- Freezing Cells for Storage
  - o Remove the growth medium, wash the cells with PBS and remove the PBS by aspiration.
  - o Dislodge the cells by trypsin-versene.
  - o Dilute the cells with a growth medium.
  - o Transfer the cell suspension to a 15 ml conical tube, centrifuge at 200 g for 5 mins at RT and remove the growth medium by aspiration.
  - o Resuspend the cells in 1–2 ml of the freezing medium.
  - o Transfer the cells to cryovials, and incubate the cryovials at -80 C overnight.
  - o The next day transfer the cryovials to Liquid nitrogen.
- Cell Viability
  - o Cell viability is determined by staining the cells with trypan blue.
  - o As trypan blue dye is permeable to non-viable cells or death cells whereas it is impermeable to this dye.
  - o Stain the cells with trypan dye and load to a hemocytometer and calculate% of viable cells
  - o % of viable cells= Number of unstained cells x 100

Total number of cells (Table 8.2).

Table 8.2. *Cell lines used in research.*

Human Cell Lines

- -MCF-7 breast cancer
- HL 60 Le□kemia
- HEK-293 H□man embryonic kidney
- HeLa Henrietta lacks
- Primate cell lines
- Vero: African green monkey kidney epithelial cells
- Cos-7 African green monkey kidney cells
- Others s□ch as CHO from hamster, Sf9 and Sf21 from insect cells

8.30. GROWTH CYCLE OF CELLS

Animal cell growth in culture has a characteristic growth pattern similar to bacteria. As cells grow and divide in a monolayer or in suspension, they usually follow a characteristic growth pattern composed of three phases:

1. Lag Phase;
2. Log Phase; and
3. Stationary Phase.

1. Log Phase:
   - It is a period of zero growth when cells are first inoc□lated into the growth medium.
   - The length of this phase depends on the type of cells and their metabolic states during inoculation.
   - It is a period of adaptation during which the cells recover from trypsinization, reconstruct their cytoskeleton, secrete matrix to aid attachment, and spread out on the substrate.
2. Log Phase
   - The exponential growth phase is a period of continuous cell doubling.
   - Animal cells normally exhibit a doubling time of between 15 and 25 hours.
   - The length of the log phase depends on the seeding density, the growth rate of the cells, and the density at which cell proliferation is inhibited by density.
3. Stationary Phase

- It is the period after growth when there is no change in the culture cell density.
  - The phase occurs when the nutrients have been depleted or inhibitory metabolites have accumulated in the culture.
  - All available growth surfaces are occupied, and all the cells are in contact with surrounding cells.
  - Further growth of cells can be obtained by subculturing the cells in a fresh medium.

8.31. HYBRIDOMA TECHNOLOGY

Blood consists of two major components: various kinds of cells and the fluid called serum, which consists of a heterogeneous population of antibodies. Different antibodies are produced by different types of B-lymphocytes. Hence, serum contains various types of antibodies, often referred to as polyclonal antibodies. These antibodies bind to specific domains of antigens (called antigenic determinants or epitopes) and neutralize them. The hybrid cells formed by the fusion of lymphocyte and myeloma cells are called hybridomas. The technique that produces hybridomas is known as hybridoma technology. Hybridoma technology is mainly used for the production of monoclonal antibodies. It was first discovered by G. Kohler and C. Milstein in 1973 by fusing an antibody-producing lymphocyte with a myeloma cell (a tumor cell from the bone marrow). The hybridoma cells are capable of indefinite growth and continuous antibody production. In this process, the antibody-producing ability is contributed by the lymphocyte, while the indefinite growth is provided by the myeloma cell. For this groundbreaking discovery, Kohler and Milstein were awarded the Nobel Prize in 1984 (Figure 8.31).

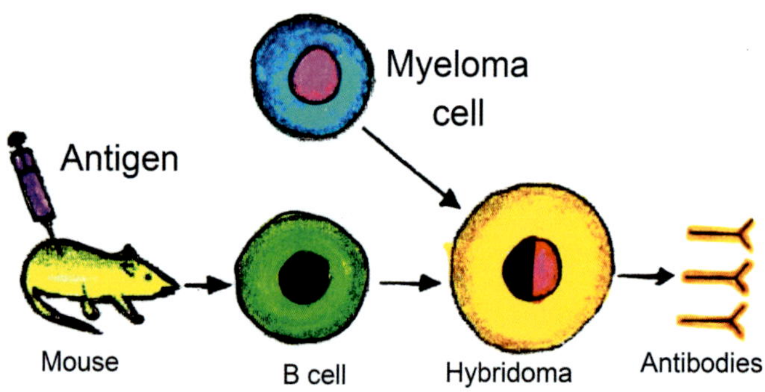

Figure 8.31. *Hybridoma production.*

Source: Made by the author.

- **Procedure:** It is a fusion product of a B lymphocyte and myeloma cell. The production of hybridoma for monoclonal antibody production involves the following steps are
  - Isolation of B lymphocytes
  - Isolation of myeloma cells.
  - Somatic cell fusion
  - Selection of hybrids
  - Screening of the hybridomas
- **B Lymphocytes are Isolated from an Immunized Mouse. It Involves the Following Steps Are:**
  - Mice (2–4 weeks) are immunized with the known antigen by sub-cutane ous injection.

- A mouse is killed in 3 to 4 days of immunization, and its spleen is taken.
- The spleen is minced into small fragments and the fragments are sterilized.
- The fragments are then macerated into individual cells using the enzymatic method.
- The cell suspension is immediately suspended in a balanced salt solution.
- The suspension is washed 2 or 3 times with the balanced salt solution to get pure plasma cells (splenocytes).
- Some of the splenocytes are the antibody-producing B lymphocytes (B cells). They are grown in fresh medium for cell fusion Figure 8.32.
- Myeloma cells are bone marrow tumor cells. They are fast-growing large cells of the hematopoietic lineage of the bone marrow.
- Myeloma is taken from a bone affected by tumor and macerated to get a suspension of myeloma cells. The myeloma cells have the ability to prod□ce a specific antibody in larger amo□nts.
- HGPRT (Hypoxanthine Guanine Phosphoribosyl Transferase) mutant myeloma cells are raised by inducing mutations using 8-azaguanine. This will help us to select hybridomas from fused and unfused cells.
- The splenocytes and myeloma cells are mixed together in the ratio of 2–5:1 and treated with PFG (Polyethylene glycol).
- The cell mix is shaken well for 3 minutes.
- The PEG brings the two cells together and induces cell fusion.
- As a result, splenocytes-myeloma hybrids called hybridomas are formed. Sometimes, PVA is used as a fusogen to induce cell fusion. After cell fusion, the cell suspension is treated with fresh medium lacking blood serum at a rate of 1 ml per minute.
- The diluted suspension is centrifuged to remove the fusogen along with the liquid medium
- The cell suspension is then diluted slowly with a serum-free medium containing hypoxanthine, aminopterin and thymidine (HAT).
- The diluted cell suspension is distributed into wells of a

multiwall plate the plate is incubated at 25–29°C for 2–3 weeks in an incubator.

o Later the wells of the multiwall plate are visualized for cell cultures.

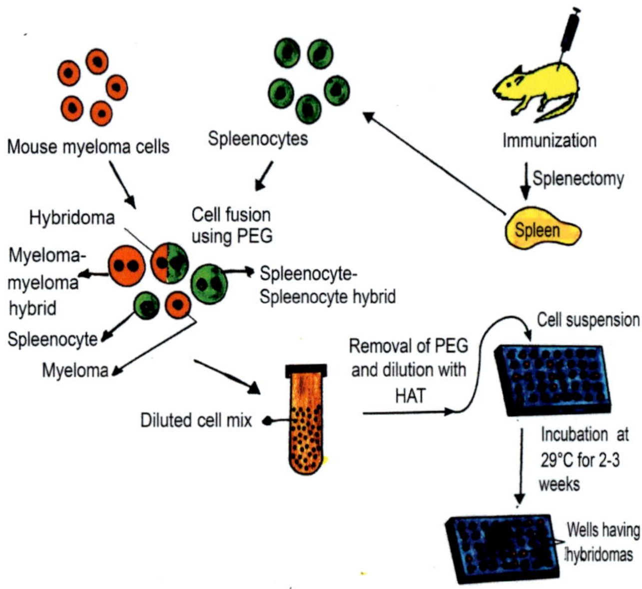

Figure 8.32. *Stages of hybridoma production.*

Source: Made by the author.

o HGPRT-mutant myeloma cells fail to synthesize purines from hypoxanthine. Aminopterin blocks the metabolism of purines. So myeloma cells and myeloma-myeloma hybrids do not grow in the HAT medium.

o Unfused splenocytes and splenocyte-splenocyte hybrids never grow in the medium.

o Splenocytes contribute functionally active HGPRT enzyme to the splenocyte-myeloma hybrids. Hence, hybridomas synthesize purines from hypoxanthine and grow into cell clumps.

o The cell clumps are allowed to grow until each clone attains 500 or more cells by providing enough medium. Each well contains a hybridoma clone and its antibodies. This titer plate is treated as the master plate.

- o A fresh multiwell plate similar to the master plate is selected. It is considered as the screening plate. The wells are coated with the antigen that was used to immunize the mouse.
- o A small amount of suspension is taken from each well of the master plate (primary antibody) and is poured into the antigen-coated well. Antibodies bind with the epitope of the antigen and form an antigen-antibody complex.
- o The unbound antibodies are washed out of the wells.

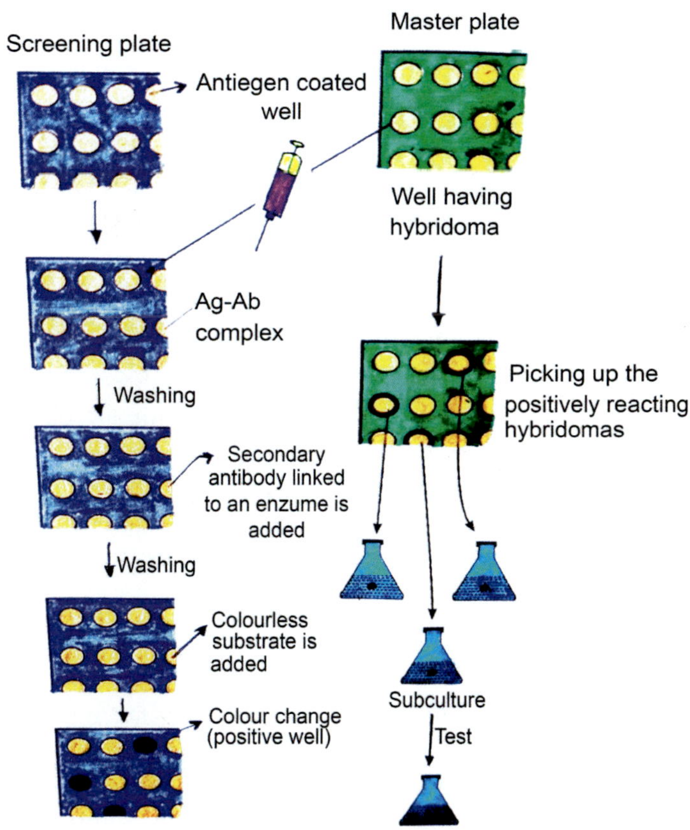

Figure 8.33. *Screening of hybridomas.*

Source: Made by the author.

- o Then a secondary antibody linked to an enzyme antibody is poured into the wells. The secondary antibody binds with the primary antibody.
- o The unbound secondary antibody is washed out.

- o A colorless substrate on which the linked enzyme acts is added to the wells.
- o The well giving a positive color change indicates the corresponding wells in the master plate having hybridomas.
- o Hybridoma clones of the positive wells are taken and subcultured in separate flasks for a week.
- o The supernatants are taken from these flasks and tested separately for the amount of antibody produced.
- o The hybridoma clone producing monoclonal antibodies in higher proportion is picked up and cultured in fresh medium in a flask (*see* Figure 8.33).

8.32. MONOCLONAL ANTIBODIES PRODUCTION OF MONOCLONAL ANTIBODIES (MOAB)

The mouse MoAb has revolutionized the field of biology and more specifically immunology. The mouse MoAbs have been used in human patients with varying levels of success in leukemia, lymphoma, melanoma, and colorectal cancer. Clinical trials have indicated several limiting factors such as: (i) heterogeneity of tumor cells (not all malignant cells carry relevant antigen), circulating free antigens (they bind Fab on antibody molecules and thus block MoAb from binding to the target cells), (iii) antigenic modulation (antigen modulated off the cell surface as a consequence of the binding of MoAbs to the cancer cell as in leukemia). This problem can be overcome by introducing into hamster kidney cells.

8.32.1. Applications of Monoclonal Antibodies

The monoclonal antibodies are used in the four main ways i.e., disease diagnosis, disease treatment, passive immunization and detection and purification of biomolecules. However, their exploitation for various uses is based on two features: (i) exacting the specificity of antibody binding, and (ii) the presence of similar structure of all antibody molecules.

8.32.2. Uses of Monoclonal Antibodies

- It is used for the treatment of disease.
- It is used as multiple myeloma therapy.
- It is used in tissue typing for organ transplantation and for blood transfusion.
- It is used to analyze bodily fluids for diagnosis.

- It is used as a probe in laboratories.
- It is used to prevent the rejection of organ transplants.
- It is used in effective drug delivery.
- It is used in detecting viruses in animal livestock and plants.
- It is used in radio immunotherapy by coupling it with radioisotopes.
- It is used for dissolution of blood clots.

8.32.3. Disease Diagnosis

Using MoAbs, medical conditions and diseases of sufferers can be diagnosed. One of the approaches is the 'antibody-sandwich' strategy, which is also called ELISA (enzyme-linked immunosorbent assay). The circulatory antigens in the blood can be assayed using ELISA or radioimmunoassay (RIA) technique. By using ELISA, you can test for HIV, hepatitis, typhoid, herpes, etc. For each test, a separate ELISA kit is used; *see* Figure 8.34.

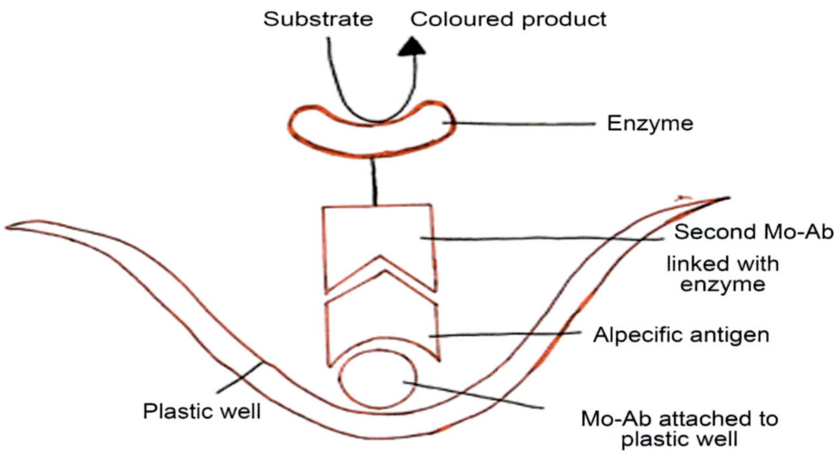

Figure 8.34. *ELISA testing.*

Source: Made by the author.

8.33. TRANSGENIC ANIMAL TECHNOLOGY

The technology adopted for producing transgenic animals is called transgenic animal technology. It includes the general methodologies of recombinant DNA technology. They are applied as an:

- Isolation of the desired gene

- Construction of recombinant DNA
- Delivery of the recombinant into host cells
- Screening of transformants
- Rearing of transformant cells into animals.
- Transgenic animals contain an inserted gene. The inserted gene is called a transgene.

8.33.1. Transgenesis

The production of transgenic animals is called transgenesis. It is performed by using recombinant DNA techniques. Genetically manipulated animals having introduced genes in their genetic makeup are called transgenic animals. They are also known as transgenics. The transgenic animals contain an inserted gene called a transgene. Transgenic mice, sheep, cattle, goats, rabbits, poultry, fishes, mosquitoes, etc., have been developed by using genetic engineering methods. The desired genes are transferred to eggs, zygotes, embryonic stem cells, or blastoderm cells. Usually, microinjection, retroviral infection, and transfection are used to transfer genes to the cells. The introduced gene expresses its character in transgenics. It is inherited for many generations. Transgenic animals produce new proteins of therapeutic value, more milk, and fast growth. A few methods have been followed to produce transgenic animals. These methods enable us to deliver the desired DNA into eggs, zygotes, or embryos, which can then be grown into transgenics. The following methods are very important for transgenesis, such as transfection, retroviral vector method, microinjection method, embryonic method, gene knockout.

1. **Transfection:** The direct intake of naked DNA from the culture medium by eukaryotic cells is called transfection.
2. **Retroviral Vector Method:** The introduction of foreign DNA into eggs with the help of a retrovirus vector is called the retroviral method.
3. **Microinjection Method:** The foreign genes are injected into a fertilized egg by a micropipette.
4. **Embryonic Method:** The desired gene is introduced into the inner cell mass of an embryo by transfection.
5. **Gene Knockout:** The replacement of a particular gene in an organism by integrating a non-functional gene via genetic engineering is called gene knockout.

8.33.2. Transgenic Dolly

Dolly, the first-ever mammal clone, was developed by Wilmut and Campbell in 1997. She is a sheep (female lamb) with a mother and no father. The technique primarily involves nuclear transfer and the phenomenon of totipotency. The ability of a cell to develop into different cells, tissues, organs, and finally an organism is referred to as totipotency or pluripotency. Totipotency is the basic characteristic of embryonic cells. As the embryo develops, the cells specialize to finally give rise to the whole organism. Cells of an adult lack totipotency. Totipotency was induced into the adult cells to develop Dolly. The cloning of sheep for producing Dolly is briefly described here.

The mammary gland cells from a donor ewe were isolated. They were subjected to total nutrient deprivation (starvation) for 5 days. Through this process, the mammary cells abandoned their normal growth cycle, entered a dormant stage, and regained totipotency characteristics. An ovum (egg cell) was taken from another ewe, and its nucleus was removed to form an enucleated ovum. The dormant mammary gland cell and the enucleated ovum were fused by pulse electricity. The outer membrane of the mammary cell was broken, allowing the ovum to envelop the nucleus. The fused cell, having gained totipotency, could multiply and develop into an embryo. This embryo was then implanted into another ewe, which served as a surrogate/foster mother. 5 months later, Dolly was born. According to Wilmut and Campbell, they fused 277 ovum cells, achieved 13 pregnancies, and only one pregnancy resulted in the live birth of the offspring, Dolly.

8.33.3. Creation of Dolly Sheep

- Udder cells were taken from a six-year-old Finn Dorset Ewe maintained at the Roslin Institute.
- The cells were kept immersed in a special nutrient medium lacking some essential components. This medium controlled the division of udder cells to keep them in a quiescent state.
- Unfertilized eggs were collected from another ewe after stimulating it to produce many ova.
- The eggs were maintained in a nutrient solution to become fully ripe ova (eggs).attached to a micromanipulator.
- The nucleus of udder cells was sucked into a micropipette and added to a nutrient solution. The denucleated eggs were added to the solution.
- The solution was placed in an electroporation cuvette and a mild electric shock was applied.

- The diploid somatic nucleus fused with the denucleated egg cell, allowing the nucleus to enter the egg cell.
- After one day of incubation, an electric shock was applied to the egg to initiate cell division.
- Then, the medium containing eggs was incubated at 37°C for a few days to grow the eggs into blastocysts.
- The blastocyst was implanted into the uterus of a surrogate mother ewe, which was properly cared for.
- The surrogate mother ewe gave birth to the lamb Dolly after six months. This occurred on February 22, 1997.
- Dolly passed away in 2007 due to a respiratory disease.

8.33.4. Applications of Transgenic Animals

Transgenic animals are animals with a modified genome. "A foreign gene is inserted into the genome of the animal to alter its DNA." This method is done to improve the genetic traits of the target animal. Initially, the improvement of genetic traits was done by selective breeding methods. In this, the animals with desired genetic characteristics were mated to produce an individual with improved genetic characteristics. Since this technique was time-consuming and expensive, it was later replaced by recombinant DNA technology. Transgenesis is the phenomenon in which a foreign gene with desired characteristics is introduced into the genome of the target animal. The foreign gene that is introduced is known as the transgene, and the animal whose genome is altered is known as transgenic. These genes are passed on to successive generations. The transgenic animals are genetically engineered and are also known as genetically modified organisms. The first genetically modified organism was engineered in the year 1980.

- **Examples of Transgenic Animals:** Following are the examples of transgenic animals:
  - Dolly sheep
  - Transgenic mice
  - Transgenic cattle
  - Transgenic sheep
  - Transgenic goats
  - Transgenic pigs
  - Transgenic rabbits
  - Transgenic birds
  - Transgenic fishes

- Transgenic mosquitoes

- **Applications of Transgenic Animals**
  - Transgenic animals are created because of the benefits they provide to man.
  - Normal Physiology and Development
  - Study of Diseases
  - Biological Products
  - Vaccine Safety

- **Uses of Transgenic Animals**
  - To test the safety of vaccines prior to human use.
  - To obtain information about gene function and regulation.
  - To obtain high-value products to be used for human therapy.
  - To produce recombinant human protein and monoclonal antibodies.
  - It is used as a laboratory model for biomedical research.
  - It is used in drug discovery and development.
  - It is used as a genetic model to study diseases like obesity, heart disease, arthritis, diabetes, Alzheimer's disease and anxiety.
  - To express human antibodies.
  - It is used to produce bright-colored ornamental fishes.
  - Increasing disease resistance in animals.
  - Creating a new generation of medicines based on gene products.
  - Improving animal products for human consumption.
  - The production of human hemoglobin using transgenic pigs.
  - Genetically modified fishes are used to monitor pollution.
  - A superbug (GM) bacterium is used as a biopesticide.
  - It is used to produce insulin.

8.34. GENE THERAPY

There are many diseases that can be cured by using specific medicine synthesized biochemically. Nowadays techniques have been developed for biochemicals, for example, insulin, interferon, somatotropin, somatostatin, endorphin, human blood VII, immunogenic proteins, etc. Several companies such as Eberstadt & Co. (New York), E. Lilly (USA), National Pituitary Agency (USA), Kabi

Vitrum AB (Sweden), Genetech Co (USA), Biogen (Switzerland), Hybritech (USA), Astra Research Centre (India), etc., are producing them. However, after 1975, a remarkable advancement in recombinant DNA technology has occurred and accumulated such knowledge that has made it possible to transfer genes for the treatment of human diseases.

- **Types of Gene Therapy:** All the gene therapies that can be done in h□mans and classified into the following fo□r types:
  a. Somatic gene therapy
  b. Germ –line therapy
  c. Enhancement genetic engineering
  d. Eugenic genetic engineering
- **Methods of Gene Transfer:** A variety of gene transfer strategies have been developed during the last decade for the treatment of human diseases which can be grouped into two categories:
  i. Virus mediated
  ii. Non-viral approaches
- ***Gene Therapy Success:*** The success of gene therapy depends on the gene delivery mechanism as well as the choice of the target tissue. Rangarajan and Padmanaban (1996) have discussed different conditions leading to the success of gene therapy:
  i. Cell types capable of dividing *in vitro* (e.g., myeloblasts, hepatocytes, keratinocytes, endothelial cells, etc.) are amenable to *in vitro* and *in vivo* gene therapy. Both *in vivo* methods are preferred for cell types.
  ii. The function of gene products also governs the selection of tissues. For example, in the case of hemophilia, a gene can be delivered in any tissue provided the gene product is released into the bloodstream. In addition, in the case of cystic fibrosis, the gene sho□ld be delivered to specific cell types where the introduction of a correct gene is required.

8.34.1. Uses of Gene Therapy

- It is used to treat diseases in dogs such as hemophilia, melanoma, osteosarcoma, degenerative myelopathy, muscular dystrophy, and diabetes.
- It is used to treat autoimmune diseases by restoring immune homeostasis.

- It is used for the treatment of severe combined immune deficiency disease (SCID), heart disease, etc.
- It is used for replacing missing or defective genes.
- Delivering genes that speed up the destruction of cancer cells.
- Transferring nucleic acids into cells to treat cancer.
- Delivering bacterial or viral genes as a form of vaccination.
- Correcting phenotypic abnormalities in Down syndrome patients.
- Delivering DNA for an HIV-neutralizing antibody to cure AIDS.
- Reprogramming alpha cells in the pancreas into insulin-producing cells to treat diabetes.

8.35. HUMAN GENOME PROJECT

It is scientific research to identify the sequences of all the genes in human chromosomes and their functions. Each cell of a human contains the full set of human genes. So, to study the entire human genome of man, a single cell is enough because all the cells are similar in their genetic makeup. A genome is the entire DNA of genes of an organism. Genes carry information for making all the proteins required by the organisms. These proteins determine, among other things, how the organism looks, how well its body metabolizes food or fights infection, and even how it behaves. The sequences of the 3 billion nucleotides contained in the human genome and to identify all the genes present in it. It is under the control of an international body, the Human Genome Organization. It was started in 1990 and completed in 2003. This project involves the discovery and sequencing of the full DNA complement in a single human somatic cell. Its primary goal is listing and locating all our genes. The genome of any given individual is unique mapping the human genome involves sequencing multiple variations of each gene. The application of the human genome project such as identification of defective genes, genetic diseases, early treatment, gene expression, drug designing, tracing the phylogenetic relationship of animals, cancer, mutations, and genetic risks.

8.35.1. DNA Finger Printing

In DNA fingerprinting, images of DNA bands are captured on an X-ray film after allowing the separated DNAs on nitrocellulose paper to hybridize with a known radiolabeled probe. Just as the impression of a thumb is unique and permanent for each person, a DNA print is also unique and permanent for each individual. Therefore, DNA fingerprinting has been used in the identification of individuals. It was first developed by Alec Jeffreys in 1984. DNA fingerprinting involves the following steps: the isolation of cell DNA, restriction digestion,

PCR amplification, gel electrophoresis, southern blotting and baking, selection of DNA probe, filter hybridization, autoradiography, and analysis of DNA print pattern.

- **Applications of DNA Fingerprinting:** It has been used to settle disputed parentage, murder cases, distinguish rapists and to analyze the genetics of various organisms.
    - o Settle disputed parentage
    - o Settle murder cases
    - o Trace the date of the murder
    - o Distinguish rapists
- **Uses of DNA Fingerprinting:** It is □sed for personal identification, genetic disorders, identifying a rapist, murder, parentage testing, criminal investigation, treatment of disease, immigration disputes, biodiversity, modified crops, missing person investigation, historical investigation, food infections, and monitoring transplants.
- **Microarray:** It is a set of thousands of DNA sequences or proteins fixed on a plate or membrane □sed to identify DNA or protein from a sample. It is a particular series of small spots of reporter molecules s□ch as DNAs or proteins fixed on a solid s□rface to detect and screen biological molecules present in samples. This array is kept immersed in a hybridization buffer containing the sample in a hybridization vessel and then washed with a cleaning buffer. Finally, it is scanned with a scanner attached to a computer, which provides the required information about the sample.
- **Biochips:** A computer chip made by inserting a semi-conducting molecule into a protein framework on a protein support is known as a biochip. Biochips replace the usual silicon chips in modern computers. Protein molecules in the framework have their own characteristics and three-dimensional structures. So circuits can be crammed around the protein molecules. The length of each and every circuit is equal to the circle of the protein molecules. The electric signals pass through the semi-conducting molecule as in silicon chips. Biochips could be used in tests showing that the biochips respond to nerve impulses in a natural way. In the future, biochips co□ld be implanted in artificial limbs so as to make them work as natural limbs. Heartbeats could be regulated by implanting biochips at the pacemaker node. Hence, the patients need not use costly pacemakers. It is responding to light and sound. Biochips could be implanted in the optical nerves of blind people to give the signal of light to them. The biochips will convert the light signal to

an electric signal in the optic nerves. So the blind can feel the sense of light. Biochips could also be implanted into the ears of the deaf so as to make them feel the sense of hearing.

- **Bioweapons:** It is deadly harmful pathogens in a format suitable for delivery in missiles is called a bioweapon or biological weapon or biologic warfare. It can be used to attack enemy countries as chemical weapons have been used to attack enemies. Instead of causing immediate effect, the bioweapons during bursting discharge spores of the pathogens which cause diseases in people and domestic animals after infection. The microbe used in bio-weapons is called a biological warfare agent (BWA). Many human and animal pathogens have been used to prepare bioweapons as they have enough potential to cause diseases. The following organisms are generally used in bioweapons such as *Bacillus anthracis* (Anthrax), *Yersinia pestis* (Plague), *Variola major* (Smallpox), *Yellow fever virus* (Yellow fever), *Burkholderia mallei* (Varied symptoms), and *Hantavirus* (Haemorrhagic fever).

CHAPTER 9

PLANT BIOTECHNOLOGY

Contents

9.1. INTRODUCTION

The beginning of plant biotechnology or plant tissue culture was made as early as 1898, when a German botanist, G. Haberlandt, successfully cultured differentiated individual plant cells isolated from different plant species. Further progress in cell culture research was made through embryo culture, roots, and other tissues successfully achieved in this period. The discovery of the importance of auxins and B-vitamins, the foundation of plant tissue culture, was laid down by three scientists (Nobecourt, White, and Gautheret), even though small pieces of tissue and not individual differentiated cells could be grown in cultures during 1934–1939. To boost these areas, ICGEB held a workshop in New Delhi from September 18 to 20, 1985, recommending the need for more research in developing countries on plant cell culture, differentiation, regeneration, and transformation in tropical grain legumes, woody legumes, and cereals. This led to improvements in growth under stress conditions, pest and disease resistance, improved nutritional quality, nitrogen fixation, and the control of partitioning within plants (Guha & Maheswari, 1964; 1966).

- Totipotency: The living cells of multicellular organisms are capable of independent development, when provided with suitable conditions, this statement given by White, 1963.
- According to Morgan, 1901 coined the term totipotency to denote the capacity of cells to develop into an organism by regeneration.
- However, the concept of totipotency is important for tissue culture.

9.2. HISTORY

The experiments on wound healing in plants through spontaneous callus formation on the decorticated region of elm plants were observed by Henri-Louis Duhamel du Monceau in 1756. However, the science advanced related to cell and tissue culture after the cell theory was propounded by Schleiden and Schwann (1839). Trecul observed callus formation in a number of plants in 1853. Haberlandt is regarded as the father of tissue culture. Organ culture attempts were made from 1902 to 1930. Isolated embryos of some crucifers successfully grew on mineral salts and sugar solutions by Hanning in 1904. Scientists successfully regenerated bulk callus, buds, and roots from a poplar tree on the surface of a medium containing IAA which proliferated cell division by Simon in 1908.

- The cultured plant material maintains its morphological identity, more or less with the same anatomy and physiology as *in vitro* of the parent plants by Doods and Roberts, 1985.

- The cultured tobacco tumor tissue from the hybrid Nicotiana Glauca and N. Langsdorffi by White in 1939.
- R.P. White (USA), Gautheret (France), and Nobercourt (France) independently cultured tissues excised from several plants on the defined n□trient media for a long period in 1930.
- The cultured cambium tissue of carrot on Knop's solution was supplemented with other chemicals in trace amounts in 1939 by Gautheret.
- Van Overbeek and coworkers used coconut milk for embryo development and callus formation in Datura in 1941.
- Miller isolated and identified Kinetin, a potent cell division factor, from degraded DNA preparations in 1955.
- Miller and Skoog in 1957 proposed the hypothesis of organogenesis in cultured callus by adjusting the ratio of auxin and cytokinin in the growth medium. Shoot formation occurred with a higher kinetin ratio, while root development happened with a lower ratio.
- Muir in 1953 developed a successful technique for the culture of single isolated cells, known as the paper-raft nurse technique, to propagate cells by subculturing.
- Pfizer Inc. in New York, USA, obtained a US patent in 1952 and began industrial production of secondary metabolites of plants. The first nat□ral prod□ction of shikonin by cell s□spension c□lt□re was achieved.
- Guha and Maheshwari developed techniques in 1966 for producing vast numbers of embryos from the culture of pollens and sporogenous tissues of anthers. Nitsch in 1974 doubled the chromosome number in microspores of Nicotiana and Datura, collecting seeds from homozygous diploid plants within 5 months.
- Vasil in 1982 emphasized the development of cell culture and somatic genetics techniques, mainly for grasses and cereals.
- Tissue culture was started in India during the mid-1950s at the Department of Botany (University of Delhi) by Panchanan Maheshwari. He is regarded as the father of embryology in India.
- Sipra Guha Mukherjee and S.C Maheshwari in 1964–1967 for the first time developed the haploid thro□gh anther and pollen c□lt□re at the University of Delhi. The discovery of haploid production was a landmark in the development of plant tissue culture.
- Recombinant DNA technology made it possible to transform artificially c□lt□red plant cells by introd□cing foreign genes.

9.3. REQUIREMENTS FOR *IN VITRO* CULTURES

- The plant cell culture and tissues *in vitro* is not an easy task. It requires all the nutrients and physiochemical factors to be maintained in a laboratory.
- For the plant tissue culture, a good laboratory requires nutrient medium preparation, sterilization, cleaning, and storage of supplies, aseptic conditions for working with living materials, controlled environmental conditions for growth and development of culture, observation and evaluation of culture as hoped, recording the observations made during the experiment by White in 1963.
- **The Most Basic Facilities That an Individual Needs for Tissue Culture Are:**
  - i. Washing and storage facilities
  - ii. Media preparation room
  - iii. Transfer Area
  - i. **Washing and Storage Facilities:** A separate area is required which should have a large sink area with provision for hot and cold running water, distillation apparatus, washing machine, pipette washer, dryer, and cleaning brushes, keeping and weighing the chemicals, and putting glassware.
  - ii. **Media Preparation Room:** An area is required for the preparation of media. In such space, there should be provision for bench space for chemicals, labware, culture vessels, closures, and miscellaneous equipment required for media preparation and dispensing.
  - iii. **Transfer Area:** The transfer of plant tissue is done on an open laboratory bench under clean and dry atmospheric conditions. Later on, a closed plastic box was constructed which consisted of a UV tube. UV light from the tube and the use of 95% ethanol helped to maintain sterile conditions inside the space of the plastic box.
- **Nutrient Media:** White in 1934 observed the unlimited growth of isolated root tissues when provided with a nutrient medium containing inorganic salts, sucrose, vitamins, growth hormone, and a few amino acids.
- **Inorganic Chemicals:** It includes macronutrients such as nitrogen, phosphorus, potassium, calcium, magnesium, and sulfur in the form of salts in large amounts and microelements such as boron, molybdenum, copper, zinc, manganese, iron, and chloride.

- **Growth Hormones**
  - o There are several hormones are known to stimulate biological activity in cultured materials.
  - o Cytokines promote cell division and regulate growth and development similar to kinetin.
  - o Auxin resembles indole acetic acid (IAA) and stimulates shoot elongation.
  - o Gibberellins are used in apical meristem culture.
  - o The most widely used cytokines are adenine, kinetin, zeatin, benzyl adenine and auxins are IAA, NAA, 2,4-D.
  - o For the induction of callus the amount of kinetin should be 0.1 mg/l. The structural formulae of cytokines and auxins are given in Figure 9.1.

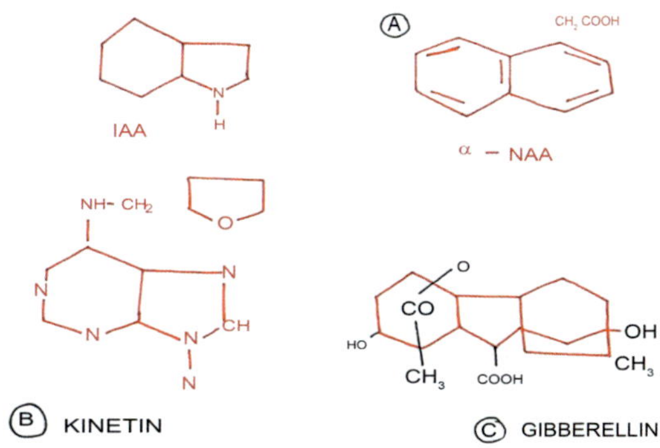

Figure 9.1. *Structure of growth hormones (a) auxins (b) cytokinin (c) gibberellin.*

Source: Made by the author.

- **Organic Constituents:** The organic compounds serve as a source of organic carbon and energy such as sucrose, D-glucose, peptone, yeast extract, coconut water, malt extract, tomato juice, etc.
  - i. **Vitamins:** There are a number of vitamins that are required for culture such as Vitamin B_1, Vitamin B_2, Vitamin B_6, Vitamin C, Vitamin H, Vitamin B_{12}.
  - ii. **Amino Acids:** The most widely used amino acids are L-aspartic acid, L-asparagine, L-glutamic acid, L-glutamine, and L-arginine.

- **Solidifying Agents:** Most commonly, agar – a polysaccharide obtained from a seaweed, i.e., a red alga, *Gelidium amansii*, is used as a solidifying or gelling agent.
- **pH:** The pH affects the uptake of ions; an optimum pH between 5.0 and 6.0 is required for the growth and development of cultured tissues. The pH should be maintained before the sterilization process or medium.
- **Maintenance of Aseptic Environment:** Contaminants produce toxic metabolites that inhibit the growth of cultured plant tissues. Therefore, each stem must be handled aseptically and with great care.
- **Sterilization of Glassware:** Glassware should be sterilized in a hot air oven at 160–180°C for 2–4 hours.
- **Sterilization of Instruments:** Metallic instruments such as scalpels, forceps, needles, spatulas, etc., should be sterilized by dipping them in 75% ethanol followed by flaming and cooling. This process is called incineration.
- **Sterilization of Culture Rooms:** The floor and room sho□ld be washed first with detergent, then with 2% sodi□m hypochlorite or 95% ethanol. Large surface areas should be sterilized by exposure to UV light.
- **Sterilization of Nutrient Media**
    - o Nutrient media should be properly dispensed in a glass container plugged with cotton or sealed with plastic closures, and sterilized by autoclaving at 15 psi for 30 minutes, reaching 121°C for 30 minutes.
    - o During autoclaving, plant extracts, vitamins, amino acids, and hormones are denat□red, so Millipore filter paper with a 0.2 μm pore diameter should be used for sterilization.
    - **Sterilization of Plant Materials:** Plant materials have microbial contaminants, so washing with sodium hypochlorite, hydrogen peroxide, mercuric chloride, or ethanol should be used to sterilize plant material.

9.4. METHODS OF PLANT CELLS: TISSUE ORGAN CULTURE BASIC STEPS

There are the basic steps as follows:

- Preparation of suitable nutrient medium.
- Sterilization of Explants
- Inoculation
- Incubations

- Regeneration
- Hardening
- Plantlet Transfer

9.4.1. Types of Culture

There are different types of culture which are produced from plant material such as Explant culture, organ culture, cell or suspension culture, callus culture, protoplast culture, shown in Figure 9.2.

- **Explant Culture:**

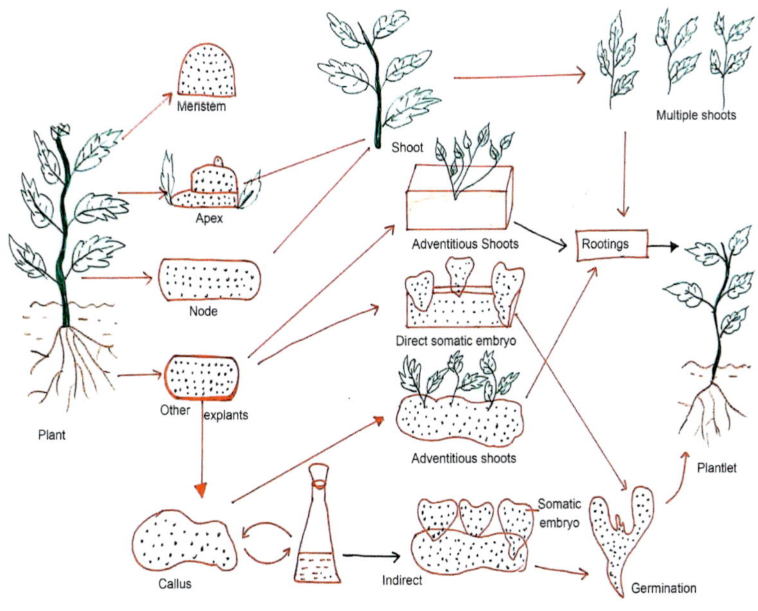

Figure 9.2. *Regeneration of a plant through plant tissue culture.*

Source: Made by the author.

- o There are different varieties of seeds plants such as trees, herbs, grasses which exhibit the basic morphological units such as root, stem and leaves.
- o There are various stages and differences in cells, tissues and their topography.
- o Parenchyma is the most versatile of all types of tissue.
- o They are capable of division and growth. However, the development of any tissue is characterized by three types of cell

growth: cell division, cell elongation and cell differentiation.

- o The healthy part of the plant should be used for culture purposes.
- o Explant cultures are the cultures of plant material. Any part of a plant may be explant such as young and healthy pieces of stem, leaf, stem hypocotyl, cotyledons, etc.
- o Explant cultures are generally used for induction of callus or regeneration of the plant.

- **Callus Formation:**
  - o The callus develops an infection of microorganisms from wounds due to stimulation by endogenous growth hormones, the auxins and cytokinins.
  - o However, it has been artificially developed by adopting tiss□e culture techniques.
  - o It is governed by the source of the explant, nutritional composition of medium and environmental factors.
  - o Explants of meristematic tissues developed cells more rapidly than thin three- walled and lignified cells of tiss□e.
  - o The callus is formed through three developmental stages: induction, cell division and differentiation.
- **Organogenesis**
  - o Roots, shoots and leaves are the organs that are induced in plant tissue culture.
  - o Since the embryo is an independent structure and does not have a vascular supply, it is not supposed to be the plant organ.
  - o It starts with stimulation caused by the chemicals of the medium, substances carried over from the original explants and endogenous compounds produced by the culture by Tomas and Davey, 1975.
  - o Skoog and co-workers gave the concept of regulation of organogenesis by a balance between cytokinin and auxin.
  - o A high ration of auxins and cytokinin stimulated the formation of the root in tobacco callus, but a low ratio of the same induced shoot formation.
- **Root Culture**
  - o The root is cultured in a liquid medium. It has several advantages over solid media.

- o The techniques of root culture give certain important information's such as nutritional requirements, infection by Rhizobium and nodulation, physiological activities for e.g., production of alkaloids, nicotine, etc., by given Dodds and Roberts, 1985.
- o White in 1934 reported the first time s□ccessf□l organ c□lt□re e.g., potentially unlimited growth of excised tomato roots.
- o Subsequently, the roots of several species of gymnosperms and angiosperms have been successfully cultured.

- **Shoot Culture**
  - o In the shoot culture, the apical meristem is cultured (the region of the shoot apex lying to leaf primordium). It is clearly different from the shoot apex by having shoot apex and a few leaf primordia. The culture of relatively large tips (5–10 mm long sections), this technique is also known as meristem culture, Meristemming, and mericlones (Murashige, 1974).
- **Micropropagation:** Murashige in 1974 described the procedure of development of micropropagation into three different development stages such as:
  - o Establishment of explant aseptically
  - o Multiplication of propagules by repeated subcultures on a specific n□trient medi□m.
  - o Rooting and hardening of plantlets and planting into soil.
- **Cell-Suspension Culture:**
  - o It is prepared by transferring a fragment of callus (500 mg) to the liquid medium (500 ml) and agitating them aseptically to make them cells free.
  - o It is diffic□lt to have a s□spension of single cells. However, the suspension includes single cell, cell aggregates residual inoculum and dead cells (Dodds & Roberts, 1985).
  - o King (1980) has described that a good suspension consists of a high proportion of single cells than a small cluster of cells.
  - o It is more diffic□lt to have a good s□spension than to find o□t the optimum environmental factors for cell separation.
- **Benefits of Cell Culture**
  - o Suspension can be pipetted.
  - o They are less heterogeneous and cell differentiation is less pronounced.

- They can be cultured in volumes up to 1,500 liters.
- They can be subjected to more stringent environmental controls.
- They can be manipulated for the production of natural products by feeding precursors by Kurz and Constable, 1979.

- **Somatic Embryogenesis**
  - It can be initiated in two ways.
  - Inducing embryogenic cells within the preformed callus.
  - Directly from the pre-embryonic determined cell (without callus).
  - Which are ready to differentiate into embryoids by Sharp et al., 1980.
  - In the first case, embryoids are initiated in call□s from s□perficial cell aggregates where cells contain a large vac□ole, dense cytoplasm, large starch granules, and nucleus by McWilliam et al., 1974.
  - Two nutritional media of different compositions are required to obtain embryoids.
  - The first medi□m contains a□xins to initiate embryogenic cells.
  - The second medium lacks auxin or a reduced level of auxin is needed for subsequent development of the embryonic cells into embryoids and plantlets.
  - In both cases, a reduced amount of nitrogen is required (Ammirato, 1983).
  - The embryogenic cells pass through 3 different stages such as globular, heart-shaped, and torpedo-shaped to form embryoids.
  - These embryoids can be separated and isolated mechanically by using glass beads. When embryoids reach the torpedo stage, they are transferred to a filter paper bridge and s□crose on which Whatman no.1.
  - Filter paper is placed to make a bridge, as described by Dodds and Roberts, 1985.
- **Somaclonal Variation**
  - The genetic heterogeneity of cells in a population represents a continuity of genotypes, whereas phenotypically the population is represented as a discrete sum of subclones.
  - After cloning a single cell from the population of the strain

of *Dioscorea deltoidea*, there developed a subpopulation that differed in their growth rate and sapogenin content.

- o The mass occurrence of somaclonal variants, increase in resistance, productivity, and vital force of the plant has been explained by Carlson in 1983.
- o This would be due to the dominance of nonlethal mutations that lead to heterozygosity with a wild-type allele, which is thus phenotypically expressed as hybrid heterosis.
- o There are some somaclonal variations in plants such as *Brassica spp., Nicotiana tabacum, Oryza sativa, Saccharum officinarum,* and *Zea mays.*

- **Protoplast Culture**
  - o A protoplast is the biologically active and most significant material of cells.
  - o When the cell wall is mechanically or enzymatically removed, the isolated protoplast is known as a naked plant cell.
  - o The plant cell wall acts as a physical barrier and protects the cytoplasm from microbial invasion and environmental stress.
  - o It consists of a complex mixture of cellulose, hemicellulose, pectin, lignin, lipids, proteins, etc.
  - o For the dissolution of different components of the cell wall, it is essential to have the respective enzymes. There are a number of stages given below:
- **Isolation of Protoplasts**
  - Surface sterilization of samples (Leaf)
  - Rinsing in suitable osmoticum
  - Plasmolysis of cells
  - Peeling of the lower epidermis
  - Isolation and p□rification of protoplasts
- **Regeneration:** The protoplast regenerates a cell wall, undergo cell division and form a callus.
  - The callus can also be subcultured.
  - Embryogenesis begins from the callus when it is placed on a nutrient medium lacking mannitol and auxin.
  - The embryo developed into seedlings and finally, mat□re plants, shown in Figure 9.3.

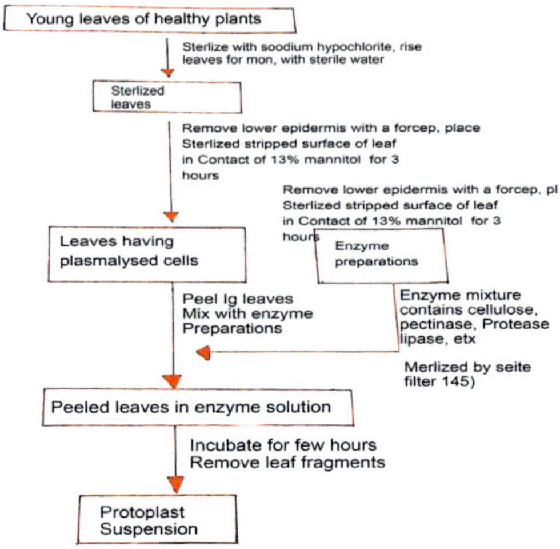

Figure 9.3. *Isolation of protoplasts.*

Source: Made by the author.

9.5. PROTOPLAST FUSION AND SOMATIC HYBRIDIZATION

- **Schieder in 1982 Discussed Four Major Aspects of Protoplast Fusion**
  - Production of fertile amphidiploid somatic hybrids of sexually incompatible species is achieved. The induced fusion of protoplasts from two genetically different lines of species must result in a variety of homo as well as heterokaryotic fusion products.
  - The selection of a few somatic hybrid colonies from the mixed population of regenerated protoplasts is a key step in a successful somatic hybridization technique.
  - Production of heterozygous lines within one plant species which normally will be propagated only vegetatively, e.g., potato.
  - Transfer of only a part of genetic information from one species to another using the phenomenon of chromosome elimination.
  - The transfer of cytoplasmic genetic information from one to a second line or species. It has been possible to transfer useful genes, e.g., nif genes, disease resistance genes, rapid

growth genes from one species to another, thereby widening the genetic base for plant breeding, as shown in Figure 9.4.

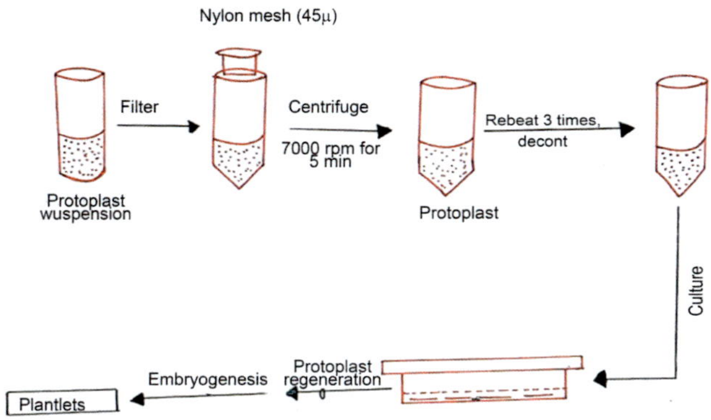

Figure 9.4. *Purification, culture and regeneration of protoplasts.*

Source: Made by the author.

9.5.1. Fusion Product

- The fusion of the cytoplasms of two protoplasts results in the fusion of cytoplasms. The nuclei of two protoplasts may or may not fuse together even after the fusion of cytoplasms.
- The binucleate cells are known as heterokaryon or heterocyte when nuclei are fused. The cells are known as hybrid or synkaryocyte when only cytoplasm fuses and genetic information from one of the two nuclei is lost, known as cybrid, i.e., cytoplasmic hybrid or heteroplasts (Doods & Roberts, 1985), as shown in Figure 9.5.

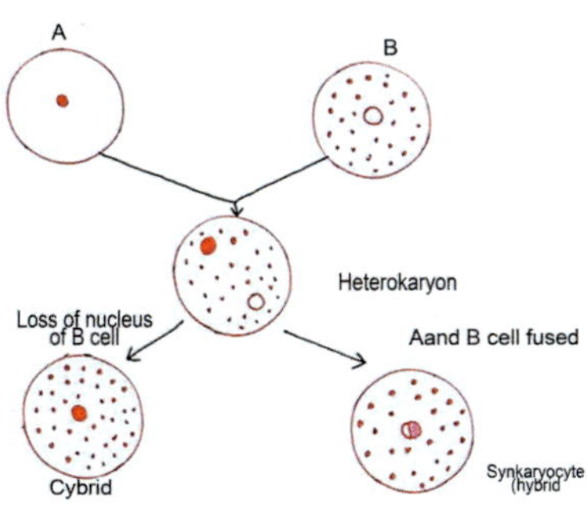

Figure 9.5. *Cybrid/hybrid productions.*

Source: Made by the author.

9.5.2. Methods of Somatic Hybridization

- Isolation of protoplasts from suitable plants.
- Mixing of protoplasts in a centrifuge tube containing fusigenic chemicals, i.e., chemicals promoting protoplast fusion such as polyethylene glycol (PEG) 20% W/V, sodium nitrate ($NaNO_3$); maintenance of high pH of 10.5 and temperature at 37°C.
- Wall regeneration by heterokaryotic cells.
- Fusion of nuclei of heterokaryon to produce hybrid cells.
- Plating and production of colonies of hybrid cells.
- Selection of hybrids, subculture, and induction of organogenesis in the hybrid colonies.
- Transfer of mature plants from the regenerated callus.

9.5.3. Selection of Somatic Hybrids and Cybrids

- After the fusion, protoplasts of media regenerate cell walls and undergo mitosis.
- This results in a mixed population of parent cells, homokaryotic fusion products and hybrids.
- Hybrid cells should be differentiated from other cells.
- There are various selection methods used for the selection of fusion products. Selection methods are dependent on:
  - Physical properties of fused cells.
  - Biological properties of fused cells.
  - Biological properties of colonies from fused cells.
  - The somatic hybrids cannot be identified.
- For the first time, P.S. Carlson and co-workers prod□ced the first interspecies somatic hybrids between *Nicotiana glauca* and *N. langsdorfii* in 1972.
- The somatic hybridization experiment's selection procedure includes the f□ll chlorophyll-deficient (non-green) chloroplasts of one parent with the protoplasts of the other parent.
- This helps in the visual selection of heterokaryon under a microscope.

9.5.4. Anther and Pollen Culture

- A male reproductive organ is haploid in chromosome numbers.
- As a result of microsporogenesis, tetrads of microspores are formed from a single spore mother cell.

- They are known as pollen grains after release from tetrads by Bhojwani and Bhatnagar in 1974. Haploid plants are very useful in the direct screening of recessive mutations because in diploid or polyploid screening of recessive mutations is not possible. The development of homozygous diploid plants following chromosome doubling of haploid plant cells is shown in Figure 9.6.
- In China, the most widely grown wheat is a doubled haploid produced through homozygous diploid lines.
- Anther cultures of rice are also successfully grown.
- Haploid plants have been produced in tobacco, wheat, and rice through pollen culture.

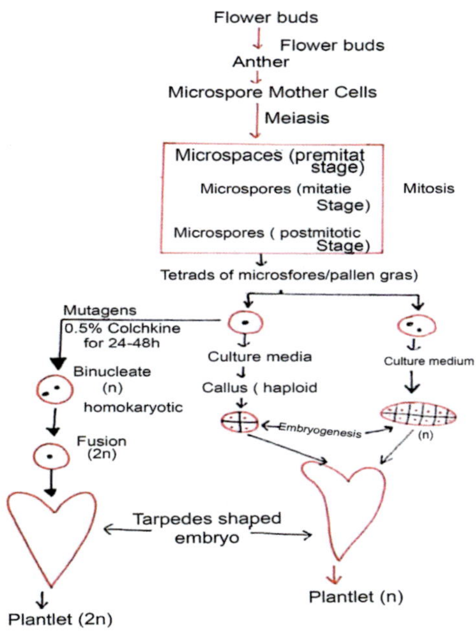

Figure 9.6. *Anther culture.*

Source: Made by the author.

- More than 247 plant species and hybrids belonging to 38 genera and 34 families of dicots and monocots have been regenerated using the anther culture technique.
- They include economically important crops and trees such as rice, wheat, coconut, rubber trees, etc., by Maheswari et al., 1983.

9.5.5. Culturing Techniques

- Anthers are s□perficially sterilized and washed with do□ble distilled sterile water.
- They are excised from the flower b□ds and their proper developmental stages are determined under a microscope.
- On confirmation of stages s□ch as (a) anthers are directly transferred onto nutrient agar or liquid medium where induction of embryogenesis occurs, (b) the pollen grains are aseptically removed from the anthers and cultured on liquid medium (Reinert and Bajaj, 1977).
- Methods of anther culture are given below.

9.5.5.1. In Vitro Androgenesis

- *In vitro* androgenesis is the formation of a sporophyte from the male gametophyte on an artificial medi□m.
- It is most commonly found in families *Solanaceae and Poaceae* (Graminae).
- The success of *in vitro* androgenesis is based on adjustment of the development stage of pollen, minerals of culture medium and growth regulators as well as thermal shock or other treatments.
- Pollen grains are isolated from excised anthers and extracted in a beaker.
- Well-collected pollens are washed properly, centrifuged, and decanted. They are inoculated in a liquid medium, then subcultured in a solid MS medium.
- In culture, pollen grains can be induced to produce callus or embryos from which whole plants are regenerated in one month.
- This technique is most successful in Brassica, Datura, petunia, etc. Mainly there are two ways of *in vitro* morphogenesis of pollen grains: direct and indirect.

9.5.5.2. Direct Androgenesis

- It is also called pollen-derived embryogenesis. Here, pollen directly acts as a zygote and therefore passes through various embryogenic stages similar to zygotic embryogenesis.
- Pollen grains reach the globular stage of the embryo, then the plants.
- Direct androgenesis is very common in many plants of the family Solanaceae and Brassicaceae, as discussed by Prakash and Giles in 1987.

9.5.5.3. Indirect Androgenesis

- In indirect androgenesis, the pollen grains, instead of normal embryogenesis, divide erratically to develop a callus. Indirect androgenesis has been found in barley, wheat, Vitis, Coffea, etc.
- The possibility of pollen morphology from the dividing pollen varies.
- The haploid callus, embryo, or plantlets may originate from the continued division of the vegetative cell of the pollen, multiple divisions of generative cells, division products of both generative and vegetative cells as described by Sangwan in 1981.

9.6. MENTOR POLLEN TECHNOLOGY

- Pollination is a natural phenomenon that occurs in many plant species.
- However, there are certain plants where pollen grains remain viable but unable to set seed on their own pistil, i.e., they are self-incompatible.
- Similarly, in several other plants, besides being viable, they even fail to germinate on stigmas of another species, i.e., they are interspecific incompatible.
- After the birth of recent mentor pollen technology, the incompatibility problem of pollen grains can be overcome by altering the fertilization ability.
- The pollen which has been purposefully treated is called mentor pollen.
- Hence, the mentor pollen is defined as the compatible pollen which has been treated in many ways to inhibit its fertilization ability and retain its power to stimulate incompatible pollen to accomplish fertilization.
- The mentor pollen itself may or may not germinate on stigmas.
- As a result of this treatment, hybridization between closely related species or genera and pollination in cross -pollinating plants could be achieved.

9.7. EMBRYO CULTURE

- Embryo culture has also been done for the production of haploid plants. It is used for the recovery of plants from distinct crosses.

- It is useful when embryos fail to develop due to the generation of embryonic tissues.
- It is being used extensively in the extraction of haploid barley *Hordeum vulgare* from the crosses *H. vulgare* × *H. bulbosum*. Das and Barman (1992) developed the method of regeneration of tea shoots from embryo callus.
- It produced somatic embryoids within 8 weeks of culture in the second medium which differentiated into buds after 2 weeks. Several shoots with 4–6 leaves developed after 16 weeks of culture.

9.7.1. Embryo Rescue

- Viable hybrids are produced as a result of sexual crosses between two varieties of the species. However, if sexual crosses are done between the species of the same genus or between two different genera, production of hybrid is rather difficult because of several barriers arising either during pollination, fertilization, or embryogenesis.
- It has been observed that in some cases, in spite of successful pollination and fertilization, the embryo does not develop.
- This is due to inherent deficiencies or incompatibility between the developing embryo or endosperms.
- In such cases, immature embryos are dissected out from the fruit (seed) and grown artificially on a medium which differentiates into shoots, roots, and plantlets.
- This technique of growing immature embryos is termed as "embryo rescue."
- Embryo rescue technique is very useful in wide hybridization, complete growth of the embryo in a plant, breaking dormancy of certain seeds where the dormancy period is very long. By using the embryo rescue technique, wild varieties can be crossed with cultivars.
- Compared with cultivars, the wild species have greater resistance to pests and pathogens and produce grains of better quality.

9.7.2. Triploid Production

- The triploid primary endosperm (fusion of the second male gamete with two polar nuclei).
- Some examples of triploid plants raised from endosperm cultures such as Rice (*Oryza sativa*), Maize (*Zea mays*), and Barley (*Hordeum vulgare*).

- The triploids of poplar (*Populus tremuloides*) have better quality pulpwood. Therefore, it is important to the forest industry.

9.7.3. Protoplast Fusion in Fungi

- The protoplast fusion in fungi is the improvement of strains to be used for commercial purposes.
- For example, two strains of *Cephalosporium acremonium* were crossed and *Cephalosporin C* was improved.
- Attempts to improve the citric acid production in *A. Niger* were unsuccessful (Peberdy, 1989).
- **Protoplast Techniques Are Applied in the Following Four Major Areas to Achieve the Following Crosses**
    - Between different species and genera.
    - Between apparently incompatible strains of the same species.
    - Between different strains or isolates of a species primarily for breeding.
    - Between isogenic strains which may provide an opportunity for genetic mapping.
- **Intraspecific Protoplast Fusion**
    - It is the cross between the same species in an individual which involves isogenic strains or non-isogenic ones, e.g., *Absidia glauca, Pleurotus ostreatus, Fusarium graminearum*, and *Treesi,* etc.
- **Interspecific Protoplast Fusion**
    - It is the cross between two different species. It is important in areas where new products are to be produced, e.g., *S. cerevisiae* × *S. fermentali, Pleurotus ostreatus* × *P. florida*, etc.
- **Application in Agriculture**
    - There are a number of applications in agriculture which has solved the problem of incompatibility of plants and scope of production of new verities within a short period of time.
    - Improvement of Hybrids
    - Production of Encapsulated Seeds
    - Production of Diseases Resistant Plants
    - Transfer of the *nif* gene to Eukaryotes

9.8. BIOETHICS IN PLANT GENETIC ENGINEERING

There are new varieties of production of a plant due to mutation. Cultivation of genetically modified crops by farmers is increasing fast throughout the world. It will support healthcare and industry and provide food, feed and fiber security on a global basis. The major concern about genetically modified plants or crops foods is given below such as:

- Genetically modified crops may bring abo□t changes in evol□tionary patterns.
- Effect of genetically modified crops on biodiversity and the environment.
- Transfer of transgene from genetically modified food to pathogenic microbes.
- Risk of change in fundamental vegetable nature of plants.
- Effect of genetically modified on non-target and beneficial insects and microbes.
- Pollen transfer from genetically modified plants.
- The risk of transfer of allergies (Table 9.1).

Table 9.1. *Some Important Contributions through Scientists*

Year	Author
1902	Plant Tissue culture (Haberlandt)
1904	Embryo culture of selected crucifers (Hannig)
1922	Asymbiotic germination of orchid seeds *in vitro* (Knudson)
1922	*In vitro* culture of root tips (Robbins)
1925	Use of embryo c□lt□re techniq□e in interspecific crosses of Linum (Laibach).
1934	Successful culture of tomato roots (White).
1939	Successful establishment of continuously growing callus cultures (Gautheret, Nobecourt and White).
1941	Use of coconut milk containing a cell division factor for the first time in Dat□ra (Van Overbeek).
1944	*In vitro* adventitious shoot formation in tobacco (Skoog).

1946	Raising whole plants of *Lupinus and Tropaeolum* by shoot tip culture (Ball).
1952	Use of meristem culture to obtain virus- free dahlias (Morel and Martin).
1952	First application of micrografting (Morel and Martin).
1952	Got the US Patent on their claim that plant cells could be grown in a liquid nutrient medium and had the potential to produce useful compounds (Routien and Nickel of Pfizer and Co. Inc., New York).
1953	Production of haploid callus of the gymnosperm *Ginkgo biloba* from pollen (Tulecke).
1954	First plant from a single cell (Muir et al.,).
1955	Discovery of kinetin a cell division hormone (Miller et al.,).
1957	Discovery of the regulation of organ formation by changing the ratio of auxin: cytokinin (Skoog and Miller).
1958	Regeneration of somatic embryos *in vitro* from the nucellus of citrus ovules (Maheshwari and Rangaswamy).
1959	Regeneration of embryos from callus clumps and cell suspensions of Daucus carota (Reinert, Steward).
1959	P□blication of the first handbook on plant tiss□e c□lture (Gautheret).
1960	First successful test tube fertilization in *Papaver rheas* (Kanta).
1960	Use of the microculture method for growing single cells in hanging drops in a conditioned medium (Jones et al.).
1960	Enzymatic degradation of cell walls to obtain a large number of protoplasts (Cocking).
1960	Filtration of cell suspensions and isolation of single cells by plating (Bergmann).
1962	Development of Murashige and Skoog nutrition medium (Murashige and Skoog).

1964	Prod□ction of first haploid embryos from pollen grains of *Datura innoxia* (Guha and Maheshwari).
1967	Androgenic haploid plants of *Nicotiana* were produced (Bourgin and Nitsch).
1968	Restriction endonuclease term coined to a class of enzymes involved in cleaving DNA (Meselson and Yuan).
1970	Selection of biochemical mutants *in vitro* by the use of tissue culture derived variation (Carlson).
1970	The first achievement of protoplast f□sion (Power et al.).
1970	Discovery of first restriction endon□clease from *Haemophillus influenzae* It was later p□rified and named HindII (Kelly and Smith).
1970	In certain cancer -causing animal virus genetic information flow in reverse form. Discovery of reverse transcriptase (Temin).
1971	The first restriction map □sing HindII enzyme to c□t circ□lar DNA of SV 40 into 11 specific fragments was prepared (Nathans).
1971	Regeneration of first plants from protoplasts (Takebe et al.)
1972	Interspecific hybridization thro□gh protoplast f□sion in two species of *Nicotiana* (Carlson et al.).
1972	First recombinant DNA molecule produced using restriction enzymes (Jackson et al.).
1972	Joining of two restriction fragments regardless of their origin produced by the same restriction enzyme by the action DNA ligase (Mertz and Davis).
1974	Ti plasmid is the tumor- inducing principle of *Agrobacterium* (Zaenen et al., Larebeke et al.,).
1975	Positive *in vitro* selection of maize callus cultures resistant to T toxin of *Helminthosporium maydis* (Gengenbach and Green).
1975	High resolution 2-Dimensional gel electrophoresis procedure developed that led to the development of proteomics (O'Farrel).

1976	Shoot initiation from cryopreserved shoot apices of carnation (Seibert).
1977	Successful integration of the Ti plasmid DNA from *A. tumefaciens* in plants (Chilton et al.,).
1977	Gene sequencing based on the degradation of the DNA chain (Maxam and Gilbert).
1977	Discovery of split genes (Sharp and Roberts)
1978	Somatic hybridization of tomato and potato resulting in potato (Melchers et al.).
1979	A co-cultivation procedure developed for the transformation of plant protoplasts with *Agrobacterium* (Marton et al.,)
1981	Somaclonal variation (Larkin and Scowcroft).
1982	Incorporation of naked DNA by protoplasts resulting in the transformation with isolated DNA (Krens et al.)
1984	Transformation of tobacco with *Agrobacterium* transgenic plants developed (De Block et al., Horsch et al.).
1986	A genetic fingerprinting techniq□e developed for identifying individuals by analyzing polymorphism at the DNA Sequence level (Jeffreys).
1990	Random amplified polymorphic DNA techniq□e developed (Williams et al., Welsh and McClelland).
1991	DNA microarray system using light directed chemical synthesis system developed (Fodor).
1995	DNA fingerprinting by amplified fragment length polymorphism (AFLP) technique developed (Vos et al.,).
1997	Sequencing of *E. coli* genome
2002	The draft sequence of the rice genome (*Oryza sativa L.ssp. japonica*)
2003	The draft sequence of the rice genome (*Oryza sativa L.ssp. indica*) (Yu et al., 2002).
2005	Map- based sequence of the rice genome (International Rice genome sequencing project).

9.9. PLANT GENETIC MATERIAL

- The cell is the basic unit of life. All living organisms, including bacteria, plants, and animals, are built from cells. Generally, cells are very small, with diameters much less than 1 mm, so they are invisible to the naked eye.
- In the simplest cells, bacteria, a cell wall surrounds a very thin fatty acid-containing outer plasma membrane, which in turn surrounds a s□perficially □nstr□ct□red inner region located in the bacterial DNA that carries the genomic information.
- These nucleus-free bacteria and close relatives are known as prokaryotes. In prokaryotes, nuclear material is not separated from the cytoplasm by a discrete membrane.
- These also lack an extensive cellular architecture: membranous organelles are absent, and the genetic material is not enclosed in a distinct structure.
- In all the cells other than bacteria, the inner cellular mass is partitioned into a membrane-bound spherical body called a nucleus and an outer surrounding cytoplasm.
- In the nucleus is located the cellular DNA in the form of coiled rods known as chromosomes.
- Cells that contain a nucleus along with other distinctive membranous organelles, such as mitochondria, vesicles, and Golgi bodies, are referred to as eukaryotes.
- In the cells of eukaryotes, the nuclear membrane separates the genetic material from the cytoplasm, which is then further subdivided by other distinct membranous structures.
- Eukaryotes can be unicellular and multicellular and comprise all macroscopic forms of life such as plants, animals, and fungi.
- Plant cells differ from animal cells due to the presence of an outer cell wall surrounding the cell membrane and chloroplasts, whereas animal cells contain centrosomes, which are involved in cell division and lysosomes.
- The essence of cells is the ability to grow and divide to produce progeny of cells that are capable of generating new cellular molecules and replicating themselves.
- Thus, cells are in effect tiny factories that grow by taking up simple molecular building blocks. Considering the complexity of an organism, one is struck by the relatively small number of organic components used in the formation of substances.

- Cells contain molecules of great variety, but living cells are composed of specific sets of elements, six of which (C, H, N, O, P, and S) make up more than 99% of their weight.
- Certain simple combinations of atoms such as methyl ($-CH_3$), hydroxyl (OH), carboxyl (-COOH), and amino groups ($-NH_2$) occur repeatedly in biological molecules.
- Each such group has distinct chemical and physical properties. Molecules can be broadly grouped into two categories.
- **Small Organic Molecules**
  - These are compounds with a molecular weight in the 100–1000 range containing up to 30 or so carbon atoms.
  - Molecules of this kind are usually found free in solution in the cytoplasm where they form a pool of intermediates from which large molecules are made.
  - Some of the small organic molecules are sugars, amino acids, and fatty acids. At least 750 different types of small molecules are found.
- **Macromolecules**
  - These are invariably polymeric molecules that are formed by joining together specific types of small molec□les.
  - A large polymeric molecule or more than one polymer associated with large molecules can be referred to as macromolecules.
  - Some of the macromolecules such as carbohydrates, proteins, and nucleic acids are most important. Three simple molecules such as sugar, amino acids, and nucleotides which form macromolec□les have been briefly described.
- **Sugars:** There are three types of sugar molecules which involved different role in organisms.
  - i. Monosaccharide $(CH_2O)_n$
  - ii. Disaccharide $(C_{12}H_{22}O_{11})$
  - iii. Polysaccharide $(C_6H_{10}O_5)_n$
- **Amino Acids**
  - Amino acids are chemically varied but they all contain a carboxyl acid group and an amino group both linked to a single carbon atom.
  - Proteins are long linear polymers of amino acids joined head to tail by a peptide bond between the carboxyl acid group of

one amino acid and the amino group of the next amino acid.

- There are 20 amino acids in proteins, each with a different side chain attached to the α-carbon atom.
- All amino acids contain a nitrogen atom; therefore, the protein represents a class of nitrogenous compounds. The properties of amino acid side chains in aggregates determine the properties of proteins they constitute.

- **There Are Different Levels of Protein Structure Such As:**
  - **Primary Structure:** The sequence of amino acids linked by peptide bonds along the polypeptide chain is the primary structure.
  - **Secondary Structure:** The structure of a protein in which the polypeptide chain is folded in a regular manner with a helical or pleated structure. Much of the folding is the result of linking carbonyl and amide groups of the peptide chain backbone by hydrogen bonds. In many proteins, the hydrogen bonding prod□ces a reg□lar coiled arrangement called the α-helix.
  - **Tertiary Structure:** The final str□ct□re of a protein in three-dimensional form from a single polypeptide chain. The final confirmation of a protein from a single polypeptide which is bent or tightly folded and compact is determined by a variety of interactions including hydrogen bonds, salt bonds, hydrophobic or nonpolar bonds, and van der Waals forces. It is by virtue of their tertiary structure that proteins adopt a globular shape.
  - **Quaternary Structure:** It contains more than a single polypeptide chain to indicate the spatial relationships among the separate chains or subunits. It means how two or more chains are arranged in relation to each other. Most large proteins contain two or more polypeptide chains, and in between, there are no covalent linkages. Proteins with two or more polypeptide chains are known as oligomeric proteins. A well-known example of an oligomeric protein is hemoglobin.

9.10. NUCLEOTIDES

Nucleotides have 3 basic parts:

1. A 5-carbon sugar.
2. A phosphate group.

3. A nitrogenous (nitrogen-containing) base, as shown in Figure 9.7.

9.10.1. Nucleic Acids

- Nucleic acids contain instructions that cells need to carry out all the functions of life.
- Nucleic acid – very large organic molecules made up of carbon, oxygen, hydrogen, nitrogen, and phosphorus (Figure 9.8).

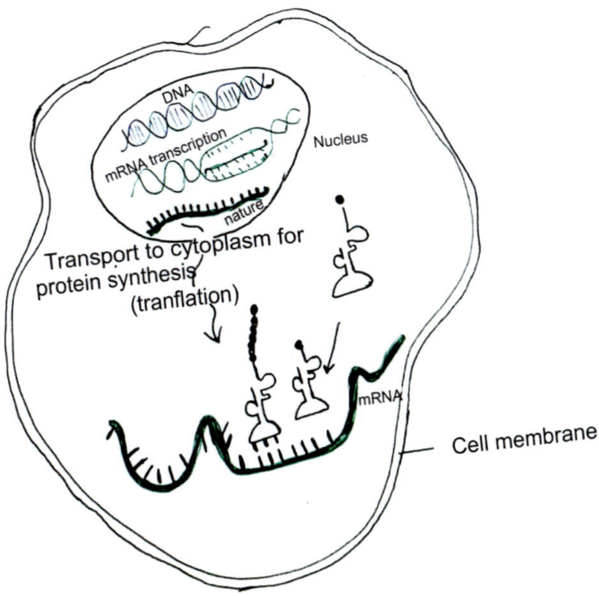

Figure 9.7. *Cell structure.*

Source: Made by the author.

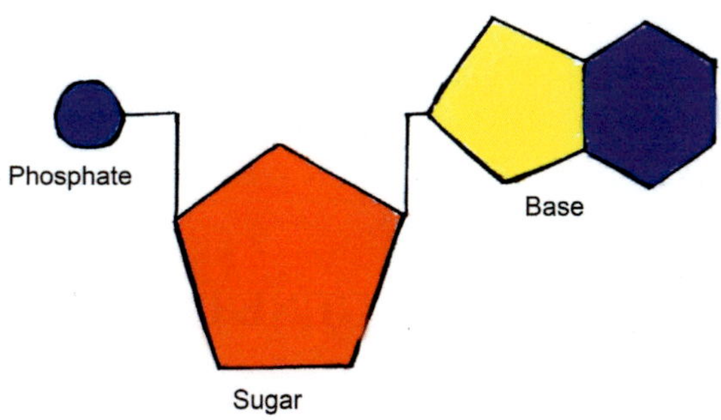

Figure 9.8. *Nucleotides.*

Source: Made by the author.

9.10.2. Types of Nucleic Acid

- DNA (deoxyribonucleic acid)- the genetic material that carries information about an organism. It is passed from parent to offspring.
- DNA directs all cellular activities and is only found in the nucleus of a cell.
- RNA (ribonucleic acid) – a nucleic acid found in the nucleus and cytoplasm of cells that plays an important role in the production of proteins; *see* Figure 9.9.

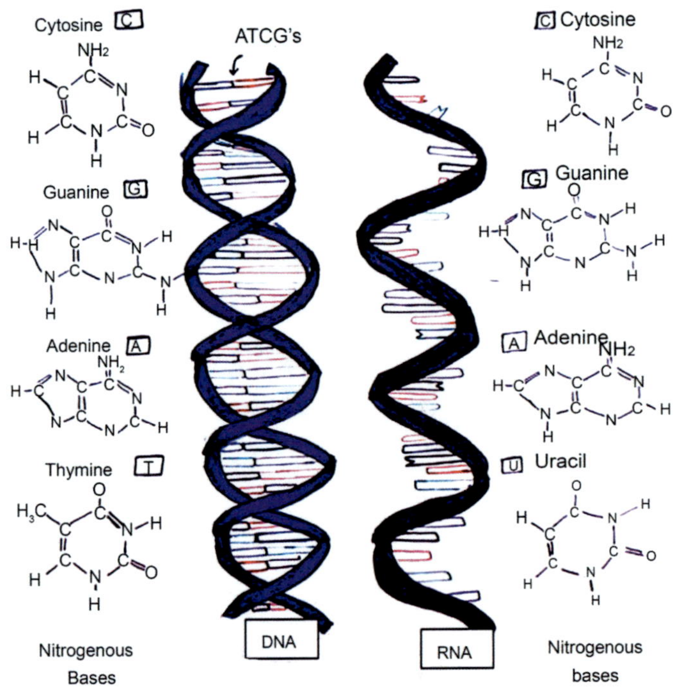

Figure 9.9. *DNA and RNA sugars.*

Source: Made by the author.

9.11. NUCLEOTIDES AND GENETIC ORGANIZATION

Nucleic acids are made up of nucleotides.

- Nucleotides have three basic parts:

1. A 5-carbon sugar.

2. A phosphate group.
3. A nitrogenous (nitrogen-containing) base (Table 9.2).

Table 9.2. *Levels of genetic organization*

Nucleotide	Gene	DNA	Chromosomes	Genome
The smallest building block of gene	Code for traits	Made of many genes	Tightly coiled strands of DNA	The entire genetic make-up of an organism

- **Composition of DNA:** Major parts of a DNA nucleotide:
  1. A 5-carbon sugar called deoxyribose.
  2. Phosphates
  3. Four different nitrogenous bases (rungs of the ladder); *see* Figure 9.10:
     i. Adenine (A);
     ii. Thymine (T);
     iii. Guanine (G); and
     iv. Cytosine (C).

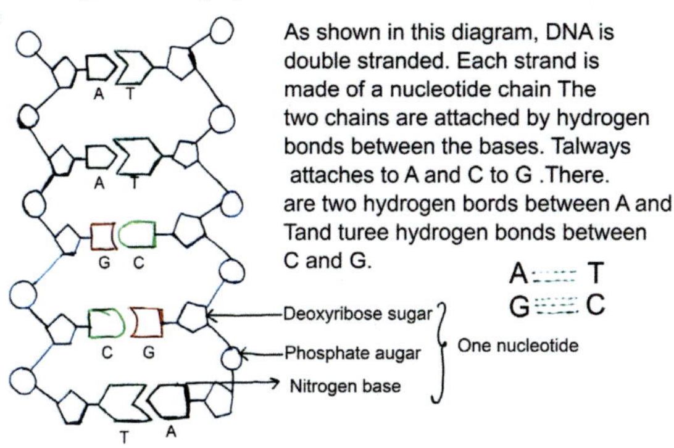

Figure 9.10. *Nucleotide chain.*

Source: Made by the author.

9.11.1. Pairing the Nitrogenous Bases in DNA

- Adenine always pairs (matches up) with thymine and guanine always pairs (is complementary to) with cytosine.
- These nitrogenous bases make up four different possible nucleotides in DNA.
- Many nucleotides (basic unit of nucleic acids) are placed together to make one DNA molecule; *see* Figures 9.9 and 9.10.

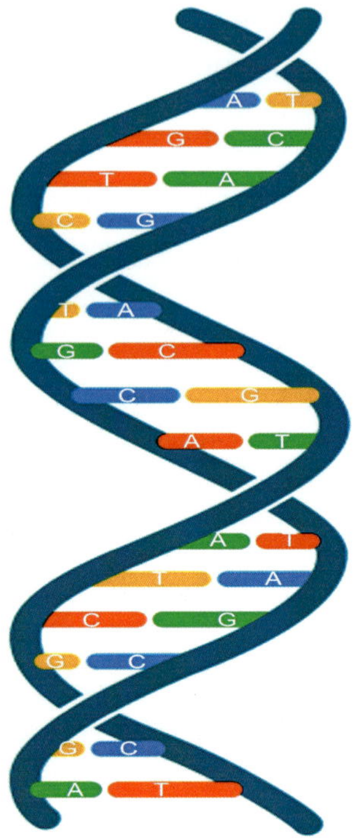

Figure 9.11. *DNA structure.*

Source: Made by the author.

9.11.2. Structure of DNA

- James Watson & Francis Crick – American biologist & British physicist b□ilt the first
- accurate structural model of DNA (Nobel Prize in 1962).
- Watson & Crick's model of DNA is a double helix (twisted ladder), in which 2 strands were wound around each other.
- The outside/sides of the ladder are made of deoxyribose sugar and phosphates.
- The rungs of the ladder are the nitrogenous bases (connect to deoxyribose sugar in the outside/side of the ladder) Figure 9.11.
- Identifying Base Pairs
  - DNA strands are *complementary* to one another (match up with each other). A & T and G & C.
  - Given the following seq□ences of DNA, find the complementary strands:

AGT TCA	CAG GTC
CTT GAA	AGC TCG

- DNA Replication
  - Before cells divide, they make copies of their DNA. DNA replication ensures that each daughter cell will have all of the genetic information it needs to carry out its activities.
  - DNA replication is semiconservative, meaning that when DNA copies, each new strand has ½ of the original strand.
- Major Steps in DNA Replication
  - DNA unwinds (untwists) exposing the nitrogenous bases.
  - DNA unzips – weak hydrogen bonds between nitrogenous bases break.
  - Hydrogen bonds are the weakest type of bond that occurs in molecules.
  - New complementary strands form as nitrogenous bases floating in the nucleus of the cell pair up with the bases on each half of the DNA molecule.
  - Two new DNA molecules that are exactly alike are formed. Both new strands are identical to the original strand of DNA.
- The Function of Genes in DNA
  - Proteins determine the size, shape, and many other traits of an organism, for example: eye color, hair color, type of hairline, etc.
  - Each gene in a DNA molecule codes for the production of a particular protein.
  - Gene – a small portion of the DNA molecule that codes for a particular trait/protein.
- Order of Nitrogenous Bases
  - The order of the nitrogenous bases along a gene forms a genetic code that specifies what type of protein will be produced. The order of the nitrogenous bases tells the cell which amino acids to put together to make a protein.
  - If one amino acid is changed or out of place, a different protein is made. Since DNA cannot leave the nucleus to make proteins at the ribosomes, RNA directs protein synthesis; *see* Figures 9.12 and 9.13.

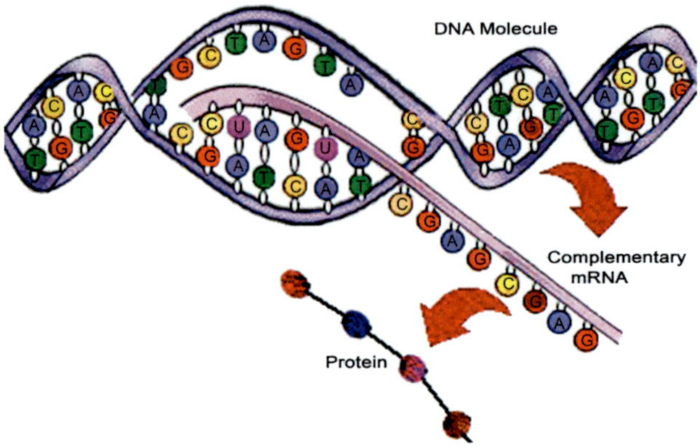

Figure 9.12. *Nitrogenous bases involved to make protein.*

Source: Made by the author.

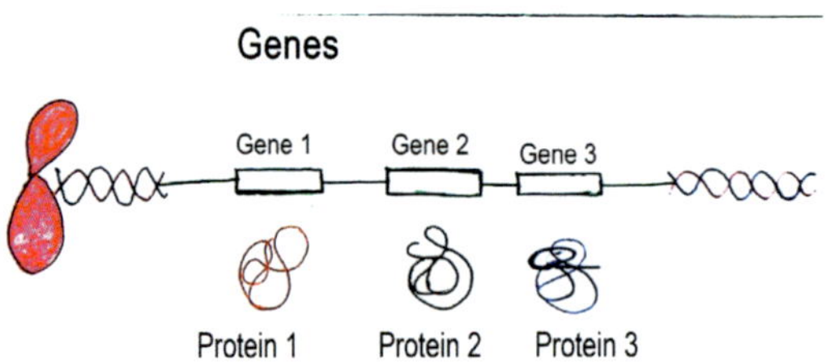

Figure 9.13. *Sequence of genes.*

Source: Made by the author.

- **It All Started with Mendel:** The beginning of our thinking of the possibility of genetic material begins with Mendel. He described the genetics of pea crosses, although he did not know what genetic material he was referring to. He used the term "factors" to describe the things that gave his pea plants their physical characteristics. While Mendel's work provided a logical description of heredity, it was largely ignored for many decades. His work was finally reproduced in the early 1900s and was thereafter regarded as a fundamental contribution to genetics.
- **Dots and Slot Blots**
  - Another variety of Southern or Northern is dot blots or slot blots.
  - In dot blots, cloned or pure extracted DNAs to be tested are spotted adjacent to each other on a nitrocellulose membrane.

- DNA blots thus produced arc immobilized and denatured so that they are bound on the membrane as single-stranded DNA blots.
- "The membrane is then hybridized with a radioactively labeled probe (DNA, or RNA). The dot representing the sequence related to the probe will light up autoradiography.
- The intensity of the dot will indicate the relative concentration of sequences related to the probe in the DNA sample used for a particular dot.
- No steps involving digestion with an enzyme concentration of a sequence related to the probe in the DNA electrophoresis or transfer from the gel to the membrane are needed in this technique."
- Therefore, it is considered much more convenient, when o restriction sites are needed to be studied.
- Sometimes, an apparatus is used for placing spots on the membrane through slots made in this equipment.
- The spots made thus are in the form of oblong slots rather than round blots.
- These slots are used just like dot blots and are described as slot blots.

• Applications of Molecular Probes
  - Molecular probes can be used for both basic and applied studies in the field of molecular biology and biotechnology. These uses will be discussed in detail in the different subsequent chapters of this book as appropriate but will be described briefly in this section.

9.12. USE OF MOLECULAR PROBES IN RESTRICTION FRAGMENT LENGTH POLYMORPHISMS (RFLPS) AND RELATED ANALYSES

1. **What are RFLPs?:** When genomic DNAs from each of several individuals belonging to one or more species are digested separately with restriction enzymes, electrophoresed, blotted on a membrane, and probed with a labeled DNA clone, polymorphism the hybridization pattern is sometimes revealed and attributed to sequence differences between the individuals. Such variation has

been termed as "restriction fragment length polymorphism" (RFLP). A variation in one DNA fragment obtained with a specific enzyme is treated as RFLP.

2. **Detection of RFLPs:** An RFLP can be demonstrated using the following steps:
   - Extract and purify DNA from several individuals which may differ in some respects among themselves.
   - Digest with a restriction endonuclease.
   - Subject each "DNA digest to gel electrophoresis separately but on the same gel slab.
   - The DNA fragments of different lengths resulting from digestion can be separated by gel electrophoresis, but the differences in the distribution patterns of fragments in several DNA samples due to digestion by several endonucleases cannot be detected directly. This is because the number of fragments is large and the range in size is rather continuous, such that they form a continuous smear on the gel. Electrophoresis is, therefore, followed by the following steps.
   - Fragments are transferred from the gel to nitrocell□lose filters using the technique of Southern blotting.
   - The filters with small fragments, f□lly or partially homologo□s to the probe, can be detected by autoradiography after hybridization.

Since repetitive DNA may hybridize with many fragments, they will give a smear on autoradiography. Therefore, unique DNA sequences are generally used as probes for detecting RFLPs. The DNA fragments hybridizing with the probe can be visualized by autoradiography and will be called RFLPs at the phenotypic level, where RFLPs are detected by using four plants (which differ by point mutation as well as by insertions) and three enzymes individually and in combination.

3. **RFLPs for Evolutionary Studies:** The restriction fragment length polymorphisms (RFLPs) can be studied in a set of related species, □sing a random or a specific DNA probe. The similarities and differences can be used to infer phylogenetic relationships. This has actually been done in a number of cases, both in plants and in animals.

4. **RFLP Maps and Linkage of RFLPs with Specific Genes:** RFLPs have been used to prepare chromosome maps in humans, mice, fruit fly, and in Mendelian markers for genetic mapping is sometimes

limited due to the non-availability of mutants.

- The list of markers can be extended by using molecular markers, which are examined in the form of RFLPs in plants including maize, tomato, lettuce, and rice.
- The use of Mendelian markers for genetic mapping is sometimes limited due to the non-availability of mutants.
- The list of markers can be extended by using molecular markers, which may be examined in the form of RFLPs.
- "Once a large number of RFLPs are available in a species, the parents, F, and F2's can be used to study their inheritance and linkage relationship and genetic linkage maps can be prepared.
- They have been assigned to specific chromosomes □sing monosomies in maize and efforts are being madc to rclatc them with morphological and economic traits so that they can be used for practical plant breeding.
- A difference in restriction maps between two individuals or an RFLP can be used as a genetic marker in much the same way as any other phenotypic marker. The polymorphism at the molecular level may be due to the presence or absence of a restriction site.
- It shows how the presence of an additional site will produce two fragments and its absence will produce only one fragment.
- Although there may be a linkage between a molecular marker and a phenotypic trait, the change in the molecular map (restriction map) may not affect the phenotype.
- In order to study the linkage relation between a molecular marker and the DNA present in the parents that differ morphologically as well as for molecular markers, we may examine the restriction fragment patterns and the phenotypes by making crosses between two such parents. We can estimate the recombination frequencies between molecular markers and phenotypic markers using a test cross.
- We used a hypothetical example showing the linkage of red and white eyes with different molecular markers in *Drosophila*.
- "It can be seen that as a result of recombination in the test cross, 15% of the progeny show the association of the red eye with a molecular marker associated with the white eye in the parent and vice versa.
- Therefore, it is concluded that the restriction marker is 15 map

units away from the marker eye color although it has no causal relation with eye color.

- Such a linkage between the molecular marker and phenotypic marker allows the identification of genetic loci at the molec□lar level.
- A tight linkage of such a molecular marker with disease resistance may also be recommended as a tool for identification of disease- resistant cultivars in many of our crop plants and will be increasingly utilized in the future. Molecular markers are also being used for the selection of plants at the seedling stage for plant breeding purposes."

9.13. USE OF MOLECULAR PROBES IN MOLECULAR CYTOGENETICS

9.13.1. Isolation of Gene or a Related Sequence

This can be facilitated through the use of a probe carrying a known sequence.

For this purpose sequences may be hybridized following any one of the following techniques:

- Southern blots can be prepared (from genomic DNA or cDNA) to isolate DNA corresponding to a band that hybridizes with the probe.
- This DNA can then be used for cloning experiments to get a partial genomic library or cDNA library (see later in this chapter for details).
- The genomic or cDNA library can be screened through colony hybridization or plaque hybridization to identify and isolate the clone carrying the desired sequence.
- Sometimes the step (ii) is not used for complete genomic/cDNA libraries can be prepared and used in this step.
- The identified clone sho□ld carry the gene or seq□ence of interest, that can be multiplied and the sequence or gene retrieved according to a standard procedure.

9.13.2. In Situ Hybridization (ISH) for the Location of Sequence Chromosomes

- A molecular probe can also be used for *in situ* hybridization where chromosomes in mitotic or meiotic preparation are used for hybridization with the probe (without extracting DNA).
- In a number of cases, rye chromosomes in the wheat background have been identified □sing this techniq□e.

- Sat-chromosomes with NORs (nucleolar organizing regions) are recognized using probes for ribosomal DNA as done in many cases in wheat and can be identified using a D genome-specific probe.
- Telomeres have been identified in human eukaryotes using a telomeric sequence as a molecular probe, showing all chromosomes carrying the same sequence.
- In D genome telomere-related species, chromosomes of a genome-specific probe (pASI separated from Aegilops) can be cited where the use of molecular probe in hybridization served a very useful purpose.

9.13.3. Use of Molecular Probes in DNA Fingerprinting

With the help of probes prepared from human minisatellite DNA dispersed throughout the human genome, it has become possible to identify individual humans, as the possibility of any two humans (except identical twins) showing the same pattern of fingerprinting is remote. This aspect is already playing an important role in identifying criminals like murderers and rapists. This subject will also receive detailed treatment.

9.13.3.1. Construction and Screening of Genomic and cDNA Libraries

- In order to isolate one or more related genes from a genome, we prepare a mixture of clones each carrying DNA derived either from the genomic DNA or from cDNA (derived from the mRNA isolated from a specific metabolically active tissue of an organism).
- This mixture may contain thousands of clones, which when derived directly from the genomic DNA are collectively called a genomic library.
- Similarly, when these clones are derived from cDNA, they are collectively called a cDNA library construction and the use of these libraries will be discussed in this section.

9.13.3.2. Genomic Library by Shotgun Experiment

- Cloning an entire genome in the form of a library of random genomic clones (without identifying them) is often called a shotgun experiment.
- "This experiment, genomic DNA is extracted, broken into fragments reasonable size by a restriction endonuclease and then inserted into a cloning vector to generate a population of chimeric vector molecules.

- A set of fragments cloned in this manner is called a genome library.
- Once such a library is available, then clones can be perpetuated indefinitely in a plasmid vector and retrieved whenever needed for a variety of purposes, including identification and isolation of a gene, when a specific probe is available.
- Genomic libraries can be prepared by using a number of restriction endonucleases, one at a time so that fragments of varying sizes having different places in the genome will be available.
- However, this may lead to cuts at inconvenient places, including sites within a gene, so that is having complete genes will be difficult to obtain.
- In order to overcome this difficulty, we use the following strategy in the shotgun experiment:
  – We use restriction endonucleases, which have short (4 bp) recognition sequences, so that such a sequence may be frequently distributed.
  – Conditions are used that give only partial digests, so that a particular restriction site is only occasionally cleaved, and long fragments without having any breaks on recognition sites available within a gene can be easily obtained.
- This technique of shotgun experiment leads to the construction of a random genomic library, in which all fragments have the same ends, thus helping retrieval of a fragment from the vector with the same enzyme cuts.
- The number of fragments representing every sequence of the genome increases with genome size.
- For instance, for a probability level of 99 that all sequences are present in our library of a species, we may need 1,500 cloned fragments for *E. coli*, 4,600 for yeast, 48,000 for *Drosophila melanogaster,* and 800,000 for a mammal like the human being. Libraries reaching these desired limits have been prepared in all these cases.

9.13.3.3. cDNA Library from mRNAs

- Complementary DNA (cDNA) libraries can also be prepared by isolating from tissues that are actively synthesizing proteins, such as roots in plants, ovaries, or reticulocytes in mammals.
- The mRNA is copied into cDNA using reverse transcriptase. The mRNA is used to copy it into cDNA through the use of cDNA molecules, which can be made double-stranded and cloned.

- However, cDNA clones will differ from genomic clones in lacking the introns present in split genes, but the advantage of being capable of being expressed in bacteria, which do not have the machinery to process HnRNA (obtained cDNA clones from split genes) mRNA.

9.13.3.4. Colony (Plaque) Hybridization for the Screening of Libraries

- Once a genomic library or cDNA library is available, we may want to use it for the isolation of a gene sequence.
- This can be achieved by the colony hybridization technique. In this technique, bacterial-transformed chimeric vectors are grown into colonies, which are lysed on nitrocellulose filters.
- Their DNA is denatured in situ and fixed on the filter, which is hybridized with a radioactively labeled probe carrying a sequence related to the gene to be isolated (usually a cloned cDNA for the screening of a genomic library).
- Colonies carrying this sequence will be identified by dark spots after autoradiography so that the original chimeric vector carrying the desired gene sequence can be recovered from one or more colonies in the original master plate and used for further experiments. This technique is described as colony hybridization.
- It is possible that the probe may identify more than one clone or that a gene is fragmented in the library.
- In such a case, one needs to reconstruct the desired sequence using several overlapping sequences available in the library.
- This is a very routine exercise whenever we like to isolate specific DNA sequences from the genome of a species, or from cDNA derived from mRNA of a specific tissue of species.
- Sometimes the library may be available not in the form of bacteria is formed with chimeric DNA molecules but in the form of chimeric particles carrying the cloned segments.
- In such a situation, a bacterial is infected with a mixture of chimeric phage particles (i.e., the library) and a large number of plaques develop overnight.
- These plaques can be treated just like the colonies in colony hybridization to identify and isolate the chimeric phage particle carrying the gene of interest. This technique is then described as plaque hybridization.

9.13.3.5. Chromosome Walking and Characterization of Chromosome Segments

- When a probe is used for the identification of a gene sequence in a genomic library, the probe may hybridize with a number of clones, each carrying a part of a large gene fragment during the preparation of a genomic library.
- If we obtained partial digests (by digesting the DNA only partially) from the genome, different genomes (from a large number of cells) may give fragments which have overlapping sequences, because sites cleaved in different genomes of the same organism will differ randomly.
- Since one of these fragments may have the entire sequence represented in the probe, overlapping sequences may be used to construct the original genome sequence. Identification of fragments with overlapping sequences may be key to the reconstruction or characterization of large chromosome regions.
- This is achieved by the technique popularly described as chromosome walking. The technique of chromosome walking involves the following:
  - From the genomic library, select a clone of interest (identified by a probe) and subclone a small fragment from one end of the clone (there is a technique available to subclone a fragment from the end).
  - The subcloned fragment of the selected clone may be hybridized with other clones in the library, and a second clone hybridizing with the subclone of the first clone is identified due to the presence of an overlapping region.
  - The end of the second clone is then subcloned and used for hybridization with other clones to identify a third clone having an overlapping region with the subcloned end of the second clone.
  - The third clone identified as above is also subcloned and hybridized with clones in the same manner, and the procedure may be continued.
  - A restriction map of each selected clone may be prepared and compared to know the region of overlapping so that identification of a few overlapping restriction sites will amount to walking along the chromosome, or along a long chromosome segment.

- Regions of chromosomes approaching 1000 kb have been mapped following the above technique.
- Restriction maps of entire chromosomes can be prepared in this manner following the technique of chromosome walking.

9.13.3.6. Reverse Genetics and Chromosome Jumping (or Hopping Libraries)

- In most cases, we identify the genes by their protein products, which help in the isolation and cloning of a specific gene.
- However, for many human disorders which are each controlled by a single gene (e.g., H□ntington's disease, cystic fibrosis, D□chenne muscular dystrophy, etc.), the gene product is not known despite intensive investigations.
- S□ch genes are cloned processes often identified primarily by their map position.
- Once the gene is cloned, it can be utilized for getting information about the encoded protein, the gene mutation, and the metabolic defect that may be associated with the disease.
- Since the information about the is coming in the reverse order, this is popularly described as reverse genetics approach involves the following two steps: (i) locate the gene to a particular chromosome through RFLP linkage analysis; (ii) map the position of the gene with respect to molecular marker, both through recombinational analysis (in terms of recombination cM=centiMorgan units) and by physical mapping (in terms of kilobase pairs) with p□lsed-field gel electrophoresis (PFGE) and preparation of restriction maps.
- Even if the distance is 1 cM, which is the limit of resolution, this will be equivalent to 1,000 kb.
- Therefore, if the segment carrying the gene is to be cloned using the linked molecular marker such a segment will be too large to be cloned.
- In order to bring the molecular marker close to the gene of interest, the 'chromosome jumping' approach has been utilized.
- The technique of 'chromosome jumping' is based on the following steps depending upon the distance between the gene and the marker, decide about the distance of jumps' or 'hopsize' (e.g., 100 kb or 200 kb): (i) genomic DNA molecules in the range of selected size (say 80 kb-130 kb in case of 100 kb 'hop size are 16)-240 kb for 'hopsize' of 200 kb) are selected thro□gh p□lsed-field gel elcctrophoresis (ii)

for circularization of DNA segments, ligation been two ends of each long linear DNA molecule was allowed using T4 ligase in the presence of $supF^+$ (iv) DNA circles obtained in (ii) above are digested with *EcoRI;* (v) the vector iCh3A A lac, an amber mutated phage vector (supF is also cut with *EcoRI* and used for cloning mal DNA fragments representing the junctions of the circularized genomic molecules and carrying sup F (vi) the cloned DNA fragments and in(v) above represent the jumping library, which can be plated on the host and screened through the technique of plaque hybridization described earlier.

- The above technique of chromosome jumping will help to narrow the gene and available molecular markers.
- After several cycles, the jumping is followed by cloning the regions that are very close to the desired gene and cloning it. It is thus obvio□s that in the field of 'reverse genetics' □sing the techniq□e of chromosome jumping, genes will be characterized, whose gene products are unknown.
- The gene for cystic fibrosis has recently been isolated □sing this techniq□e. It has also been shown that the 'cystic fibrosis' disease is caused by a single base substitution in this gene. This technique of 'reverse genetics' will become a very active area of research, particularly in human genetics in the coming years.
- The term 'reverse genetics' has, however, been redefined recently. Classical genetics, sometimes described as 'forward genetics,' involves a study that starts from the phenotype and follows through the identification of the gene, and finally concl□des with the isolation and sequencing of the gene.
- In view of this, if 'reverse genetics' has to use steps in the reverse order for a genetic study, then it should start with DNA molecules or clones with unknown effects and should conclude with the determination of the phenotype which it controls.
- This can be achieved by preparing random genomic DNA clones, which can subsequently be utilized for a genetic study in the reverse order. This will then be described as 'reverse genetics.'
- This definition of 'reverse genetics' differs from the earlier conventional meaning of this term, where 'reverse genetics' was used to describe cases of genes with unknown products, which are isolated (with the help of phenotype and its linkage with molecular markers) and then studied in some detail.

9.14. ISOLATION, SEQUENCING AND SYNTHESIS OF GENES

- Molec□lar manip□lation of specific genes is part of biotechnology research.
- These genes, isolated or artificially synthesized before they are manipulated and used or transformation heading to the production of transgenic animals and plants that are discussed respectively, genes are also sometimes sequenced for a better understanding of its structure that needs to be manipulated.
- We, therefore, discuss the gene one of the very popular areas that sho□ld be the isolation, seq□encing, and synthesis of specific genes in this chapter.

9.14.1. Isolation of Genes

- D□ring the last two decades, significant progress has been made in the techniques for isolation of a variety of genes, including those for (ribosomal RNA; (ii) specific protein prod□cts; (iii) phenotypic traits with an unknown product and those for (iv) regulatory functions e.g., promoter genes, etc.
- Different techniques have been used for the isolation of different types of genes and will be briefly described in this section.
- Early attempts for isolation of ribosomal RNA and proteins. As we know, ribosomes consist of ribosomal RNA (rRNA) and proteins. Ribosomal RNA (rRNA) can be of four sizes namely, 28S, 16S, 58S, and 5S (S stands for Svedberg units, a measure of speed with which a molecule sediments on centrifugation).
- This ribosomal RNA makes up 80% of the cellular RNA and is synthesized on ribosomal genes, which could be isolated.
- Isolation of ribosomal genes was considered convenient due to their following three characteristic features (i) availability of homogeneous rRNA; (ii) differences between ribosomal RNA genes and other genes due to relatively high G + C content in rRNA (rRNA has 45–60% G + C while the remaining RNA has only 40% G + C); (iii) ribosomal genes are present in multiple copies, their number within several thousand. In view of the above, ribosomal RNA genes were isolated for the first time (in 1965) in an amphibian named *Xenopus*, by Hugh Wallace and Max L. Birnstiel, working at the University of Edinburgh (UK).
  - o Following steps were involved in this isolation: (i) RNA was

isolated from ribosomes of *Xenopus* and made radioactively labeled, due to its replication in a medium containing tritiated uridine; (ii) ribosomal DNA was isolated by density gradient centrifugation followed by its denaturation (G +C content of rDNA differs from that of bulk DNA and helps in separation by centrif□gation); (iii) single-stranded DNA was fixed on filter paper; (iv) labeled rRNA was added on filter paper carrying single-stranded DNA; (v) DNA-RNA hybridization was allowed to take place (vi) excess labeled RNA was washed (vii) radioactivity was measured and duplex hybrids isolated, which on denaturation, gave single-stranded DNA, which could be made double-stranded.

- Isolation of ribosomal DNA allowed the first characterization of ribosomal genes, which was followed by the isolation of 5S genes by Donald Brown.
- For a detailed account of isolation of ribosomal genes in Xenopus as originally done, readers are referred to an article "Isolation of Genes" in Scientific American (A□g□st 1973).

9.14.2. Isolation of Genes Coding for Specific Proteins

- Isolation of genes for specific proteins became possible only after the discovery of the reverse transcriptase enzyme in 1970.
- This enzyme can be easily used for the synthesis of copy DNA or complementary DNA (cDNA) from mRNA.
- This complementary DNA can then be used for the isolation of the corresponding gene from genomic DNA.
- It is, therefore, obvio□s that for the isolation of a specific gene, techniq□es sho□ld first be available for the isolation of specific mRNA.
- For this p□rpose, antibodies are prod□ced against a specific protein for which the gene is to be isolated. Therefore, isolation of a gene coding for a specific protein involves the following basic steps.
- The p□rification of the protein prod□ct of the gene; (ii) prod□ction of antibodies against this protein product by immunizing animals like rabbits or mice; (iii) precipitation of polysomes engaged in synthesizing the specific protein with the help of the antibody raised in step (ii); (iv) isolation and p□rification of mRNA from the polysome fraction; (v) using mRNA for synthesizing cDNA with the help of reverse transcriptase; (vi) cloning of cDNA thus synthesized for the preparation of a cDNA library; (vii) immunological and

electrophoretic analysis of translation products of cDNA clones to identify the specific cDNA meant for the specific protein in mind; (viii) use of the specific DNA probes selected in step (vii) for the identification and isolation of the gene from genomic DNA through screening a complete or partial genomic library.

- In step (vi) above for the preparation of the cDNA library, cDNA may be cloned in expression vectors (vectors in which the gene can be expressed due to the presence of promoter sequences for RNA polymerase enzyme) like ågtll, which can accept gene insertions into the β-galactosidase gene.
- The chimeric vector will then produce the hybrid protein if correctly expressed. The hybrid protein can be identified using an antibody as above. Plaques belonging to the chimeric vector carrying the desired gene can also be identified by the reaction of the antibody attached to a radioactive protein.
- This procedure was used to clone a number of genes including a gene regulating zein synthesis in maize.
- The cDNA clones may also be used directly for gene manipulation I transformation experiments.
- However, when a cDNA clone is used, it can be described as a synthesis of the gene rather than as isolation of a gene, cDNA has been artificially synthesized, and not isolated from the organism.
- The above technique for isolation of gene has now been successfully utilized both in plants and animals.
- However, the first genes isolated were those which existed in multiple copies and were expressed at high levels in specific tissues, e.g., ovalbumin gene in chicks, globin and immunoglobulin genes in mouse, genes for storage proteins in cereals and legumes, amylase genes in barley, actin genes in some legumes, etc.

9.14.3. Isolation of Genes That are Tissue- Specific in Expression

- It is much easier to isolate genes that are expressed in specific tissues. For instance, genes for storage proteins are expressed only in developing seeds, the ovalbumin gene is expressed in the oviduct, or the globin gene is expressed.
- Such genes can be easily isolated because mRNA extracted from these specific tissues will either exclusively belong to the gene of interest or it will be rich in this species of mRNA.

- Other mRNA molecules in minor quantities can be eliminated since these can be identified through their isolation from tissues where this gene is silent.
- This strategy, which was actually followed by isolation and cloning carrot genes expressed during the development of the somatic embryos, and also for the isolation of several genes for storage proteins in a crop, is outlined.

9.14.4. Isolation of Genes using DNA or RNA Probes

- Specific molecular probes (whether DNA or RNA probes), if available, can be used for the isolation of specific genes.
- These probes may be available either from another species for the same gene or may be artificially synthesized using a part of the amino acid sequence of the protein product of the gene.

9.14.4.1. Use of Heterologous Probes

- The probes obtained from one species are heterologous probes. These have been found to be effective in identifying blots.
- For instance, the gene for chalcone synthase or on Southern clones during colony hybridization or plaque hybridization or on Southern blots synthase has been isolated from *Anuirhinum majus* and *Petunia hybrida* heterologous cDNA probe from parsley.
- Similarly, heterologous probes from maize were used for the isolation of barley genes in a cDNA library and not with the genomic library since the vector (e.g., pGEM-4 blue) can also be used for getting the waxy gene) and Al (aleurone gene).
- These heterologous clones if available have been found to be more sensitive and efficient.

9.14.4.2. Use of cDNA or Synthetic Probes

- We have already discussed the cDNA probes above. However, if a protein purified using the technique of two-dimensional gel electrophoresis is used for 5–15 consecutive amino acids, this information can be used for the synthesis of oligonucleotides (using automated DNA synthesizers).
- These oligonucleotides may then be directly utilized for screening cDNA genomic libraries for isolation of specific genes.
- These can also be used for designing primers for the amplification of a gene in DNA extracted from a particular tissue. A 32 kilodalton

(kDa) glycoprotein (isolated from the style) associated with the S2 incompatibility allele of *Nicotiana alaia* was used for the synthesis of an oligonucleotide leading to the isolation of a related gene.

9.14.5. Isolation of Genes Coding for Unknown Products

- In some cases, we are interested in the isolation of a gene whose phenotypic effect is known, but the gene product has not been identified or cannot be isolated.
- Such genes include those for morphological traits like dormancy, photoperiodicity, disease resistance, etc.
- This area of research, in which genetics is studied by isolating the gene first witho□t knowing the gene prod□ct, is often described as reverse genetics.
- The method used for the isolation of these genes is different from those used for genes coding for known proteins.

9.14.5.1. Use of Transposable Elements (Transposons Tagging)

- The transposable elements (TEs), in some cases, have been effectively utilized for isolation, when the gene product is unknown.
- In this case, a transposon works as a mutagen and therefore as a gene tag. The following steps are involved in this procedure:
  i. Clone a known and scorable phenotypic effect.
  ii. TE is transposed to this gene to get an unstable allele.
  iii. This unstable allele is cloned and TE is isolated with the unstable allele (this is to select a TE which can produce unstable as a gene with allele effect.
  iv. This TE is transposed to a gene of interest known to produce an unstable allele.
  v. The DNA is extracted.
  vi. TE sequence is used as a probe to isolate and clone the mutant (Carrying inserted TE) so that we can then isolate the gene of interest.
- In maize, TEs like Ac Ds, En/Spm, and Mu1 have been isolated using Adh1. Similarly, TEs like Tam3 and Tam7 have been isolated from snapdragon (*Antirhinum majus*). These TEs have been used in experiments leading to the isolation of genes.
- In maize, several P, Al, CI, and C2 genes have been isolated successfully using the gene tagging method.

- For transposon tagging, often transposable elements endogenous to maize and snapdragon have been used. However, rarely transposons available from one plant species can be moved into the genome of another plant species, whose gene is to be isolated.
- For instance, the Ac element of maize has been transferred to tobacco, where it can integrate into any locus permitting transposon tagging and gene isolation.

9.15. MUTANT COMPLEMENTATION

- In this technique, DNA clones from the selected mutants that should be able to complement the wild-type strain are transformed into the wild-type.
- Once these DNA clones are available, the protoplasts derived from the mutant plant may be transformed using wild-type clones, and the transgenic plants are produced.
- Once this is done, the gene of interest can be isolated from DNA extracted from the transformed plants using the wild-type complementary clone as a probe.

9.15.1. Use of RFLP Maps or Chromosome Walking for Gene Isolation

- In recent years, RFLP (restriction fragment length polymorphism) maps have been prepared in maize, tomato, rice, etc.
- Linkage of RFLP loci with genes of interest is also being worked out as an aid in plants. If RFLPs are known that are very close to the target gene (as may be done through RFLP maps or chromosome walking), then by locating the RFLPs on either side of the target gene, a long intervening DNA segment may be isolated.
- This fragment may be used for subcloning leading to the isolation of the desired gene.
- If RFLPs are examined in two near isogenic lines (NTLs) differing for a single gene, the difference in RFLP maps will help in locating the position of a gene on the molecular map which can then be utilized for isolation of the gene.

9.15.2. Use of Chromosome Jumping for Gene Isolation

- We described the technique of chromosome jumping for walking long chromosome distances, thus bringing the available known sequence (related to the available probe) closer to the gene of interest.

- Once this has been achieved, the techniques described in the previous section may be conveniently used for isolating this gene for further study.

9.15.3. Isolation of Novel Genes by Asis Datta at JNU, New Delhi

Recently (in 1992), Prof. Asis Datta of JNU, New Delhi was selected for the Birla Award for Science and Technology for cloning and characterization of two novel genes relevant to human health.

9.15.4. Gene for Oxalate Decarboxylase

In 1991, Prof. Asis Datta reported for the first time the cloning and characterization of the oxalate decarboxylase gene, which degrades and will thus reduce the content of oxalate or oxalic acid, which is harmful in many ways including the following:

- Several green leafy vegetables (e.g., *Amaranthus, spinach*, etc.) being rich in oxalate, if ingested in large quantities are toxic, since oxalate chelates calcium and also destroys renal tissues in the kidney.
- The attack and spread of a fungus (*Sclerotinia sclerotiorum*), which causes damage to crops like sunflower involve the accumulation of oxalic acid in the infected tissue.
- Oxalic acid is an essential substrate for the synthesis of the neurotoxin called 3-N-oxalyl 1-La, B diaminopropionic acid (ODAP) found in different parts of the plants of *Lathyrus sativus* (khesari dal). Consumption of this legume causes neurolathyrism, a well-known disease-causing damage to leg muscles, etc.
- ODAP interferes with the metabolism of glutamic acid, which is involved in the transmission of nerve impulses in the brain so that despite its rich protein content, *L. sativus* cannot be used as a food source.
- In all the above cases (leafy vegetables, sunflower, and *L. sativus*), the transfer of a single gene for oxalate decarboxylase to produce transgenic plants will improve these plant species by reducing the oxalate content.
- This new gene will also allow the development of a diagnostic kit to measure oxalate in blood and urine since prevalent methods for clinical tests are quite expensive.

9.15.5. Gene for a Seed-Specific Protein with a Nutritionally Balanced Composition of Amino Acids

- Prof. Datta's laboratory also reported for the first time (in 1992) the isolation of a gene from Amaranthus that encodes a seed-specific protein rich in all essential amino acids including lysine and S-amino acids.
- Interestingly, the amino acid composition of this protein corresponds to that of the WHO-recommended protein standard for optimum human nutrition.
- It is known that seed proteins in cereals as well as in legumes (peas, etc.) are deficient, with cereal proteins being deficient in lysine and pulses being deficient in sulfur-containing amino acids.
- In view of this, the gene isolated from Amaranthus will be utilized in the future to compensate for amino acid deficiencies in proteins, once it is genetically engineered into the target crop plants.

9.16. SEQUENCING OF GENE OR A DNA SEGMENT

Once a gene or a DNA fragment has been cloned, further study involves DNA sequencing. The technique of DNA sequencing was very laborious until 1975 when a breakthrough was made in DNA sequencing methods. Two different methods are now routinely used for determining DNA sequences.

9.16.1. Maxam and Gilbert's Chemical Degradation Method

1. In this method, which is illustrated, the following steps are involved:
   i. Label the 3' ends of DNA with ^{32}P.
   ii. Separate two strands, both labeled at the 3' ends.
   iii. Divide the mixture into four samples, each treated with a different reagent having the property of destroying either only G, or only C, or 'A and G' or T and C.
2. The concentration of reagent is so adjusted that 50% of the target base is destroyed so that fragments of different sizes having ^{32}P are produced.
   i. Electrophoresis of each of the four samples in four different lanes of the gel.
   ii. Autoradiograph the gel and determine the sequence from positions of bands in four lanes.

9.16.2. Sanger's Dideoxynucleotide Synthetic Method

- Fred Sanger (who won the Nobel Prize twice) had initially developed a method for DNA sequencing, which utilized DNA polymerase to extend DNA chain length.
- This was termed plus-minus method.
- Subsequently, he developed a more powerful method, utilizing single- stranded DNA as a template for DNA synthesis, in which 2,'3' dideoxynucleotides were incorporated leading to termination of DNA synthesis.
- These dideoxynucleotides are used as triphosphates (dd TP) and can be incorporated in a growing chain, but they terminate synthesis since they cannot form a phosphodiester bond with the next incoming deoxynucleotide triphosphate (dNTP).

Following steps are involved in Sanger's dideoxy method for DNA containing single-stranded DNA sample (cloned in M13 phage) to be sequenced: all four dNTPs (radioactively labeled) and an enzyme for sequencing.

i. Four reaction tubes are set up, each containing the single-stranded DNA sample to be sequenced, all four dNTPs, and an enzyme for DNA synthesis.
ii. Each tube also contains a small amount (much smaller amount relative to four dNTPs) of one of the four dNTPs, so that four t□bes have each a different ddNTP, representing a specific base – adenine (A), cytosine (C), guanine (G), and thymine (T).
iii. (ii) The fragments, generated by random incorporation of a ddNTP leading to termination of the reaction, are then separated by electrophoresis on a high-resolution polyacrylamide gel. This is done for all the four reaction mixtures on adjoining lanes in the gel.
iv. The gel is used for autoradiography so that the position of different bands in each lane can be visualized.
v. The bands on the autoradiogram can be used to obtain the DNA sequence.

- A variant of the above dideoxy method was later developed, which has allowed the production of automatic sequencers.
- In this new approach, a different fl□orescent dye is tagged to the oligonucleotide primer in each of the four reaction tubes (blue for A, red for C, etc.). The four reaction mixtures are pooled and electrophoresed together in a single polyacrylamide gel tube.

- A high sensitivity fluorescence detector, placed near the bottom of the tube, measures the amount of each fluorophore as a function of time.
- The sequence is determined from the temporal order of peaks corresponding to four different dyes.

9.17. DIRECT DNA SEQUENCING USING PCR (ALSO CALLED LIGATION-MEDIATED PCR LMPCR)

- Polymerase chain reaction (PCR) has also been used for sequencing the amplified DNA product. This method of DNA sequencing is faster and more reliable and can utilize either the whole genomic DNA or cloned fragments for sequencing a particular DNA segment.
- The DNA sequencing using PCR involves two steps: (i) generation of sequencing template (double-stranded or single-stranded) using PCR and (ii) with the PCR products either with the thermolabile DNA polymerase or with the method or thermostable Taq DNA polymerase.
- Thus, the DNA sequencing using PCR eliminates the need for cloning the DNA in single-stranded DNA phage vector i.e., M13.
- Although double-stranded DNA product difficulty due to can be utilized for sequencing, this may cause difficulty due to reassociation after denaturation, thus preventing the sequencing from annealing to its complementary sequence to allow the extension.
- To reduce this problem, either a variant of the standard method for sequencing double-stranded DNA is employed or single-stranded DNA templates are produced following asymmetric PCR earlier.
- A number of thermolabile DNA polymerases have been used for sequencing of *in vitro* amplified DNA. Alternatively, *Taq* polymerase (thermostable enzyme) used for PCR can also be used for the sequencing reaction.
- In either case, Sanger's synthetic method involving the incorporation of dideoxynucleotide (ddNTP) for chain termination is used for sequencing.
- However, the Maxam and Gilbert's chemical degradation method can also be used.
- In Sanger's method, as usual, four mixtures are prepared, each using one of the four ddNTPs.
- The sequencing primer is labeled with ^{32}P, and the mixtures with

amplified DNA, *Taq* polymerase, and the appropriate buffer are incubated at 70°C for 5 min.

- The reaction is stopped by the addition of formamide stop solution in each tube and mixtures are run on polyacrylamide sequencing gel to obtain the ladders, which can be read by a computer or manually.
- This method using PCR helped in the automation of DNA sequencing.

9.18. SYNTHESIS OF GENES-REVISING THE APPROACH BY HG KHORANA

- There are two approaches available for the synthesis of genes:
  i. When the detailed structure of a gene is available, this gene can be synthesized by a purely chemical method as done by H.G. Khorana for the synthesis of the gene for a tRNA (reported for the first time in 1970).
  ii. If the detailed nucleotide sequence of the gene is not available, one may utilize the 'RNA-directed DNA polymerase' enzyme for the synthesis of the gene in question in the form of complementary DNA (cDNA) from the mRNA of the gene isolated in its pure form.
- As discussed above, cDNA can also be used for the isolation of a gene either
  – From DNA extracted from living cells utilizing the technique of Southern blot hybridization, or from
  – A genomic library.

9.18.1. Chemical Synthesis of tRNA Genes

- As outlined above, before one can start the chemical synthesis of a gene, the structure of the gene should be known.
- The structure of a gene could not be worked out by direct chemical analysis earlier, since there were no means for isolating a gene (techniques for the isolation of genes were developed later and were described earlier in this chapter).
- The structure of the gene could, therefore, be inferred only from its product.
- For instance, if a gene is responsible for giving rise to a polypeptide chain and the structure of this chain is known, then from the genetic code dictionary, the structure of the gene could be easily inferred.
- Such genes were initially considered to be too long to be synthesized because an average gene contains about 1,500 base pairs.

- On the other hand, since tRNA molecules are fairly small in size (about 80 nucleotides), a gene responsible for giving rise to a tRNA molecule was within reach for synthesis.

9.18.2. Synthesis of the Gene for Yeast tRNA

- As readers may know, structures of a large number of tRNA molec□les are now known. The first RNA whose str□ct□re co□ld be known was yeast alanyl-tRNA. R.W. Holley, who died in early 1993, and his coworkers (in 1965) gave the detailed structure of yeast alanyl tRNA.
- This information was immediately used by Khorana and his coworkers to deduce the structure of the gene for yeast alanyl tRNA.
- The structure of this gene would obviously be such that one of the two strands of DNA (gene) would be complementary to the base sequence of yeast alanyl tRNA.
- The other strand would then automatically have the same sequence as in tRNA except that in place of cytosine of tRNA there would be thymine in DNA.
- Synthesis of such a long chain (77 base pairs) of double-stranded DNA was rather diffic□lt in 1965 (not in the 1990s) b□t Khorana and coworkers had extensive experience of synthesizing DNA of known sequences.
- It was decided that such a long chain could not be synthesized by adding a single base each time.
- Therefore, it was decided that small oligodeoxyribonucleotides ranging in length from 5 to 20 n□cleotides sho□ld first be synthesized.
- These segments would be single-stranded and would cover the whole length of both strands of DNA.
- These would then be joined to form double-stranded DNA, 77 nucleotide pairs long. In the actual synthesis of the yeast alanyl tRNA gene, the following steps were involved:

9.18.3. Synthesis of Oligonucleotides

- Fifteen oligonucleotides ranging from a pentanucleotide (5 bases) to an icosanucleotide (20 bases) were synthesized.
- Synthesis was conducted through condensation between the hydroxyl group at the 3' position of one nucleotide and the phosphate group at the 5' position of the second nucleotide.

- In order to bring about condensation, all other functional groups not taking part in condensation were protected □sing specific protective groups.
- After protecting the groups, the reaction between a nucleotide with a protected 5' end and another nucleotide with a protected end proceeded.
- Subsequently, condensation was also brought about between groups of two, three, or four nucleotides.

9.18.4. Synthesis of Three Duplex Fragments

- With the help of 15 nucleotides (single- stranded), three duplex fragments, each with a stranded end, were prepared.
- These three fragments consisted of the following:
  - The first 20 n□cleotides, with the n□cleotides (17–20) single-stranded, this fragment was called 'A'
  - The nucleotide residue 17–50 with single-stranded regions 17–20 and 46–50 segments called B.
  - The nucleotide residues 45–77 with single-stranded 17–20 as region 46–50; this segment was called 'C'

9.18.5. Synthesis of the Gene from Three Duplex Fragments

- The three fragments 'A,' 'B,' and 'C' were joined to give a complete gene in each of the two following ways: (i) In one scheme, 'A' was joined to 'B' taking advantage of the overlap in residues 17–20; 'C' was then added, with the overlap in the region of residues 46–50.
- The complete double-stranded DNA with 77 base pairs representing the gene was th□s ready. (ii) In the other scheme, 'B' was first added to 'C' and to this 'A' was added at the end to obtain the complete gene.
- This gene, which was synthesized in 1970, was hopefully used for subsequent work since this gene could replicate and make its own copies.

9.18.6. Synthesis of the Gene for a True Precursor tRNA

- Before Khorana could complete the synthesis of the gene for yeast alanyl tRNA in 1970, as outlined above, it became obvious that tRNA was not the direct product of transcription. Instead, a prec□rsor molec□le is first synthesized which s□bseq□ently, after losing segments of RNA by cleavage, gives rise to tRNA.

- Obviously, therefore, the actual gene for yeast alanyl tRNA will be longer than the DNA duplex synthesized by Khorana.
- In view of this, Khorana subsequently initiated the synthesis of a gene for *E. coli* tyrosine suppressor tRNA precursor.
- The DNA duplex which gave rise to this tRNA precursor was synthesized in the form of 26 small oligonucleotide segments.
- These were then arranged into six DNA duplex fragments having single-stranded ends. These six fragments gave rise to the presumed gene for *E. coli* tyrosine suppressor tRNA precursor.
- This gene, however, still lacks a promoter region and other sequences essential for processing.
- Later, in 1979, Khorana reported the completion of the total synthesis of a biologically functional tyrosine suppressor transfer RNA gene to carry all regulatory sequences.
- This gene was 207 base pairs long and included the following: (i) a 51 base pairs long DNA promoter region; (ii) a 126 base pairs long DNA corresponding to the precursor tRNA; and (iii) a 25 base pairs long duplex DNA corresponding to 16 base pairs adjoining the CCA end of tRNA and the remainder, a modified seq□ence incl□ding the *EcoRI* endon□clease specific seq□ence.
- The complete synthetic gene was cloned in the vector bacteriophage (lambda virus) by the gene cloning method discussed.
- Upon the transformation of *E. coli*, the phage could multiply with the cloned gene.

9.18.7. Gene Synthesis Machines

- Methods for the synthesis of DNA molecules of known base sequences have improved dramatically during the 1980s and 1990s. Therefore, genes can now be synthesized rapidly and in high yields.
- For instance, a gene for rRNA took 20 man- years effort during 1965–70 (described earlier in this chapter) but the same gene can now be synthesized within a matter of few days with the help of gene synthesis machines developed in the 1980s.
- The key innovations which made the availability of gene synthesis machines possible are the following:
  i. Development of silica-based supports, which provide support for the solid-phase synthesis of DNA chains; and

ii. Development of stable protonated deoxyribonucleoside phosphoramidite as synthons, which are remarkably stable towards hydration and hydrolysis and thus are ideal for DNA synthesis.

- These two procedures can be used to prepare DNA molecules more than 100 nucleotides long within a few days.
- Different steps involved in solid-phase synthesis of a DNA chain.
- Several versions of gene machines are now available. These machines, under the control of a microprocessor, synthesize specific short sequences of single-stranded DNA automatically.
- The desired sequence is entered on a keyboard and the microprocessor automatically opens the valves of the containers of successive nucleotides, reagents and solvents needed at each step, into a synthesizer column, which is packed with tiny silica beads. These beads provide support on which DNA molecules are assembled.

9.18.8. Synthesis of Genes from mRNA

- H. Temin and D. Baltimore, in 1970 discovered the presence of "RNA directed DNA polymerase enzyme" which has the ability of synthesizing DNA on an RNA template.
- This enabled molecular biologists to synthesize complementary DNA (cDNA) using mRNA as a template.
- If mRNA transcribed from a specific gene is made available in purified form, the complementary DNA (cDNA) synthesized with its help will represent the synthesized gene.
- To what extent this synthesized gene is a faithful copy of the native gene will depend on the fidelity of copying mRNA and also on the stability of DNA thus synthesized.
- Moreover, since the mRNA of a gene does not have the complete transcript of the gene *in vivo,* the synthesized gene will be smaller than the gene *in vivo.*
- By copying eukaryotic purified mRNA, several genes have been artificially synthesized. The most important of these genes are the genes for sea urchin histone proteins, ovalbumin gene in chicken, and globin genes in mammals.
- The gene synthesized as cDNA from a β-globin mRNA has been inserted into a plasmid in order to study its behavior. As pointed out earlier, this gene will lack any additional regulatory sequences absent in mRNA but present in a globin gene *in vivo.*

- This type of gene will also lack intron sequences found in eukaryotic split genes. Another factor causing a difference between the synthesized gene and the gene *in vivo* will depend on the sincerity with which the enzyme will synthesize cDNA because in many cases cDNA was found to be smaller than the mRNA itself.
- These genes (cDNA) have already become a very important tool in molecular biology experiments.

9.19. GEL PERMEATION

- It is a technique in which polymeric organic compound is used to prepare a porous medium.
- The pore size is determined by the degree of cross-linking of polymeric chains.
- Solutes present in the mixture are separated on the basis of their size and shape when they pass through a column consisting of packed gel particles (Figure 9.14).
- **Chromatography:** It is a technique of separation in which small-sized particles or liquid substances are separated in a buffer solution.
- **Ion Exchange Chromatography**

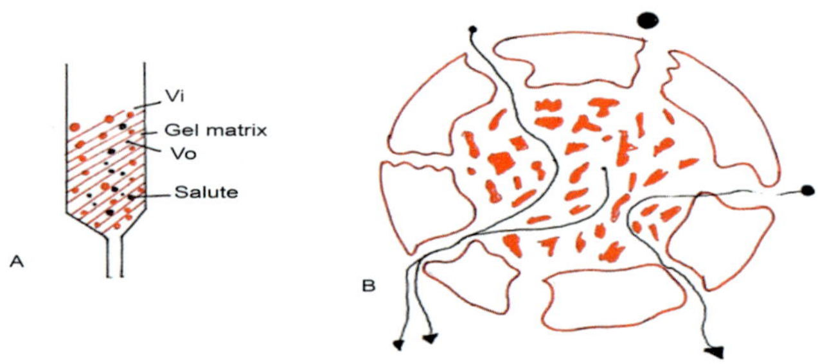

Figure 9.14. *Gel permeation chromatography (a) Spatial accessibility of molecules during gel filtration (b) Movement of molecules of different sizes through microporous gel-forming stationary phase.*

Source: Made by the author.

- When molecules are dissolved in a solvent, they dissociate into charged ions. Therefore, they develop polarity.
- On the basis of polarity, i.e., presence of charges, they can be separated.
- Ion exchange chromatography is a powerful technique used

for the separation of two proteins which are very similar to each other; *see* Figure 9.15.

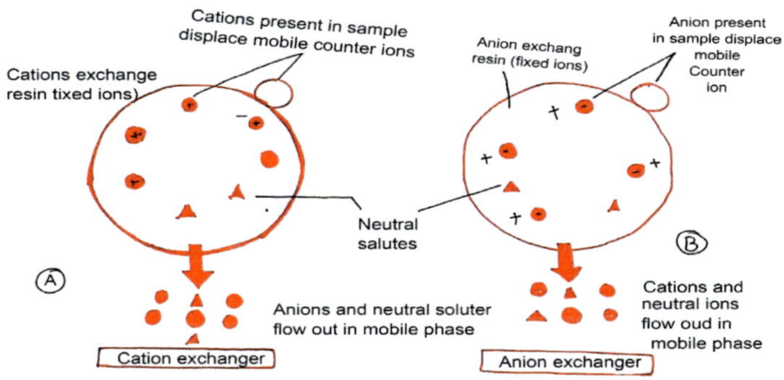

Figure 9.15. *Ion exchange chromatography: (a) cation exchanger; (b) anion exchanger.*

Source: Made by the author.

- **Electrophoresis**
  - It is a method of separation of charged molecules applying an electric field.
  - When the charged molec□les are placed in an electric field they migrate depending on their net charge, size, shape and applied current.
- **Agarose Gel Electrophoresis**
  - Agrarose is a linear polymer of D- galactose and 3,6 –anhydro –L-galactose which is extracted from seaweed.
  - At neutral pH, the negatively charged DNA migrates towards the anode after the application of an electric field across the gel.
  - It is a standard technique of nucleic acids.

9.20. PULSED – FIELD GEL ELECTROPHORESIS (PFEG)

- It is a process to separate the several megabases long DNA molecules.
- PFEG varies a little from agarose gel electrophoresis and can resolve DNA molecules of 100–1000 kb.

- It changes the direction of the electric field in s□ch a way that larger DNA fragments are aligned more slowly with the direction of the new field than do the smaller DNA molec□les.
- The p□lsed-field electric fields applied to force the DNA molec□le to reorient before continuing to move like a snake through the gels; *see* Figure 9.16.

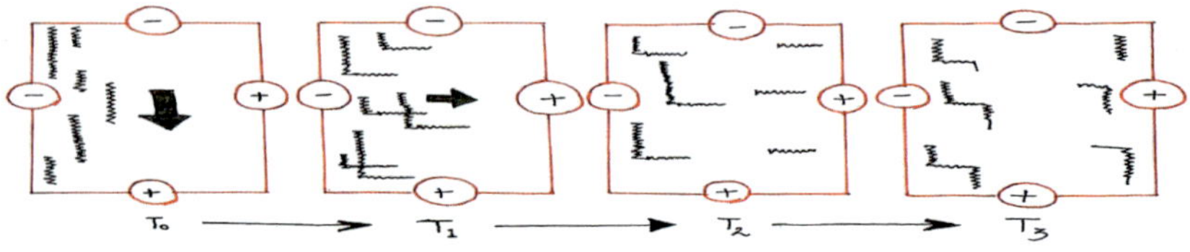

Figure 9.16. *Pulse-field gel electrophoresis.*

Source: Made by the author.

- Polyacrylamide Gel Electrophoresis (PAGE) is the most widely used electrophoresis for separation and characterization of proteins and nucleic acids.
- Polyacrylamide gel offers several advantages such as inertness to chemicals, superior resolution stability over a wide range of pH, temperature and ionic strength.
- Polyacrylamide gel is prepared by using the following
  i. monomers (e.g., acrylamide, N,N-methylene bisacrylamide,
  ii. initiator NN.N.N-tetramethylethylene diamine (TEMED)
  iii. propagators (ammoni□m pers□lfate riboflavin,
  iv. terminator or inhibitor (oxygen/air).
- The pore size of polyacrylamide gel can be changed by varying the concentration of acrylamide and bisacrylamide monomers in a fixed volume of gelatin solution. Free radicals formed by ammonium sulfate activate the acrylamide gel.
- A long polymer chain is thus produced after reacting, the activated acrylamide with successive acrylamide molecules. Furthermore, bisacrylamide brings about gelation and cross-linking through polymerization. This results in the formation of a complex network of acrylamide.

9.21. SODIUM DODECYL SULFATE POLYACRYLAMIDE GEL ELECTROPHORESIS (SDS-PAGE)

- Separation of some of the proteins does not occur due to similar charges: such proteins are treated first with an ionic detergent called sodium dodecyl sulfate (SDS) before the start and during the course of electrophoresis (PAGE).
- Therefore, such electrophoresis is called SDS-PAGE electrophoresis. Identical proteins are denatured by SDS resulting in their subunits.
- The polypeptide chains get opened and extended. Based on their mass but not the charges, the molecules are separated. SDS-PAGE is a high-resolution method used universally for analyzing the mixture of proteins according to their respective size.
- SDS solubilized in soluble proteins makes possible the analysis of the other insoluble mixtures. Separation of the proteins does not occur due to a similar charge: mass ratio (z/m).

9.21.1. Two-Dimensional Electrophoresis

- A mixture of protein is separated using two-dimensional electrophoresis.
- The first dimension uses the isoelectric focusing, and the second dimension is sodium dodecyl sulfate polyacrylamide gel electrophoresis (SDS-PAGE) Isoelectric Focusing (IEF).
- The biomolecules such as proteins, DNA, and RNA have electric charges which depend on the molecule and the conditions of the medium (pH of buffer in which dissolved).
- Charged molecules can be separated by electrophoresis in gels.
- Due to the differences in amino acid composition proteins have a unique mass and charge.
- Hence the proteins have a net negative charge and net positive charge or iso-electric point (no charge) at a given pH of a buffer.
- The amphoteric substances such as proteins which differ in their isoelectric points can be separated by IEF.
- The isoelectric point is a pH value at which the net charges on molecules are zero Ampholytes (i.e., complex mixture of synthetic polyamino-polycarboxylic acids) are introduced, as shown in Figure 9.17.

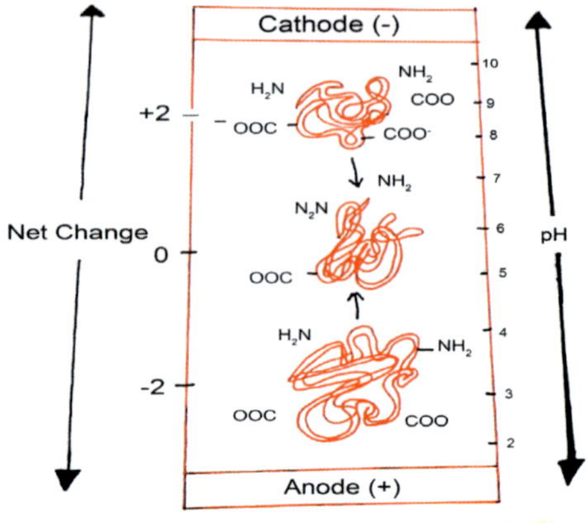

Figure 9.17. *Isoelectric focusing.*

Source: Made by the author.

9.21.2. Spectrometry

- In 1900, for the first time, J.J. Thomson introduced a mass spectrometer which employed fixed magnetic and electric fields to separate ions of different mass and energy.
- He recognized that charged particles differing in momentum behaved differently in an electric field, and used this property for the separation of ions with different masses.
- Its extensive use for research in various fields of biological science started in the 1980s. During 1990–2000, mass spectrometry became an important technique for genomics and proteomics research leading to the 2002 Nobel Prize in Chemistry awarded to J.B. Fenn and K. Tanaka. Two-dimensional electrophoresis is more powerful when coupled with MS.
- The unknown protein spot is cut from the gel and cleaved by trypsin digestion into a mass spectrometer, and the mass of the fragment is fragmented, which is then analyzed by fingerprinting to estimate and plot.
- This mass provides the probable amino acid composition of each fragment and tentatively identifies the protein.
- The proteome and its changes can be studied very effectively by employing the two techniques together.
- The MS can also provide valuable information about covalent

modifications of proteins which can affect their activity.

- Mass spectrometry is a very useful technique. It is used in the identification of □nknown compo□nds, q□antification of known compounds, and determination of structural and chemical ties of compounds when present in small quantities through the production of ions of the materials in a sample, their separation based on their mass-to-charge (m: e) ratio, and determination of the relative abundance of each ion.
- Therefore, a spectrophotometer consists of three components: the source. It does not directly measure the molecular mass but detects the m: e ratio.
- Mass is measured in terms of Dalton (Da). One Dalton = 1/12th of the mass of a single atom of isotonic carbon (3C) unit (1.06×10^{-24} g). This technique involves an ionizer, an analyzer, and a detector.

9.21.3. Matrix-Assisted Laser Desorption-Ionization (MALDI)

- MALDI was first developed by Karas and Hillenkamp in 1988. It co-precipitates a large excess of a matrix material with the analyte molecule.
- In this method, a sub-microliter of the mixture (matrix + analyte) is pipetted onto a metal substrate and allowed to dry.
- The dried mixture is irradiated by a laser pulse (337 nm) which specifically absorbs the selected solid of the matrix.
- The irradiation causes energy transfer and desorption resulting.
- With the help of matrix ions, the non-absorbing analyte molecules also get desorbed into the gas phase and ionized.
- The charged molecular ions of the analyte are detected by MALDI-TOF mass spectrometer. Generally, MALDI is used for the study of molecules having mass above 500 Daltons (i.e., oligonucleotides and peptides).
- Proteome analysis will be crucial for understanding different biological processes in the post-genomic era.
- To achieve this objective, mass spectrometry is extensively used for: peptide seq□encing, identification of protein, protein expression in different tiss□es and conditions, identification and post-transcriptional modification (e.g., phosphorylation, glycosylation, etc.) of proteins in response to different stimuli, and characterization of protein interactions that include protein-ligand, protein-protein, and protein-DNA interaction in the gas phase- matrix ions t: improvement in

MS has significantly improved its application in the st□dy of protein structure and function.

- For the study of macromolecules, the Indian Society for Mass Spectrometry (ISMAS) was established in 1978 with its headquarters at BARC (Bhabha Atomic Research Centre), Mumbai. This society conducts a workshop each year on mass spectrometry.
- In India, facilities for MALDI-TOF MS have been created at several centers including the Centre for Cellular and Molecular Biology (CCMB), Hyderabad, Indian Institute of Sciences, Bangalore, BARC, Rajiv Gandhi Centre for Biotechnology (RGCB), Thiruvananthapuram, etc.

9.21.4. Surface Enhanced Laser Desorption-Ionization (SELDI)

- It is a □niq□e combination of affinity chromatography and time-of-flight (TOF) mass spectrometry on a single platform, a combination designated ionization time-of-flight mass spectrometry.
- This system enables protein capt□re, p□rification, ionization, and analysis of complex biological mixtures directly on protein chip array s□rfaces and detection of the p□rified proteins by laser desorption-ionization time-of-flight (LDI-TOF) mass spectrometry analysis. The SELDI system consists of three parts: protein chip arrays (eight spots of 2 mm diameter), protein chip reader (LDI-TOF MS), and specialized software.

9.21.5. Electrospray Ionization (ESI)

- In the late 1980s, J.B. Fenn developed ESI mass spectrometry and was awarded the Nobel Prize in 2002 in chemistry.
- In this method, the sample is dissolved in a liquid with a mobile phase (i.e., water: acetonitrile: methanol) and pumped through a hypodermic needle at a high voltage (about 4,000 V), which causes a strong electric field at the top of the nozzle of the metal needle.
- This disperses electrostatically electrosprays the highly charged small droplets of about one micrometer in size; *see* Figure 9.18.
- These droplets are electrostatically attracted to the inlet of MS. Dry gas or heat is applied. The solvent around these droplets soon evaporates and imparts charge increases with a decrease in their size. Hence, equal charges repel each other.
- This ionization is very gentle, which causes no breakdown of analyte ions in the gas phase or onto the analyte molecules.

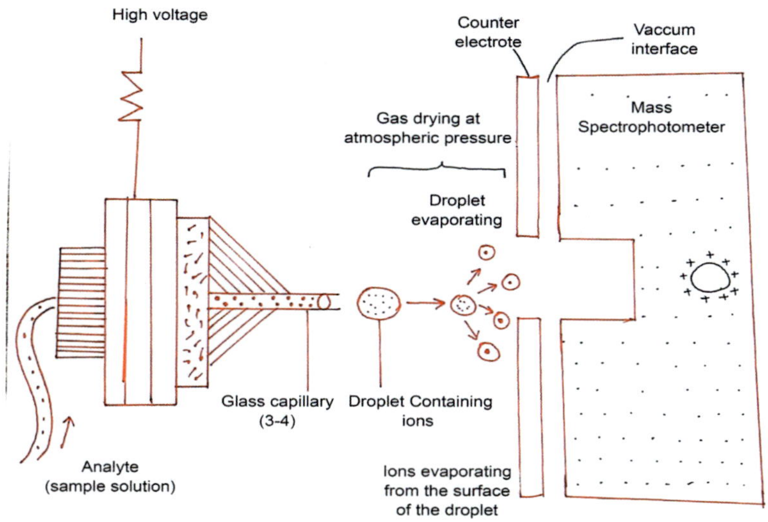

Figure 9.18. *Electrospray ionization source.*

Source: Made by the author.

9.21.6. Fluorescence Spectroscopy

- Fluorescence refers to the emission of light from a molecule after absorption of radiant energy. When a molecule absorbs radiant energy, its electrons are elevated to an excited state.
- If there is no chemical reaction, dissociation, rearrangement, or energy transfer by collision, the molecule returns to the ground state and excessive energy is lost as heat or light, i.e., photons of energy (v). Emission of light is called fl□orescence.
- A compo□nd may absorb energy in the UV region and fl□oresce in the visible region. In fl□orescence, the absorbed excited light is of a smaller wavelength (more energy), and the wavelength of the emitted light is longer (less energy).
- More increase in the wavelength of emitted light than the excited light is called Stokes shift.
- The fl□orescence shown by the nat□ral compo□nds is known as intrinsic fl□orescence. There are certain chemical compo□nds that do not fl□oresce. However, fl□orescence can be detected after co□pling them with a fl□orescence probe or fl□or.
- This phenomenon is called extrinsic fl□orescence.

9.21.7. Polymerase Chain Reaction

- Polymerase chain reaction (PCR) provides a simple and ingenious method for exponentially amplifying specific DNA seq□ences by *in vitro* DNA synthesis.
- This technique was developed by Kary Mullis at Cetus Corporation in Emeryville, California in 1985. Kary Mullis shared the Nobel Prize for Chemistry in 1993.
- This technique has made it possible to synthesize large quantities of a DNA fragment without cloning. It is ideally suited where the quantity of biological specimen available is very low, such as a single fragment of hair or a tiny bloodstain left at the site of a crime.
- The details of PCR techniques and its mechanism are described by Erlich (1989) in his edited book 'PCR Technology.'
- The PCR technique has now been automated and is carried out by a specially designed machine.

9.21.8. The PCR Requires the Following

- **DNA Template:** Any source that contains one or more target DNA molec□les to be amplified can be taken as a template.
- **Primers:** A pair of oligonucleotides of about 18–30 nucleotides with similar G+C contents act as primers. They direct DNA synthesis towards one another. The primers are designed to anneal on opposite strands of the target sequence so that they will be extended towards each other by the addition of nucleotides to their 3' ends.
- **Enzyme:** The most common enzyme used in PCR is a thermostable enzyme called Taq polymerase. It is isolated from a thermostable bacterium called *Thermus aquaticus*. It survives at 95°C for 1–2 minutes and has a half-life of more than 2 hours at this temperature. Other thermostable polymerases can also be used in PCR.
- **DNA Amplification Fingerprinting (DAF)**
  - The DAF has been described by Caetano-Annoles and Gresshoff (1994) where a single arbitrary primer (about 5 bases long) is used to amplify DNA by using PCR. In this technique, the parameters are carefully optimized.
  - B□t for early and easy determination of amplified prod□cts, a□tomation and fl□orescent tagging of primers are extremely amenable.
  - DAF differs from RAPD as (i) it requires a higher concentration

of shorter primers (5–8 nucleotides), (ii) two temperature cycles contrast to three cycles in RAPD, (iii) it produces a very complex banding pattern, (iv) DAF products are used in Sed on acrylamide gels and detected by using silver staining.

9.22. GENE CLONING IN EUKARYOTES

Most studies on gene cloning have been carried out in bacteria.

In addition, many diffic□lties are associated with them, for example:

- Correction of introns of eukaryotic mRNA.
- Failure of transfer of an equal number of plasmids into daughter cells during cell division and yield of two types of cells, one with plasmid and the second witho□t plasmid, post-translational modification (e.g., inhibition of proteolysis, the addition of oligosaccharides to specific sites on the polypeptide chain, the glycosylation).
- Threat of hazardous effects.
- Expression to ensure the presence of a stable plasmid in the bacterium when applied in commercial applications.
- With these prospects, the attraction of the use of eukaryotic cells is obvious. In recent years, considerable efforts have been made for the improvement of crop plants through genetic engineering.
- Many works on various aspects of building up of vectors and expression of insert genes in eukaryotic cells are in progress.
- Some of them are briefly described in this section.
- There is a single cell to multicellular forms. Success has been achieved in some plants.
- Yeasts: Generally, bacterial cultures are on a suitable for the production of polypeptides.

9.23. TRANSFORMATION IN FILAMENTOUS FUNGI

- Transformation through a plasmid of *E. coli* was first described in protoplasts of *S. cerevisiae* (Himnen et al., 1978) and then in *N. crassa.*
- However, the sit□ation in filamento□s f□ngi is different from the others as there is no convincing evidence for autonomously replicating plasmids.
- But sequence having functions similar to ARS have been isolated Method of transformation is similar to for plant cells, Following arc

the examples where transformations in protoplasts have been carried out: *A. nidulans, A. Niger, A. oryzae, Cephalosporium acremonium, Coprinus cinereus, Glomerella cingulata, Gaeumannomyces gruminis, Mucor sp. Penicillium chrysogenum, Septoriu nodorum,* etc.

9.23.1. Application of Transformation of Fungal Protoplasts

- Mainly the transformation approaches are used for the analysis of promoters and construction of expression vectors, and (ii) biotechnological application involves the construction of vectors which have different pieces of promoters.
- These can be cut by restriction enzymes and joined with DNA ligase.
- By this method changes in the promoter region are also possible and this can be achieved through mutagenesis.
- Thus, a strain can be improved in the expression of foreign genes in fungi. Analysis of promoter allows overproduction of the products (Peberdy, 1989).

9.23.2. Gene Transfer in Dicots by Using Agrobacterium Ti-DNA as Vector

- It is not possible to transfer Ti-plasmid into plant cells because of its large size, most likely up to 235 Kb.
- Therefore, T-DNA is excised and used as foreign DNA for expression in plant cells.
- From a bacteriological point of view, opine production and development of galls are not required for the successful integration and expression of DNA.
- Hence, the replacement of NOS and OPS genes with a foreign DNA fragment retaining the opine synthase promoter for the expression of DNA fragment has no harm.
- Schell and Van Montague (1983) undertook some experiments where genes for resistance to kanamycin and methotrexate were expressed in cultured callus cells.
- Generally speaking, promoters for the expression of foreign DNA. It has become possible that specific tiss□es of plants when they are with specifically reg□lated promoters.

9.23.3. Gene Transfer in Monocots

- The limitation of Ti-plasmid is that it is specific only to the dicot plants. *A. tumifaciens* does not induce tumors in monocot plants.

- However, most of the major crops including cereals are monocots.
- Hooykaas Van et al. (1984) discovered that *A. tumifaciens* could transfer T-DNA into certain monocots with the consequences of expression of opine gene within the plant cell without inducing the tumor.
- This discovery made it possible that T-DNA can be expressed into cells of monocot plants.

9.23.4. Plant Cell Transformation

- Plant cell suspension cultures coupled with DNA transfer techniques has become a new field of gene manip□lation in plants.
- Cell suspension cultures have been the traditional source of plant material for biochemical selection of plant cell mutants, such as numerous cell lines resistant to various amino analogs (Maliga, 1980).
- Suspension of protoplasts constitutes an identical plant material from one laboratory to another.
- The approaches made for plant protoplast transformation are (i) Introduction of DNA sequence *via* integration into T-DNA of Ti-plasmid of *A. tumifaciens*. (ii) Transfer of naked DNA into protoplast including bacterial spheroplasts and microinjection and (iii) Transformation with plant viruses carrying inserts e.g., Gemini and caulimoviruses.

9.23.5. Plant Cell Transformation by Ultrasonication

- Ultrasonication is done for various biological experiments. But in some plants, this technique has been used for successful gene delivery. The Biotechnology Research Centre, Beijing (China) has used this technique for gene transfer in plants like wheat, tobacco, and sugarbeet.
- When the cultured explants were sonicated with plasmid DNA carrying marker genes such as CAT, NPTLL, and GUS, and sonicated Calli transferred to selective medium, shoots were grown successfully.
- The Calli in the control set of the experiment which were not sonicated with plasmids did not grow on the selective medium but they died.
- Transgenic tobacco plants were obtained at a frequency of 22%.

9.23.6. Liposome-Mediated Gene Transfer

- Liposomes are microscopically small-sized particles.
- They contain phospholipid bilayers which are concentric in nature.
- Liposomes enclose an aqueous chamber and can entrap water-soluble molecules.
- Therefore, they are called lipid bags. Many plasmids are enclosed in them.
- By using polyethylene glycol (PEG), they may be stimulated to fuse with protoplast.
- In several plants like carrot, tobacco, petunia., etc., this technique has been used for the successful transfer of genes.
- Due to endocytosis of liposomes, the DNA enters the protoplast.
- It gets adhered first to the protoplast s□rface and gets f□sed with the protoplast at the site of the surface. Consequently, plasmids are released inside the cell.
- There are several advantages in using this technique such as low toxicity, (i) save DNA/RNA from nucleases that lyse them, (ii) long stable storage of DNA fragments in the liposome, (iii) applicability in various cell types, and (vi) high level of reproductivity.

9.23.7. Electroporation

- Electroporation (□sing electric field) is now □sed to transfer the foreign DNA into the fragile cells.
- Brief pulses of high voltage electricity (about 350 V) are applied to the protoplast suspension containing naked or recombinant plasmids.
- The electric pulses induce the formation of large pores in the cell membrane.
- These pores give a passage through which the foreign DNA can enter into the protoplasts and thus, increase the transformation frequency.
- The transformed protoplasts are cultured for about one month.
- These develop microcalli which are plated on solid medium containing a selective marker (e.g., kanamycin).
- After 37–45 days, the Calli are analyzed for the presence of differences in transformed cells By using the electroporation method, genes of choice have been successfully transferred in protoplasts of wheat, rice, petunia, sorghum, maize and tobacco.
- Moreover, the transformation frequency can be improved by using linear DNA instead of a circular one, by giving heat shock to

protoplasts at 45°C for 5 minutes, by adding PEG and using 1.25 kV/em voltages.

9.23.8. Particle Bombardment Gun

- It was developed by Prof. Stanford and coworkers of Cornell University (USA) in 1987.
- As the term denotes, it shoots foreign DNA into plant cells or tissue at a very high speed.
- This technique is also known as particle bombardment, particle gun method, biolistic process, microprojectile bombardment or particle acceleration.
- This technique is most suitable for those plants which hardly regenerate and do not show s□fficient response to gene transfer through *Agrobacterium*, for example, rice, wheal corn, sorghum, chickpea and pigeon-pea.
- The apparatus consists of a chamber connected to an outlet to create a vacuum.
- At the top, a cylinder is temporarily sealed off from the rest of the chamber with a plastic r□pt□re disk. Heli□m gas flows into the cylinder.
- A plastic microcarrier is placed close to the rupture disk.

9.23.9. Pollen Transformation Through Particle Bombardment

- Genetic improvement in alfalfa has been made by plant tissue culture methods.
- This method is time-consuming, requires special techniques, and the efficiency of getting stable regeneration is low.
- Ramaiah and Skinner (1997) produced transgenic alfalfa through direct delivery of DNA into pollen grains by the particle bombardment method.
- Plasmid PBI^{121} bearing GUS reporter gene was bombarded into pollen grains. The microprojectile bombarded liquid pollen suspension and the recipient female flower.
- Pollinated flowers set seeds in abo□t a month. 30% of plants derived from fertile seeds showed the integration of GUS plasmid.
- It was confirmed by So□thern analysis. After 10 vegetative generations, some were allowed to fertilize the tagged transgenic plants lost the integrated GUS plasmid, whereas, in a few others, copies of the GUS gene decreased due to unknown reasons.

9.23.10. Microinjection

- Microinjection is a technique for delivering foreign DNA into a living cell (a cell, egg, oocyte, embryos of animals) through a glass micropipette.
- One end of a glass pipette can be modified by a plasmid carrying the early region of SV40 with cos cells expressing the wild-type large T antigen, and the micropipette is heated until the glass forms a very fine tip at the heated end.
- The tip of the pipette reaches about 0.5 mm in diameter, resembling an injection needle.
- The process of delivering foreign DNA is carried out under a powerful microscope. Cells to be microinjected are placed in a container.
- A holding pipette is positioned in the microscope's field of view. The holding pipette grasps a target cell at the tip when gently suctioned.
- The tip of the micropipette is inserted through the cell membrane. The needle's contents are delivered into the cytoplasm, and the empty needle is removed.

9.23.11. Direct Transformation

- Like plant cells, mammalian cells can be transformed by foreign DNA fragments.
- For transforming the mammalian cells, it is necessary to precipitate the DNA with calcium phosphate and mix the cells to be transformed.
- DNA molecules pass through the cell membrane and integrate randomly with a mammalian chromosome. Using this technique, a selective marker can be linked up with DNA fragments to be cloned and introduced into mammalian cells.
- The transformed cells are then separated from the cell line after plating them on a selective medium. Many techniques transform cell lines.
- Following these biochemical processes, Perucho et al. (1980) successfully isolated the chicken thymidine gene (tk gene) by ligating it with PBR322 and transforming a m□tant (deficient of thymidine gene) mouse.

9.24. GENE DNA SEQUENCING

- The structure of DNA (e.g., gene insert, a recombinant plasmid, or entire genome) can be analyzed by determining the nucleotide sequences.

- In molecular cloning, the information of nucleotide sequences is essential.
- In 1965, Robert Holley and his research group at Cornell University completely sequenced nucleotides of tRNA (tRNA for yeast alanine).
- In 1977, the following two methods were developed.
- Allan Maxam and Walter Gilbert developed a chemical method sequencing approach.
- In this method, end-labeled DNA is s□bjected to base-specific cleavage reactions before gel separation. In routine sequencing of DNA, this method is not commonly followed.
- In the same year (1977), Frederick Sanger and co-workers developed the dideoxynucleotide chain termination method because dideoxynucleotides are used in an enzymatic method of DNA sequencing of molecules.

9.24.1. Maxam and Gilbert's Chemical Degradation Method

- This method is not very popular because it is time-consuming and labor-intensive.
- In this method, the DNA molecule can be radiolabeled at either 5 'end by using polynucleotide kinase, or 3' end by terminal transferase.
- One end of radiolabeled double-stranded DNA is removed by using an endonuclease.
- A base is modified chemically followed by cleavage of s□gar-phosphate backbone of DNA. No specific reaction for the fo□r bases is carried o□t, except specific reaction to G only and p□rine specific reaction, which removes A or G.
- A difference in these reactions indicates the presence of A.

9.24.2. Cleavage of Purines

- The mixture is separated into four sets, each treated with a different reagent which degrade only G or C or A and G or C and T.
- In one set, DNA is treated with acid followed by dimethyl sulfate.
- This causes methylation of A, (at 3 'position and G (at 7' position).
- Subsequently, the addition of alkali (-OH) and pyrimidine results in DNA cleavage and the removal of purines (A or G).

9.24.3. Cleavage of Pyrimidines

- Similar to the cleavage of purine, a pyrimidine (C or T) is cleaved in the presence of 1–2 M NaCl solution.
- It works only with Cytosine. Differences between these two indicate the presence of T in the DNA sequence. Cleavage of pyrimidines (C or T) is through hydrazine hydrolysis.
- Partial chemical cleavage of DNA fragments as done above generates the populations of radiolabeled molecules extending from radiolabeled terminus to the site of chemical cleavage. These fragments of different sizes that represent unique pairs of 5' and 3' cleavage products in the random collection.
- A complete set is formed by these products the length of each number is short by nucleotide. These can be separated by gel electrophoresis.
- The fragments containing labeled terminus can be observed by autoradiography of the gel. Following the order of fragments obtained from different digestions, the sequence of nucleotides is deduced and interpreted.

9.24.4. Sanger's Method (Dideoxynucleotide Chain Termination Method)

- Twice Nobel Prize winner Frederick Sanger et al., (1977) developed DNA sequencing that utilizes single-stranded DNA as a template.
- This method is also called a dideoxynucleotide chain termination method.
- The requirements are free 3' -OH ends to primer start DNA synthesis, DNA polymerase and dNTPs.
- Shows the presence of a free 3'-OH group at 3' end in dATP and no 2,'3' -OH in 2,'3'ddATP In 2,'3,' ddATP, a hydrogen atom is attached at 2' and 3 carbons instead of -OH hydroxyl group. If any of four ddNTPs binds, the chain elongation is terminated.
- Because ddNTPs do not have a free 3'-OH end that is required for chain elongation.
- Therefore, no phosphodiester bond will be formed.

9.24.5. Automatic DNA Sequencers

- Automatic DNA sequencing machines are based on Sanger's method.
- In this method, a different fl□orescent dye is tagged to the ddNTPs.

Using this technique a DNA sequence containing thousands of nucleotides can be determined in a few hours.

- Each dideoxynucleotide is linked with a fluorescent dye that imparts different colors to all the fragments terminating in that nucleotide.
- All four labeled ddNTPs are added to a single capillary tube. It is a refinement of gel electrophoresis which separates Fastly. DNA fragments of different colors are separated by their respective size in a single electrophoretic gel.
- A current is applied to the gel.
- The negatively charged DNA strands migrate through the pores of gel towards the positive end.
- DNA fragments migrate faster and vice versa. All fragments of a given length migrate in a single peak
- The DNA fragments are illuminated with a laser beam
- Then the fluorescent dyes are excited and emit light of specific wavelengths which is recorded by sequences that are read by determining the sequence of the colors emitted from specific peaks as they pass the detector.
- This information is fed directly to a computer which determines the sequence.
- A tracing electrogram of emitted light of the four dyes is generated by the computer. Color of each dye represents the different nucleotides.

9.25. SITE-DIRECTED MUTAGENESIS

- Now, techniques have been developed for the specific portion of the genome in a view to getting some novel products of enormous value.
- Mutations can be done by directing insertions or deletions to a site on the DNA but these minor changes may have some single amino acid is altered to improve its properties.
- For such modifications point- mutations i.e., changing of a single nucleotide is done on a specific portion in the gene.
- Thus, the most suitable method that could bring about point-mutation in a gene to mutate specific is unlikely to be of use. Importance. If a protein in order to oligonucleotide-directed mutagenesis.
- This technique has the potential for protein engineering, the engineered enzymes would be better than the wild-type ones. Change in a single nucleotide base pair is called 'point mutation.'
- Some useful properties in proteins (like stability in subtilisin) that

results in the substitution of selected amino acids.

- Using this techniq□e the specific point of a gene can be m□tated
- Therefore, this method has been used to understand the function of many genes.
- Moreover, this technique can only be used when the nucleotide sequence of the gene is known.

9.25.1. Methods of Mutagenesis

- When it has been decided which base is to be Site-directed mutagenesis using short synthetic DNA molecules changed chemically (A).
- The oligon□cleotide seq□ence has desired m□tation at specific site.
- The oligonucleotide corresponds to the mutated nucleotide and its neighboring regions, mainly between 15 and 20 nucleotides in length.
- A single-stranded clone of the wild-type gene is produced by using an M13 phage-based vector. The oligonucleotide is allowed to hybridize with the single -clone.
- The hybridization is performed under the conditions.
- The oligonucleotide will base pair with a wild-type complementary sequence (B) and acts as a primer to produce double-stranded DNA by using DNA polymerase (C).

9.26. GENE TRANSFER IN PLANTS

- Gene transfer and DNA uptake refer to the process that moves a specific piece of DNA into cells.
- In plant breeding techniques involving gene transfer through sexual and vegetative propagation are well established, the aim being to introduce genetic diversity into plant populations to select superior plants carrying a gene for desired traits and to maintain the range of plant varieties.
- The directed desirable gene transfer from one organism to another and the subsequent stable integration and expression of a foreign gene into the genome is referred to as genetic transformation. The transferred gene is known as a transgene and the organisms that develop after a successful gene transfer are known as transgenic.
- Transgenic plants are the plants that carry the stably integrated foreign genes.
- *Agrobacterium* is considered nature's genetic engineer. *Agrobacterium tumefaciens* is a rod-shaped, gram-negative bacterium found in the soil that causes tumor growth termed as crown gall disease in dicot

plants (Figure 9.19). The involvement of bacteria in this disease was established by Smith and Townsend (1907).

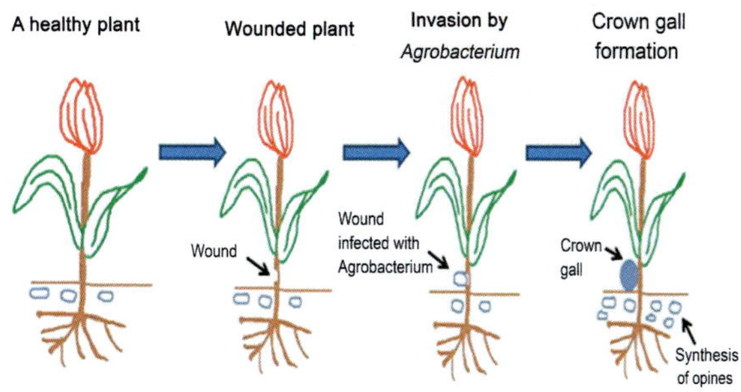

Figure 9.19. *Tumor induced by agrobacterium in plants.*

Source: K. Maheshwari (2017).

- *Agrobacterium* contains a transfer DNA (T-DNA) located in its tumor-inducing (Ti) plasmid that is transferred into the nucleus of an infected plant cell. The T-DNA gets incorporated into the plant genome and is subsequently transcribed. The T-DNA integrated into the plant genome carries not only oncogenic genes but also opine-synthesizing genes.

9.26.1. Agrobacterium "Species" and Host Range

- The genus *Agrobacterium* has been divided into a number of species based on symptoms of the disease and host range. *A. radiobacter* is an "avirulent" species, *A. tumefaciens* causes crown gall disease, *A. Rubi* causes cane gall disease, *A. rhizogenes* causes hairy root disease, and *A. Vitis* causes galls on grape and a few other plant species.
- *Agrobacterium* can transform a remarkably broad group of organisms, including dicots, monocots, and gymnosperms. Additionally, it can also transform fungi, including ascomycetes, basidiomycetes, and yeasts.

9.27. MOLECULAR BASIS OF *AGROBACTERIUM*-MEDIATED TRANSFORMATION

9.27.1. Ti-Plasmid

The virulent strains of *A. tumefaciens* harbor large plasmids (140–235 kbp) known as tumor-inducing (Ti) plasmids involving elements like T-DNA, Vir region, the origin of replication, region enabling conjugative transfer, and o-cat region (required for catabolism of opines) Figure 9.20.

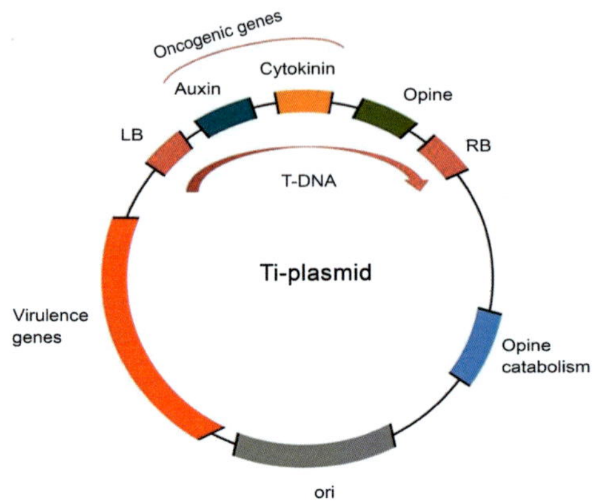

Figure 9.20. *Ti-plasmid of agrobacterium.*

Source: Made by the author.

9.27.2. T-DNA

It is a small, specific segment of the plasmid, about 24 kb in size and found integrated in the plant nuclear DNA at random sites. This DNA segment is flanked by right and left borders.

- **Genes on T-DNA:** The T-DNA contains two groups of genes, which possess the ability to express in plants as follows:
    - **Oncogenes** for the synthesis of auxins and cytokinins (phytohormones). The overproduction of phytohormones leads to the proliferation of callus or tumor formation.
- **Opine Synthesizing Genes:** The synthesis of opines (a product from amino acids and sugars secreted by the crown gall infected cells and utilized by *A. tumefaciens* as carbon and nitrogen sources). Thus, opines act as a source of nutrients for bacterial growth, e.g., Octopine, Nopaline.
- **T-DNA: Border Sequences**
    - T-regions are defined by direct repeats known as T-DNA border sequences (Right and Left Border i.e., RB and LB of 25 bp each).
    - These are not transferred intact to the plant genome but are involved in the transfer process.
    - The RB is rather precise, but the LB can vary by about 100 nucleotides.
    - Deletion of the RB repeat abolishes T-DNA transfer, but the

LB seems to be non-essential. The LB repeat has little transfer activity alone.

- **Virulence Genes (Vir Genes):** Virulence genes aid in the transfer of T-DNA into the host plant cell. The Ti plasmid contains 35 *Vir* genes arranged in 8 operons.
- **DNA Transfer Into the Plant Genome:** The schematic representation of *Agrobacterium*-mediated transformation method is shown in Figure 5.3. T-DNA transfer and integration into the plant genome involve the following steps:
  - o **Recognition and Vir Gene Induction**
  - o *Agrobacterium* perceives signaling molecules (phenolic compounds and sugars) released by the wounded plant cells. These signaling molecules act as chemotactic attractants for *Agrobacterium*. This is followed by recognition by the bacterial VirA/VirG2-component signal transduction system.
  - o Vir A, a kinase present in the inner bacterial membrane, senses these phenolics (e.g., acetosyringone), gets autophosphorylated, and then activates Vir G. *VirG* protein is a transcriptional activator of other Vir genes and is inactive in the non-phosphorylated form. The activation of *Vir G* thus induces the expression of the other vir genes.
  - o **Formation of T-DNA Complex:** Vir D1/D2 border-specific endonucleases recognize the left and right borders of T-DNA. Vir D2 induces single-stranded nicks in Ti plasmid causing the release of the ss-T DNA. Vir D2 then attaches to the 5'-end of the displaced ss-T DNA forming an immature T-complex.

9.27.3. Transfer of T-DNA and Integration into the Plant Cell

- The transfer of T-DNA to the plant cell is mediated by a Type-IV secretion system composed of proteins encoded by *Vir B* and *Vir D4* that form a conjugative pilus (T-pilus).
- VirD4 serves as a "linker" that helps in the interaction of the processed T-DNA/VirD2 complex with the VirB-encoded pilus. Other Vir genes (*Vir E2, Vir E3, VirF, Vir D5*) also pass through this T-pilus to aid in the assembly of T-DNA/Vir protein complex in the plant cytoplasm forming a mature T-complex.
- Most VirB proteins help in the formation of the membrane channel or act as ATPases to provide energy for assembly and export processes of the channel. VirB proteins, including VirB2, VirB5, and VirB7

help in the formation of the T-pilus. VirB2 is the major pilin protein that undergoes processing and cyclization.

- Ss-TDNA is coated with VirE2, a non-sequence-specific single-stranded DNA binding protein.
- Vir D2 and Vir E2 protect the ss-T strand from nucleases inside the plant cytoplasm by attaching to the 5' end. Both VirD2 and VirE2 proteins have nuclear localization signals (NLS) which serve as pilot proteins to guide the mature T-complex to the plant nucleus.
- The efficiency of transfer is enhanced by VirC2 proteins, which recognize and bind to the overdrive enhancer element.
- Some additional proteins like importins, VIP1, and VirF may interact with the T-strand, either directly or indirectly, to form larger T-complexes in the plant cell. VirF directs the proteins coating the T-complex (VIP1 and VirE2) for destruction in the proteasome.
- Inside the nucleus, ss-TDNA is converted into ds-TDNA, which gets integrated into the plant genome via a process called illegitimate recombination.

9.28. DISARMED TI-PLASMID DERIVATIVES AS PLANT VECTORS

9.28.1. Prototype Disarmed Ti Vector

- Ti plasmid is a natural vector for genetically engineering plant cells due to its ability to transfer T-DNA from the bacterium to the plant genome. But wild-type Ti plasmids are not suitable as vectors due to the presence of oncogenes in T-DNA that cause tumor growth in the recipient plant cells. For efficient plant regeneration, vectors with disarmed T-DNA are used by making it non-oncogenic by deleting all of its oncogenes. The foreign DNA is inserted between the RB and LB and then integrated into the plant genome without causing tumors; see Figure 9.21.
- For example, Zambryski et al. (1983) substituted *pBR322* sequences for almost all of the T-DNA of *pTiC58*, leaving only the left and right border regions and the *NOS* gene. The resulting construct was called *pGV3850*.

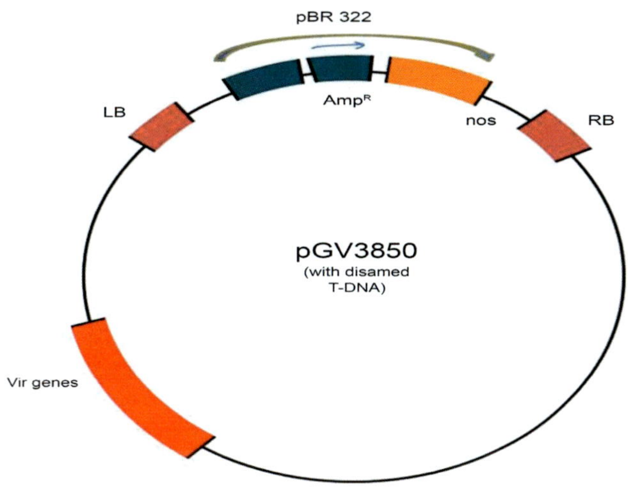

Figure 9.21. *Structure of the Ti-plasmid pGV3850 with disarmed T-DNA.*

Source: Made by the author.

- No t□mor cell formation takes place when modified T-DNA is transferred from *Agrobacterium* carrying *pGV3850* plasmid. The evidence of transfer is done by screening the cells for nopaline production.
- The creation of disarmed T-DNA is an important step forward, but the absence of tumor formation makes it necessary to use an alternative method for the identification of transformed plant cells. Opine production using *pGV3850* was exploited as a screenable phenotype, and the *ocs* and *nos* genes are now widely used as screening markers.

9.28.2. Drawbacks

Several drawbacks are associated with disarmed Ti- vector systems as discussed below;

- Necessity to carry out enzymatic assays on all potential transformants.
- Not convenient as experimental gene vectors due to large size.
- Diffic□lty in *in vitro* manipulation and
- Absence of unique restriction sites in the T-DNA.

Several approaches are employed to overcome the limitations associated with disarmed Ti vectors. The transformed plant cells can be identified based on drug or herbicide resistance by inserting selectable markers into the T-DNA.

9.28.3. Co-Integrate Vectors

Co-integrate vectors are deletion derivatives of Ti-plasmids. The DNA to be introduced into the plant transformation vector is sub-cloned in a conventional *Escherichia coli* plasmid vector for easy manipulation, producing a so-called intermediate vector. These vectors are incapable of replication in *A. tumefaciens* and also lack conjugation functions. The transfer is achieved using 'triparental mating' in which three bacterial strains are mixed together:

i. An *E. coli* strain carrying a helper plasmid able to mobilize the intermediate vector in trans;

ii. The *E. coli* strain carrying the recombinant intermediate vector;

iii. *A. tumefaciens* carrying the Ti plasmid.

iv. Conjugation between the two *E. coli* strains transfers the helper plasmid to the carrier of the intermediate vector, which in turn is mobilized and transferred to the recipient *Agrobacterium*. Homologous recombination between the T-DNA sequences of the Ti plasmid and intermediate vector forms a large co-integrate plasmid resulting in the transfer of recombinant T-DNA to the plant genome.

v. In the co-integrate vector system, maintenance of the recombinant T-DNA is dependent on recombination, which can be enhanced by extensive homology between the two plasmids, e.g., Ti plasmid *pGV3850* carrying a segment of the *pBR322* backbone in its T-DNA.

9.28.4. Binary Vector

- Binary vector was developed by Hoekma et al. (1983) and Bevan in (1984).
- It utilizes the trans-acting functions of the Vir genes of the Ti-plasmid and can act on any T-DNA sequence present in the same cell.
- The binary vector contains transfer apparatus (the Vir genes) and the disarmed T-DNA containing the transgene on separate plasmids.

9.28.5. Advantages of Binary Vector

- Small size due to the absence of border sequences needed to define the T-DNA region and Vir region.
- Ease of manipulation

9.28.6. Binary Vector System

- A plasmid carrying T-DNA containing LB and RB, called mini-Ti or micro-Ti can be sub-cloned in a small *E. coli* plasmid for ease of manipulation.

- The T-DNA of mini-Ti can be introduced into an *Agrobacterium* strain carrying a Ti plasmid from which the T-DNA has been removed but contains the vir region. The vir genes function *in trans,* causing the transfer of the recombinant T-DNA to the plant genome.
- The T-DNA plasmid can be introduced into *Agrobacterium* by triparental mating or by a more simple transformation procedure, such as electroporation.
- In a Ti-plasmid transformation system, the T-DNA is maintained on a shuttle vector with a broad host range origin of replication, such as *RK2* (that functions in both *A. tumefaciens* and *E. coli*), or separate origin for each species.
- An independently replicating vector is advantageous because the maintenance of the T-DNA does not rely on recombination. The copy number of the binary vector is not determined by the Ti plasmid, making the identification of transformants m□ch easier.
- All the conveniences of bacterial cloning plasmids are incorporated into binary vectors, such as multiple unique restriction sites in the T-DNA region to facilitate subcloning, the *lacZ* gene for blue–white screening and a λ *cos* site for preparing cosmid libraries.

Examples of Binary Vector System

- *pBIN19*- one of the first binary vectors developed in the 1980s and was widely used.
- *pGreen*- A later developed vector with more advanced features than pBIN19.
- Both vectors contain the *Lac Z* gene for blue-white screening of recombinants. The reduction in the size of *pGreen* is due to the presence of the *pSa* origin of replication. An essential replicase gene is housed on a second plasmid called *pSoup,* which functions in trans. All conjugation functions have also been removed, so this plasmid can only be introduced into *Agrobacterium*.

9.28.7. Transformation Techniques Using Agrobacterium

- The important requirements for *Agrobacterium*-mediated gene transfer in higher plants are:
- The plant explant must produce acetosyringone or other related compounds for the activation of vir genes, or *Agrobacterium* may be pre-induced with synthetic acetosyringone.

- The induced agrobacteria should have access to cells that are competent for transformation.
- Transformation-competent cells/tissues should be able to regenerate into whole plants.

CHAPTER 10

BIOSAFETY GUIDELINES

Contents

The growing concerns arising from genetically modified organisms throughout the world, the WHO/UNIDO/FAO/UNEP has built up an informal working group on Biosafety. The group prepared in 1991 a "Voluntary Code of Conduct for the Release of Organisms into the Environment." The ICGEB has also played an important role in issues related to biosafety and the environmentally sustainable use of biotechnology. ICGEB has annual workshops on biosafety and on risk assessment for the release of GMOs. It collaborates with the management of UNIDO's BINAS (Biosafety Information Network and Advisory Service), aimed at monitoring the global development in regulatory issues in biotechnology. The ICGEB has also provided an online bibliographic database on biosafety and risk assessment for the environmental release of GMOs.

10.1. GENERAL CONCERNS

- Higher cost of agriculture
- Ethical Issues
- Loss of familiarity

10.2. RISKS TO THE ENVIRONMENT

- Unpredictable gene expression
- Instability of transgene
- Increased use of chemicals in agriculture
- Resistant of targeted organisms
- Susceptibility of non-target organisms

10.3. RISKS TO HUMAN HEALTH

- Pathogens
- Drug Resistance
- Antibiotic Resistance
- Allergies
- Food Quality
- Food Safety
- Toxicity

10.4. RISKS OF INTERACTION WITH NON-TARGET ORGANISMS

- Generation of New Line viruses by recombination
- Horizontal Gene transfer
- Genetic Pollution through pollen or seed dispersal
- Transfer of foreign genes to microorganisms i.e., DNA uptake

10.5. BIOSAFETY GUIDELINES AND REGULATION

Many countries have formulated biosafety guidelines for rDNA manipulation with the aims of minimizing the probability of accidental release of GMMs and banning the deliberate release of such organisms into the environment. In India, the Department of Biotechnology (DBT) has developed the "Recombinant DNA Safety Guidelines" to exercise powers conferred through the Environmental Protection Act (1986) for the manufacture, use, import, export, and storage of hazardous microorganisms/genetically engineered organisms, cells, etc. These guidelines are implemented through three mechanisms: The Genetic Engineering Approval Committee (GEAC) of the Ministry of Environment and Forest has the authority to permit large-scale use of GMOs at a commercial level and open field trials of transgenic materials, including agricultural crops, industrial products, healthcare, etc. (Annual Report 1995–96 DBT). The Institutional Biosafety Committee (IBSCs) monitors research activities at the institutional level. The Review Committee on Genetic Manipulation (RCGM) operating within the DBT allows for risky research activities in laboratories.

- **Main Function of Institutional Biosafety Committees (IBSC)**
  - To note
  - To approve
  - To see approval of RCGM and to recommend
- **Main Function of the Review Committee on Genetic Manipulation (RCGM)**
  - To recommend generation of appropriate biosafety and agronomic data
  - To approve
  - To note
- **Main Function of Genetic Engineering Approval Committee (GEAC)**
  - Inform decision to administrative ministry to follow PVP/ Seed Act

- To approve open release to environment
- To approve for large scale use.

- **Main Function of the Indian Council of Agriculture Research (ICAR)**
  - To generate complete agronomic data on transgenics
  - To recommend suitable transgenics for commercial release
- **Main Function of MEC**
  - Recommend safe and agronomically viable transgenics
  - Inspect facility
  - Analyze data
  - Visit trial sites

To perform certain gene manipulation experiments, the workers need to acquire permission from the review committee and approval committee before commencement. For example: Toxin gene cloning, cloning of genes for vaccine production, cloning antibiotic resistance genes in pathogenic microbes, cloning oncogenes in cell cultures, cloning genes taken from tumor cells, genetic manipulation in animals, etc.

CHAPTER 11

BIOINFORMATICS

Contents

In the last few decades, advances in molecular biology and the equipment available for research in this field have increased the rapid seq□encing of large portions of the genomes of several species. In fact, to date, several bacterial genomes, more than 568, four animal genomes such as *Drosophila*, humans, and at least three plant genomes (*Arabidopsis thaliana* from dicotyledons, Rice from monocotyledons, and poplar from trees) have been sequenced. Several other genomes from microbial, animal, and plant kingdoms are actively being sequenced. The Human Genome Project, designed to sequence all of the human chromosomes, has been completed.

- Bioinformatics is a discipline emerging from a combination of Biology, Mathematics, and information technology.
- Bioinformatics is a science associated with database-like activities involving persistent sets of data that are maintained in a consistent state over essentially indefinite periods of time.
- Bioinformatics is the science of developing computer databases and algorithms for the purpose of speeding up and enhancing biological research.
- It is an integration of mathematical, statistical, and computational methods to analyze biological, biochemical, and biophysical data.
- "Biology easily has 500 years of exciting problems to work." *Donald Knuth (Stanford Professor & famous computer scientist)*; *see* Figure 11.1.

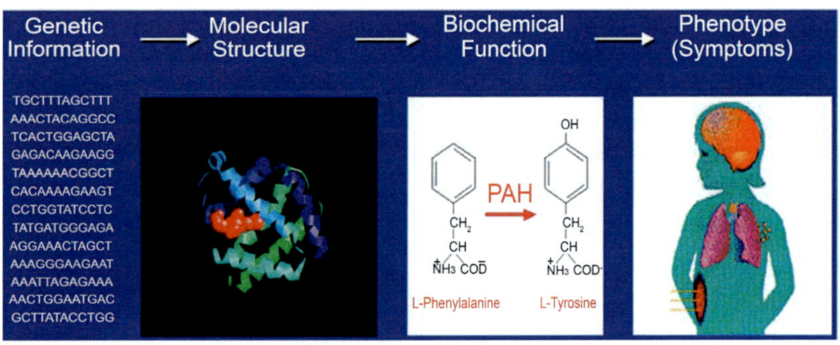

Figure 11.1. *Biology.*

Source: Adopted from introduction to bioinformatics lopresti bioS 10, November 2009.

By developing techniques for analyzing sequence data and related structures, we can attempt to understand the molecular basis of life; *see* Figure 11.2.

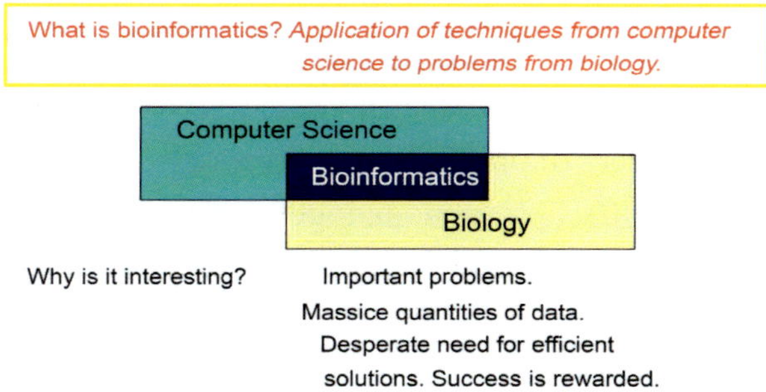

Figure 11.2. *Bioinformatics.*

Source: Adopted from introduction to bioinformatics lopresti bioS 10, November 2009.

11.1. DIFFERENT VARIETIES OF AVAILABLE BIOINFORMATICS TASKS

- Creation of a biological database
- Search similarity of sequences
- Discovery of coded genes and also their structure
- DNA and protein function and annotation
- Elucidation of protein structure
- Drug design
- Biochemical analysis (metabolic pathway).
- Tools for analysis of high-throughput molecular biology experimentation data (Microarrays, Mass spectrometry, etc.).
- Comparative genome analysis
- Building gene and protein networks
- Scientific literat□re search

11.2. COMPUTATIONAL BIOLOGY

It refers to algorithmic tools to facilitate biological analysis, whereas bioinformatics looks at the structure and comprises the entire collection of the information management system, analysis tools, and communication networks supporting biology. Computational biology involves the following points which are described below:

- Finding the genes in the DNA sequences of various organisms.

- Developing methods to predict the structure and or function of newly discovered proteins and structural RNA sequences.
- Clustering protein sequences into families of related sequences and the development of protein models.
- Aligning similar proteins and generating phylogenetic trees to examine evolutionary relationships; *see* Figure 11.3.

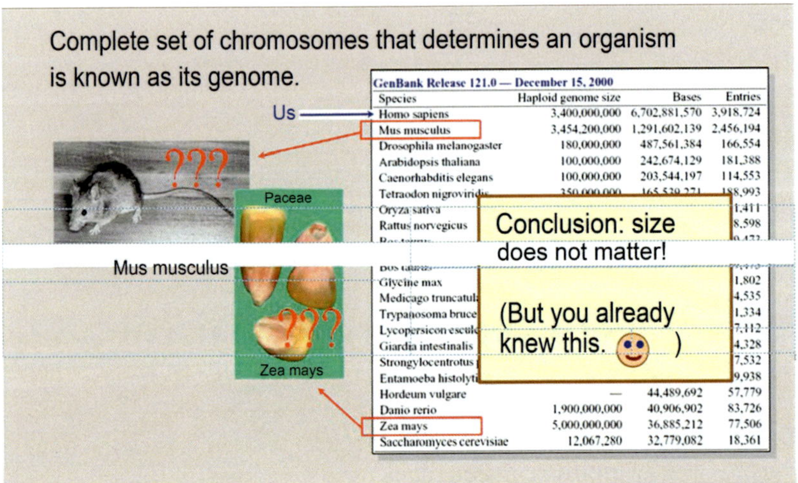

Figure 11.3. *Generating phylogenetic trees to examine evolutionary relationships.*

Source: Adopted from introduction to bioinformatics lopresti bioS 10, November 2009.

- Data Explosion
  - Genome

11.3. ALGORITHMS ARE VITAL

- An algorithm is a set of rules for calculating or problem-solving carried out by a computer program.
- An *algorithm* is a precisely-specified series of steps to solve a particular problem of interest.
- Develop model(s) for the task at hand.
- Study inherent computational complexity:
- Can the task be phrased as an optimization problem?
- If so, can it be solved efficiently? Speed, memory, etc.

- If we can't find a good algorithm, can we prove the task is "hard"? If known to be hard, is there an approximation algorithm (one that works at least some of the time or comes close to optimal).
- Conduct experimental evaluations (perhaps iterate the above steps).

11.4. SEQUENCING GENOME

Most genomes are enormous (e.g., 10^{10} base pairs in the case of a human). Current sequencing technology, on the other hand, only allows biologists to determine ~10^3 base pairs at a time. This leads to some very interesting problems in bioinformatics; *see* Figure 11.4.

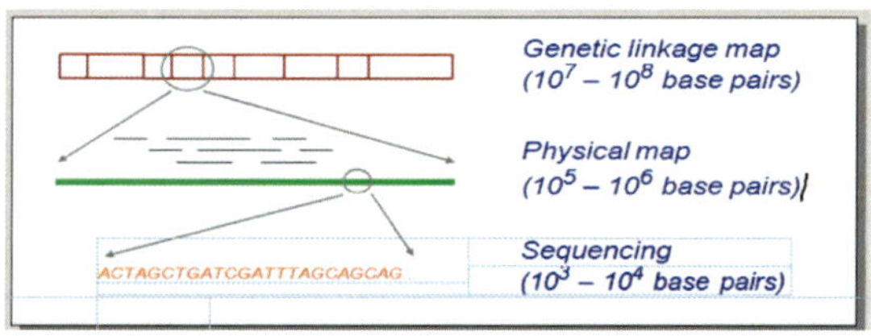

Figure 11.4. *Huge genomes.*

Source: Adopted from introduction to bioinformatics lopresti bioS 10, November 2009.

Genomes can also be determined using a technique known as shotgun sequencing.

11.4.1. Sequence Assembly

A simple model of DNA assembly is the *Shortest Super Sequence Problem*: given a set of sequences, find the shortest sequence *S* such that each of the original sequences appears as a subsequence of *S*. Look for overlap between *prefix* of one sequence and *suffix* of another*; see* Figure 11.5.

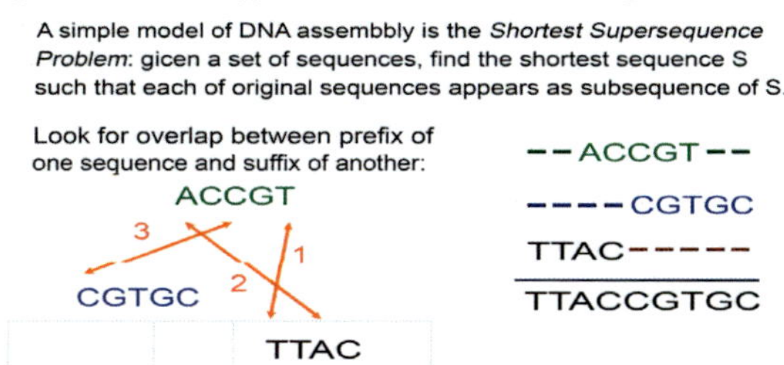

Figure 11.5. *Overlap between prefix of one sequence and suffix of another.*

Source: Adopted from introduction to bioinformatics lopresti bioS 10, November 2009.

11.5. SKETCH OF ALGORITHM

Create an overlap graph in which every node represents a fragment and edges indicate overlap. Determine which overlaps will be used in the final assembly: find an *optimal spanning forest* in the overlap graph; *see* Figure 11.6.

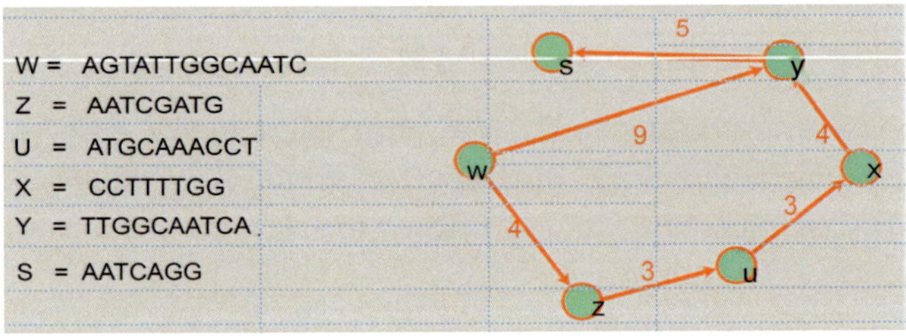

Figure 11.6. *Overlap graph construction.*

Source: Adopted from introduction to bioinformatics lopresti bioS 10, November 2009.

Look for paths of maximum weight: use a greedy algorithm to select an edge with the highest weight at every step. The selected edge must connect nodes with in- and out-degrees <= 1. One may end up with a set of paths: each corresponds to a contig fig□re; *see* Figure 11.7.

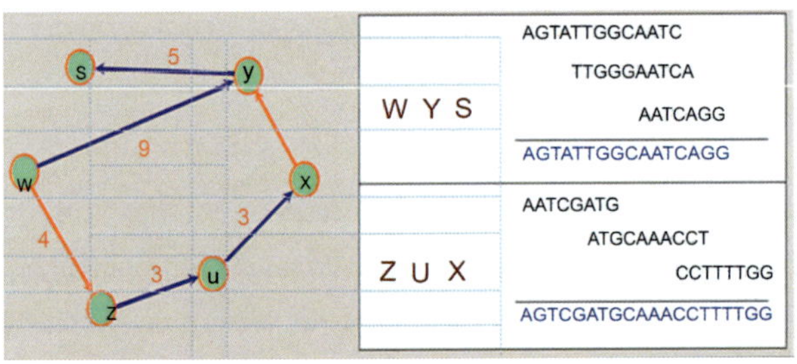

Figure 11.7. *A contig figure.*

Source: Adopted from introduction to bioinformatics lopresti bioS 10, November 2009.

Scientists build phylogenetic trees in an attempt to understand evolutionary relationships. Reversal distance is often used here.

11.6. DNA MICROARRAY

This allows the simultaneous measurement of the level of transcription for every gene in a genome (gene expression). Differential expression changes over time. The single microarray can test ~10k genes. Data obtained faster than can be processed to find genes that behave similarly. A pattern discovery problem (Green = repressed and Red = induced).

- **K-Means Clustering Is One Way to Organize This Data**
  - Given a set of *n* data points and an integer *k*.
  - We want to find a set of *k* points that minimizes the mean-squared distance from each data point to its nearest cluster center.
- **Sketch of Algorithm**
  - Choose *k* initial center points randomly and cluster data.
  - Calculate new centers for each cluster using points in the cluster. Re-cluster all data using new center points.
  - Repeat the second two steps until no data points are moved from one cluster to another or some other convergence criterion is met.

11.7. APPLICATIONS OF BIOINFORMATICS

- Data acquisition: High throughput biology requires automatic capture and conversion of biological data to symbiotic digital representation without human intervention.
- Assembly of sequence contigs: Large scale sequencing projects require sequence assembly algorithms. Bioinformatics will help in generating contigs because software has been developed.
- Predicting functional domains in genomic sequences.
- Sequence alignment
- Searching structure databases
- Determination of macromolecular structure.
- Molecular evolution.
- Bioinformatics in medicine.
- Bioinformatics in agriculture
- Questions

1. What is bioinformatics and explain its application?
2. Explain about computational Biology?
3. Give an account of Data explosion?
4. Explain how algorithms are central.
5. What is the sequence nature of biology?

6. Write an essay on sequencing genome.
7. Explain DNA microarray and its applications?
8. Explain about sketch of the Algorithm?
9. What do you mean by clustering microarray data?'
10. Define Bioinformatics applications and limitations?

- Subjective Questions

1. What is biotechnology and discuss in brief the different areas of interest in biotechnology?
2. What is the achievement of biotechnology?
3. Explain the scope of animal biotechnology?
4. What is the importance of biotechnology in different fields?
5. Write an essay on the global impact of biotechnology?
6. Give a brief note on the potential of modern biotechnology?
7. Write short notes on the following:
   a. ICGEB
   b. OECD
   c. NBTB
   d. DBT
   e. Lal Bahadur Shashtri Centre for Biotechnology
   f. Biotechnology in India
   g. Biodiversity
   h. Traditional Biotechnology
   i. Modern Biotechnology
   j. Ban on Genetic food
8. Give a brief account of the emergence of modern biotechnology?
9. Why can microbiology, genetics and biochemistry, without combining with technology, not be called biotechnology?
10. Write an essay on the history of biotechnology?
11. How did traditional biotechnology help in the evolution of biotechnology?
12. What do you mean by a gene?
13. Discuss in detail the different units of a gene?
14. Give an account of the overlapping gene.
15. Disc□ss in detail different approaches made for the artificial synthesis of genes with emphasis on the works of H.G. Khorana?

16. What is chimeric DNA?
17. Write an essay on the organization of DNA in eukaryotes?
18. Write an essay on gene regulation by the lac operon model?
19. Write a short note on Chargaff's rule of equivalence?
20. What is electrophoresis? Describe the principle and working of the method?
21. Give a detailed account of working mass spectrometry?
22. What is pulsed electrophoresis?
23. What is the polymerase chain reaction and its application?
24. What is a restriction and modification system?
25. What are the DNA enzymes?
26. Give a detailed account of DNA polymerase and terminal transferase?
27. Write short notes on alkaline phosphatase, SI nuclease, Terminal transferases, sticky ends, palindromes, Linkers and adaptors, DNA ligases and Type III restriction enzyme?
28. What are plasmids?
29. What is cloning and cloning vector?
30. Explain about bacteriophage vector?
31. What is phagemid? Give a brief account of phagemids used as vectors?
32. Discuss in detail the yeast plasmid vectors?
33. What is a cloning vector?
34. Define yeast artificial chromosomes?
35. How is the copy number of the plasmid vector and yield of the recombinant protein related to each other?
36. Can exonuclease be used while producing a recombinant DNA molecule?
37. What are the features of a plasmid being used as a cloning vector?
38. What are competent cells? What does the word "competent" refer to?
39. What do "Eco," "R" and "I" refer to in the enzyme EcoRI?
40. Why are proteases added while isolating the DNA?
41. If the "denaturation" step is missed during PCR, what would be its effect on the entire process?
42. Name a recombinant vaccine.

43. How is the Ti plasmid of *Agrobacterium tumefaciens* modified to convert it into a cloning vector?
44. Do biomolecules such as DNA, proteins exhibit biological activity in anhydrous conditions?
45. What is Biotechnology?
46. What is Genetic engineering?
47. What are the Principles of Biotechnology?
48. How is Biotechnology useful in developing food crops and in the agriculture process? What are the different types of biotechnology?
49. **What do you understand by gene cloning?**
50. What is sterilization?
51. Explain the basic equipment in the cell and tissue culture laboratory?
52. Describe the hazards and safety in the cell culture laboratory?
53. What are the basics of animal cell culture?
54. What is media preparation?
55. What is the role of important components of culture media?
56. What is recombinant DNA?
57. Define major steps involved in the technology?
58. How does rDNA work?
59. Explain the applications of RNA technology?
60. What are model organisms?
61. What is the significance of model organisms?
62. Define types of molec□lar markers?
63. Describe the classification of molec□lar markers?
64. What are the principles of nucleic acid-based methods?
65. What is the need for cryopreservation?
66. Define the steps involved in cryopreservation?
67. What are the methods of cryopreservation?
68. What is cryopreservation as a strategy to establish somatic cell banking?
69. Explain the progress in somatic cell nuclear transfer?
70. Explain the methodology of animal cloning?
71. What is embryo developmental block and reprogramming in cloning?
72. Explain about requirements of the animal cell culture laboratory?

73. What are gene therapy and its applications?
74. Define the s□bstrate □sed for animal cell c□lt□re?
75. Explain about maintenance of aseptic condition?
76. Define the cryopreservation storage of animal cells?
77. Explain polymerase chain reaction?
78. Explain enzyme immobilization and its uses?
79. Define the application of transgenic animal technology?
80. Differentiate hybridoma technology and monoclonal antibodies?
81. Explain the biosafety guidelines in India?
82. Define intellect□al property rights with some examples?
83. Write a short note on:
    - Microarray
    - Biochips,
    - Bioweapons
    - –Human Genome project

- More Subjective Questions

1. Explain the classification of carbohydrates with s□itable examples?
2. What are monosaccharides?
3. What are mucopolysaccharides? Mention Describe their biomedical importance with suitable examples?
4. Compare and contrast the structural differences between starch and glycogen?
5. Differentiate between epimerism and anomerism?
6. What is mutarotation?
7. Why is sucrose a non-reducing sugar?
8. Differentiate about the composition of sucrose, maltose and lactose?
9. Define the classification of lipids with s□itable examples?
10. Explain abo□t the essential fatty acids – definition, examples and important functions?
11. What are prostaglandins?
12. What are phospholipids – definition, types and f□nctions?
13. Define l□ng s□rfactants and respiratory distress syndrome?
14. What is the structure of cholesterol and its biological importance?
15. Define the classification of amino acids based on n□tritional requirements with suitable examples?

16. Explain the classification of amino acids based on chemical str□ct□re with suitable examples?
17. Explain the structural organization of proteins (primary, secondary, tertiary and quaternary structure)?
18. Which bonds are involved in stabilizing the structure of proteins?
19. Describe the classification of proteins based on their biological function?
20. Define the biologically important peptides (biological importance of peptides)?
21. Explain the essential amino acids?
22. Explain isoelectric pH and isoelectric precipitation of proteins?
23. Explain the denaturation of proteins and agents causing denaturation?
24. What is glutathione?
25. Define the Watson and Crick model of DNA?
26. Explain the structure and function of DNA?
27. Explain different types of RNA and their functions?
28. What is the difference between RNA and DNA?
29. Explain the structure and function of mRNA?
30. Explain the structure and function of tRNA?
31. Explain the base pairing rules and Wobble hypothesis?
32. What is DNA replication prokaryotic and eukaryotic replication, requirements and process?
33. What are inhibitors?
34. Define DNA polymerases?
35. What are Okazaki fragments?
36. Explain the Transcription-Transcriptional units, promoter regions, RNA polymerases in prokaryotes and eukaryotes, process, inhibitors?
37. Differences between prokaryotic and eukaryotic transcription?
38. Explain abo□t post-transcriptional modifications?
39. Explain about translation in eukaryotes – process and inhibitors?
40. Explain post-translational modifications?
41. Explain the genetic code and its characteristics?
42. Explain the regulation of gene expression in prokaryotes (Lac Operon concept)?
43. Define the reg□lation of gene expression in e□karyotes?

44. What is recombinant DNA technology?
45. Explain the DNA cloning process and application?
46. What is PCR – requirement, steps and application?
47. Describe the blotting techniques – southern blotting, northern blotting and western blotting?
48. What are vectors, and gene therapy?
49. What are DNA probes?
50. What is RFLP?

Objective Questions

1. Restriction enzymes were discovered by
   a. Smith and Nathans
   b. Alexander Fleming
   c. Berg
   d. None

   Answer: A
2. Bacteria protect themselves from viruses by fragmenting viral DNA with
   a. Ligase
   b. Endonuclease
   c. Exonuclease
   d. Gyrase

   Answer: B
3. Klenow fragment is derived from
   a. DNA Ligase
   b. DNA Pol-I
   c. DNA Pol-II
   d. Reverse Transcriptase

   Answer: B
4. Southern blotting is
   a. Attachment of probes to DNA fragments
   b. Transfer of DNA fragments from electrophoretic gel to a nitrocellulose sheet
   c. Comparison of DNA fragments to two sources

d. Transfer of DNA fragments to electrophoretic gel from cellulose membrane

Answer: B

5. ELISA is

a. Using radiolabeled second antibody

b. Usage of RBCs

c. Using complement-mediated cell lysis

d. Addition of substrate that is converted into a colored end product

Answer: D

6. The Golden Rice variety is rich in

a. Vitamin C

b. B-carotene and ferritin

c. Biotin

d. Lysine

Answer: B

7. The DNA fragments have sticky ends due to

a. Endonuclease

b. Unpaired bases

c. Calcium ions

d. Free methylation

Answer: B

8. Plasmids are used as cloning vectors for which of the following reasons?

a. Can be multiplied in culture

b. Self-replication in bacterial cells

c. Can be multiplied in laboratories with the help of enzymes

d. Replicate freely outside bacterial cells

Answer: B

9. The human genome project was launched in the year

a. 1980

b. 1973

c. 1990

d. 1989

Answer: C

10. The vaccines prepared through recombinant DNA technology are
    a. Third generation vaccines
    b. First-generation vaccines
    c. Second-generation vaccines
    d. None

    Answer: A

11. Which is a genetically modified crop?
    a. Bt-cotton
    b. Bt-brinjal
    c. Golden rice
    d. All

    Answer: D

12. The PCR technique was invented by
    a. Karry Mullis
    b. Boyer
    c. Sanger
    d. Cohn

    Answer: A

13. The first transgenic plant to be prod□ced is
    a. Brinjal
    b. Tobacco
    c. Rice
    d. Cotton

    Answer: B

14. RNA interference helps in
    a. Cell proliferation
    b. Micropropagation
    c. Cell defense
    d. Cell differentiation

    Answer: C

15. Which of the following is the quality of improved transgenic basmati rice?
    a. Gives high yield but no characteristic aroma
    b. Gives high yield and is rich in vitamin A

c. Does not require chemical fertilizers and growth hormones

d. Resistant to insects and diseases

Answer: B

16. The first clinical application of gene therapy over a 4-year-old girl was for

a. Adenosine deaminase deficiency

b. Adenosine deficiency

c. Growth deficiency

d. Adenine deficiency

Answer: A

17. Excision and insertion of a gene is called

a. Biotechnology

b. Genetic engineering

c. Cytogenetics

d. Gene therapy

Answer: B

18. The expression of a transgene in the target tiss□e is identified by a

a. Transgene

b. Promoter

c. Enhancer

d. Reporter

Answer: D

19. ——— is used as a vector for cloning into higher organisms

a. Retrovirus

b. Baculovirus

c. *Salmonella typhimurium*

d. *Rhizopus nigricans*

Answer: A

20. Which bacterium is used in the production of insulin by genetic engineering?

a. Saccharomyces

b. Rhizobium

c. Escherichia

d. Mycobacterium

Answer: C

21. The foundation for the development of cell culture technique was laid by
    a. Roux
    b. Arnold
    c. Ross
    d. Harrison

    Answer: A

22. Who is regarded as the father of tissue culture
    a. Harrison
    b. Arnold
    c. Ross
    d. Roux

    Answer: A

23. The limited replicative capacity of human cells in culture is called
    a. Hayflick effect
    b. Brownian effect
    c. Contact inhibition
    d. None of the above

    Answer: A

24. HAT medium used for the selection of fused hybrid cells in hybridoma culture was introduced by
    a. Littlefield and Miller
    b. Kohler and Milstein
    c. Frish and Jentoft
    d. Eagle and Karl

    Answer: A

25. Which of the following is HGPRT+ and survives in HAT medium
    a. B cells
    b. Myeloma cells
    c. Hybrid cells
    d. Both a and c

    Answer: D

26. Use of trypsin to free cells from tissue matrix for cell culture was described by
    a. Jolly and Koch

b. Beebe and Ewing

c. Sims and Stillman

d. Gottlieb Maberland

Answer: C

27. Cells used in feeder layer

a. Should have the ability to divide

b. Have the ability to metabolize

c. These properties are obtained by exposing cells to irradiation

d. All of the above

Answer: D

28. Cell culture technique became simpler only after the advent of

a. Antibiotics

b. Trypsin

c. Cell culture media

d. All of the following

Answer: D

29. Which of the following is true regarding the animal cell culture technique

a. Lactic acid is the source of carbon

b. Cells have a high requirement of L-glutamine

c. Cholin is necessary for cell adhesion and cytoskeleton

d. All of the following

Answer: D

30. Which of the following is the str□ct□ral fiber in cell c□lt□re

a. Collagen

b. Elastin

c. Fibronectin

d. Both a and b

Answer: D

31. Optimum pH required for the growth of mammalian cells is

a. 5.3–7.0

b. 6.5–7.0

c. 7.2–7.4

d. 8.1–8.9

Answer: C

32. The optimum temperature of any cell or organ is called

a. Critical temperature

b. Threshold temperature

c. Ceiling temperature

d. None of the above

Answer: C

33. For the culture of avian cells the optimum temperature requirement is

a. 37°C

b. 40°C

c. 42°C

d. 35°C

Answer: B

34. Which of the following is the oldest cell line

a. HeLacell line

b. Vero cell line

c. CHO cell line

d. BHK cell line

Answer: A

35. Which cell line is used for the production of recombinant sex hormones

a. BHK cell line

b. Vero cell line

c. HeLacell line

d. CHO cell line

Answer: D

36. The ratio of CO_2: O_2 used in cell culture system should be

a. 1:5

b. 1:13

c. 1:19

d. 1:25

Answer: C

37. Which of the following is the most commonly used cell fusing agent
    a. PEG
    b. $NaNO_3$
    c. Sendai virus
    d. Polyvinyl alcohol

    Answer: A

38. Which of the following virus is used for cell fusion
    a. Sendai virus
    b. Herpes virus
    c. Myxovirus
    d. All of the following

    Answer: D

39. Which of the following is easy and rapid method to interpret viability of cells in culture system
    a. Trypan blue dye exclusion
    b. Neutral red assay
    c. Fluorescein dye assay
    d. All of the following

    Answer: A

40. Vero cell line is obtained from Kidney cells of
    a. African green monkey
    b. Chinese hamster
    c. Calf
    d. Swine

    Answer: A

BIBLIOGRAPHY

1. Abollino, O., Giacomino, A., Malandrino, M., Mentasti, E., Aceto, M., & Barberis, R. (2006). Assessment of metal availability in a contaminated soil by sequential extraction. *Water Air Soil Pollution, 173*(1–4), 315–338.
2. Agarwal, V. K. (2000). *Molecular Biology*. New Delhi: S. Chand.
3. Agarwal, V. K., & Verma, P. S. (2000). *Concepts of Molecular Biology*. New Delhi: S. Chand.
4. Ahmadpour, P., Ahmadpour, F., Mahmud, T., Abdu, A., Soleimani, M., & Tayefeh, F. H. (2014). Phytoremediation of heavy metals: A green technology. In E. Asrari (Ed.), *Heavy Metal Contamination of Water and Soil: Analysis, Assessment, and Remediation Strategies* (p. 249). CRC Press, Taylor & Francis.
5. Alexander, M. (1977). *Introduction to Soil Microbiology*. John Wiley & Sons: New York.
6. Ali, H., Khan, E., & Sajad, M. A. (2013). Phytoremediation of heavy metals–Concepts and applications. *Chemosphere, 91*(7), 869–881.
7. Amirato, P. V. (1983). In D. A. Evans, W. R. Sharp, P. V. Amminrato, & Y. Yamada (Eds.), *Handbook of Plant Cell Culture: Vol. I, Techniques for Propagation and Breeding* (pp. 82–123). Macmillan.
8. Anand, A., et al. (2004). *Physiology of Biology*, *10*, 233–242.
9. Aneja, K. R. (2014). *Experiments in Microbiology, Plant Pathology and Biotechnology* (4th ed.). New Age International (P) Limited Publisher.
10. Aneja, K. R. (2014). *Laboratory Manual of Microbiology and Biotechnology*. Scientific International PVT. LTD MedTech.
11. Atkins, F. C. (1983). Mushroom growing in great britain *Mushroom J.*, *125*, 168–171.

12. Atkinson, G. F. (1961). *Mushrooms* (2nd ed.). Hofner Publishing Co.
13. Baker, K. F., & Snyder, W. C. (1965). *Ecology of Soil Borne Plant Pathogens–Prelude to Biological Control*. John Murray. 978-0719500596.
14. Barabasz, A., Wilkowska, A., Tracz, K., R□szczyńska, A., B□lska, E., & Mills, R. F. (2013). Expression of HvHMA2 in tobacco modifies Zn-Fe-Cd homeostasis. *Journal of Plant Physiology, 170*, 1176–1186.
15. Becking, J. H. (1982). In N. S. Subba Rao (Ed.), *Advances in Agricultural Microbiology* (pp. 89–110). Oxford & IBH Publishing Co.
16. Bergersen, F. J. (1971). *Ann. Rev. Plant Physiol., 22*, 121–140.
17. Bhaskaran, S. (1987). In S. Natesh, V. L. Chopra, & S. Ramachandan (Eds.), *Biotechnology in Agriculture* (pp. 101–108). Oxford & IBH Publishing Co.
18. Biello, D. (2006). *Tobacco Plant Transformed into Plague Vaccine Factory.*
19. Binns, A. N. (1981). In M. J. Constantin, R. R. Henke, K. W. Hughes, & B. V. Conger (Eds.), *Propagation of Higher Plants Through Tissue Culture: Emerging Technologies and Strategies* (pp. 325–332). *Environmental and Experimental Botany, 21*.
20. Boffey, S. A. (1987). In M. D. Trevan, S. Boffey, K. H. Goulding, & P. Stanbury (Eds.), *Biotechnology: The Biotechnology Principles* (pp. 11–152). Open University Press.
21. Bouwer, E. J., & Zehnder, A. J. B. (1993). *Trends in Biotechnology, 11*, 287–318.
22. Bo□wer, E. J., et al. (1998). In G. A. Lewandowski & L. J. Deflippi (Eds.), *Biological Treatment of Hazardous Wastes* (pp. 237–270).
23. Brooks, R., Lee, J., Reeves, R. D., & Jaffré, T. (1977). Detection nickeliferous rocks by analysis of herbarium specimens of indicator plants. *Journal of Geochemical Exploration, 7*, 49–57.
24. Bull, A. T., Holt, G., & Lilly, M. D. (1983). *Biotechnology: International Trends and Perspectives* (p. 84). Oxford & IBH Publishing Co. PVT. Ltd., New Delhi.
25. Carlson, P. S. (1973). *Science, 180*, 1366–1368.
26. Carlson, P. S. (1983). Cell and tissue culture techniques for cereal crop improvement. New York. (pp. 407–418)
27. Chaudhary, K., & Khan, S. (2016). Physiochemical characterization of fl□oride (F) contaminated soil and its microbe-assisted bioremediation by *Prosopis juliflora*. *Journal of Plant Biology and Soil Health, 3*(2), 8.
28. Chaudhary, K., Agarwal, S., & Khan, S. (2018). Role of phytochelatins

(PCs), metallothionines (MTs) and heavy metal ATPase (HMA) genes in heavy metal tolerance. In R. Prasad (Ed.), *Mycoremediation and Environmental Sustainability, Fungal Biology* (pp. 39–60). Springer. https://doi.org/10.1007/978-3-319-77386-5_2.

29. Chaudhary, K., Sumira, J., & Khan, S. (2016). Heavy metal ATPase (HMA2, HMA3 & HMA4) genes in hyperaccumulation mechanism of heavy metals. In *Plant Metal Interaction (Emerging Remediation Techniques)* (pp. 545–552).

30. Christov, L., & Van Driessel, B. (2003). *Indian Journal of Biotechnology, 2*, 444–450.

31. Clark, D. S. (1962). *Biotechnology and Bioengineering, 4*, 1721.

32. Cohen, S. N., Chang, A. C. Y., Boyer, H. W., & Helling, R. B. (1973). *Proc. Nat. Acad. Sci. (USA), 70*, 3240-3244.

33. Collins, H. A., & Watts, M. (1983). In D. A. Evans, W. T. Sharp, P. V. Ammirato, & Y. Yamada (Eds.), *Handbook of Plant Cell Culture: Techniques for Propagation and Breeding* (Vol. I, pp. 729–747). Macmillan.

34. Crueger, W., & Crueger, A. (1984). *Biotechnology–A Textbook of Industrial Microbiology* (pp. 207–216). Sinaver Associates Inc., Sunderland, USA.

35. Cummings, M. R., & Klug, W. S. (2014). *Concepts of Genetics*. New Delhi: Pearson Education. Benjamin-Cings Pub Co. 978-0321948472.

36.

37. Dodds, J. H., & Roberts, L. W.(1985). *Experiments in Plant Tissue Culture* (2nd ed.). Cambridge University Press.

38. Dubey, R. C. (2009). *A Textbook of Biotechnology*. S. Chand and Company LTD.

39. Ferber, D. (2004). The new science of cell hacking. *Popular Science,* 43, 44.

40. Fisher, R. A. (Statistical methods for research workers).Springer series in statistcs. 66-70.

41. Frazier, W. C., & Westhoff, D. C. (1993). *Food Microbiology*. New Delhi: Tata McGraw-Hill.

42. Garg, N., Garg, K. L., & Mukerjii, K. G. (2013). *Laboratory Manual of Food Microbiology*. I. K. International Publishing House Pvt. Ltd.

43. Gautam, N., Verma, P. K., Verma, S., Tripathi, R. D., Trivedi, P. K., Adhikari, B., & Chakrabarty, D. (2012). Genome wide identification of rice class I metallothionein gene: Tissue expression patterns and induction in response to heavy metal stress. *Functional and Integrative Genomics, 12*, 635–647.

44. Gibbs, W. (2001). Cybernetic cells. *Scientific American, August*, 54–57.
45. Greipsson, S. (2011). Phytoremediation. *National Education Knowledge, 3*(10), 7.
46. Gu, C. S., Liu, L. Q., Deng, Y. M., Zhu, X. D., Huang, S. Z., & Lu, X. Q. (2015). The heterologous expression of the Iris lacteal var. chinensis type 2 metallothionein IIMT2b gene enhances copper tolerance in *Arabidopsis thaliana. The Bulletin of Environmental Contamination and Toxicology, 94*, 247–253.
47. Guha, S., & Maheswari, S. C. (1964). *Nature (London), 204*, 297.
48. Guha, S., & Maheswari, S. C. (1966). *Nature, 212*, 97, 98.
49. Gupta, C. P., Sharma, A., Dubey, R. C., & Maheshwari, D. K. (1999). Biotechnology. *Cytobios, 99*, 183–189.
50. Haberlandt, G. (1902). Sber Akad. Wiss. Wien. Springer. 111, 69–92.
51. Hanson, E. D. (1983). *Recombinant DNA Research and the Human Prospect*. Washington, DC: American Chemical Society.
52. Henshaw, G. G. (1982). In *Plant Tissue Culture* (pp. 789–792). Tokyo.
53. Huang, G. Y., & Wang, Y. S. (2010). Expression and characterization analysis of type 2 metallothionein from grey mangrove species (*Avicennia marina*) in responses to metal stress. *Aquatic Toxicology, 99*, 86–92.
54. Jennis, P., Sharon, S., Patricia, J., & Judith, A. (2018). *Kuby Immunology*. ISBN-13 978-1319114701.
55. Kumar, H. D. (2003). *Modern Concepts of Biotechnology*. New Delhi: Vikash Publishing House.
56. Lee, S. (2014). Artificial ind□ction of cadmi□m tolerance and its f□rther enhancement via heterologous co-expression of SpHMT1 and AtPCS1 in the yeast cells. *Journal of Korean Society Applied Biological Chemistry, 57*, 307–310.
57. Lopez, D. A., Williams, R. M., & Michlke. (1994). *Enzymes: The Foundation of Life*. München: The Neville Press.
58. Ma, Y., Rajkumar, M., Luo, Y., & Freitas, H. (2011). Inoculation of endophytic bacteria on host and non-host plants–Effects on plant growth and Ni uptake. *Journal of Hazardous Materials, 195*, 230–237.
59. Maestri, E., Marmiroli, M., Visioli, G., & Marmiroli, N. (2010). Metal tolerance and hyperaccumulation: Costs and trade-offs between traits and environment. *Environmental and Experimental Botany, 68*(1), 1–13.
60. Maheswari, S. C., Rasid, A., & Tyagi, A. K. (1983). Anther/pollen culture for production of haploids and their utility*IAPTC Newsletter*, *41*, 2–9.
61. Mandal, P. K. (1988). Gene therapy. *Scientific Report, 25*, 156–157.

62. Mandels, M., & Stenberg, D. (1976). Recent advances in cellulose technology. *J. Ferment. Technol., 54*, 267–286.

63. Mason, H. S., & Arntzen, C. J. (1995). Transgenic plants as vaccine production systems. *Trends Biotechnol., 13*, 388–392.

64. Maxam, A. M., & Gilbert, W. (1977). A new method for sequencing DNA. *Proc. Natl. Acad. Sci. (USA), 74*, 560–564.

65. Memon, A. R., & Schroder, P. (2009). Implications of metal accumulation mechanisms to phytoremediation. *Environmental Science and Pollution Research, 16*(2), 162–175.

66. Miller, C. O., Skoog, F., Saltz, M., & Strong, F. M. (1955). *J. Am. Chem. Soc., 77*, 1992.

67. *Molecular Biotechnology in Life,* (2006). Author not mentioned. Retrieved from: http://www.goldenessays.com/free_essays/4/technology/molecular-biotechnology-in-life.shtml (accessed on 22 June 2024).

68. Morgan, T. H. (1901). *Regeneration.*by Thomas Hunt Morgan Macmillan: London. 1866-1945.

69. Murashige, T., & (1974). Plant propagation through tissue culture. *Annual Review Plant Physiology, 25*, 135–166.

70. Murashige, T., & Skoog, F. (1962). A revised medium for rapid growth and bioassays with tobacco tissue cultures. *Physiology Plant, 15*, 473–497.

71. Nitsch, C. (1974). A pollen culture a new techniques for mass. *C. R. Acad. Sci. (Paris), 278*, 1031–1034.

72. Peters, P. (1993). *Biotechnology–A Guide to Genetic Engineering*. WMC. Brown Publishers, Dubuque, IA.

73. Prakash, J., & Giles, K. L. (1987). Induction and growth of androgenic haploids intern. *Int. Rev. Cytol., 107*, 273–292.

74. Prescott, L. M., Harley, J. P., & Klein, D. A. (2003). *Microbiology*. McGraw-Hill Higher Education, New York.

75. Purohit, S. (2004). Biotechnology: Fundamental and Applications. Agrobios ISBN 10: 8188826510.

76. Rajkumar, M., Ae, N., Prasad, M. N. V., & Freitas, H. (2010). Potential of siderophore-producing bacteria for improving heavy metal phytoextraction. *Trends in Biotechnology, 28*(3), 142–149.

77. Reinert, J., & Bajaj, Y. P. S. (1977). In J. Reinert & Y. P. S. Bajaj (Eds.), *Applied and Fundamental Aspects of Plant Cell, Tissue and Organ Culture* (pp. 251–267). Springer-Verlag Berlin.

78. Ruschel, A. P., & Vose, P. B. (1983). In D. A. Evans, W. R. Sharp, P. V. Ammirato, & Y. Yamada (Eds.), *Handbook of Plant Cell Culture–*

Techniques for Propagation and Breeding, Vol. 1 (pp. 828–851). Macmillan Publishing Co., New York.

79. Sasson, A. (1984). *Biotechnology: Challenges and Promises*. UNESCO, Paris.
80. Schell, J., & Van Montagu, M. (1983). *Biotechnology, 1,* 175–180.
81. Schesinger, H. (1999). DNA conductor. *Popular Science,* 47.
82. Schlegel, H. S. (1986). *Genera Microbiology* (6th ed., translated by M. Kogut). Cambridge University Press, London.
83. Sharp, W. R., Sondahl, M. R., Caldas, L. S., & Maraffa, S. B. (1980) The physi- ology of in vitro asexual embryogenesis,.*Hort. Rev., 2*, 268–310.
84. Shrof, J. M. (1998). Miracle vaccines. *US News & World Report, November 23*, 157.
85. Singh, M., & Ahlawat, S. (2004). Applied genetics. *MTBiology Today,* 27–29.
86. Smith, P. H. (1966). *The microbial ecology of sludge metanogenesis developments in Industrial Microbiol., 7*, 156–161.
87. Standfield, W. D. (1969). *Schaums Outline Theory & Problems in Genetics* (2nd ed.). New York. McGraw-Hill 281.
88. Strick-Berger, M. W. (Year). *Genetics* (3rd ed.). Publisher.
89. Swaminathan, M. S. (1980). *Second Annual Day Lecture*, NBRI, Lucknow.
90. Tangahu, B. V., Sheikh Abdullah, S. R., Basri, H., Idris, M., Anuar, N., & Mukhlisin, M. (2011). A review on heavy metals (As, Pb, and Hg) uptake by plants through phytoremediation. *International Journal of Chemical Engineering, 1-31.*
91. Thomas, E., & Davey, M. R. (1975). *From Single Cell to Plants*. London. ISBN- 9780851090412.
92. Van Apeldoorn. J. H. F. (1981). Biotechnology: A Dutch Perspectives. *A Report by the Netherlands Study Centre for Technology Trends.*
93. White, P. R. (1934). Potentially unlimited growth of escised tamato root tips in a liquid medium. *Plant Physiol., 9*, 585–600.
94. Williams, J. G. (1981). In R. Williamson (Ed.), *Genetic Engineering, Vol. I* (pp. 1–59). Academic Press, London.
95. Wiseman, A. (1975). *Handbook of Enzyme Technology*. Halsted Press, New York.

INDEX

E

F

G

H

I

K

L

M

N

O

P

T

U

V

W

Y